THE GOLD SEEKERS

. . . A 200 Year History of Mining in Washington, Idaho, Montana & Lower British Columbia

By: Pauline Battien

Second printing 1992

Paperback
ISBN NO.
0-940151-16-2

Hardcover
ISBN NO.
087770-478-3

Printed by
Statesman-Examiner, Inc.
Colville, Washington

Dedication

This book is dedicated to my four grandchildren:

Douglas Moore, Susan Moore,
Michael Battien and Jennifer Battien

–Who are already participating in history making events;
may that interest continue in future years.

Mines & Companies Table of Contents

(People, Places and Things Index is on pages 244-265)

Foreward

Minus chapters that one is accustomed to, this history of the mines of Idaho, Washington, lower British Columbia, and a bit of western Montana may seem disorganized. It has been said to be confusing and lacking in continuity – a downright truth if one does not wish to follow history as it was being made. We hope the maps make it easier.

Chronology was attempted but too often it became necessary to go forward with the story while in the specific mining camp; and then hope that you, the reader would return with me to follow in the wake of the horde of prospector-miners covering this area. God's plan for the prospector must have seemed a horrible jig-saw puzzle as he would hear of new camps opening miles from his present location, and it is factual that those poor souls did move about just as the book tells. We follow many of the prospector-miners from place to place and recall suddenly where we last heard his name. Some of those hearty fellows made almost every new camp. We will become just a bit better acquainted with two centenarians of the trade, Joe Moris and Al Stiles.

Material for this book has been obtained over 27 years of research and the writing has taken two years. This makes no pretense of being perfect journalism but it seemed urgent that much of this otherwise lost history needed recorded. It is factual as nearly as any mortal today can write of activity going back 200 years – with most of the action within the past 100 years. It is not intended to be a technical work but instead is narrative style, with facts and legends of interest to the lay person.

The major reason for it's being written was to honor the prospector who searched out the minerals that make our lives just a bit better; and in some cases are vital to our daily living. It also honors those who mined those minerals, one of the hardest jobs the good Lord ever made, and sometimes one of the most heartbreaking. The attempt was to find as many names as possible who were part of each occupation and by the telling, to give credit long overdue.

As we follow our prospectors across part of Oregon Territory, the present Oregon State is not included. Lines were drawn by treaty in 1846, making Oregon Country into Oregon Territory, and a line was formed between the United States and Canada during 1857 to 1861. That line took from 1903 to 1907 for survey and marking.

The portion of Oregon Territory we cover had Colville as the center of business. The huge area contained what is now Franklin, Adams, Whitman, Spokane, Lincoln, Douglas, Ferry, Okanogan, Chelan, Stevens and Pend Oreille Counties of Washington; along with what is now the State of Idaho, and the west side of Montana. As towns and cities, and new centers of government developed, segments of this total were sliced off to form states and new counties.

By 1863, Stevens County was formed and still included Ferry, Okanogan, Spokane, Pend Oreille, Whitman and Stevens. There were many legal battles fought but as areas were amputated from Stevens, counties formed were: Whitman 1871, Spokane, 1879, Okanogan, 1888, Ferry and Chelan 1899 and Pend Oreille 1911. Idaho and Montana eventually took form on the map.

Pinkney City was named as center of business for Stevens County, the name later changed to conform in 1868 with Fort Colville, a military fort. Shortened to Colville, it was still the seat of government in 1871 when there was only 1000 population in the entire area and Richard Steele tells us that only 300 were eligible to vote.

We learn the location of many claims and mines in the entire Territorial area. As new camps opened, the rush was always to be the first there. A few locations would be made and suddenly the whole pack was searching, and digging nearby. Unfortunately we cannot tell about every single prospect and hope "yours" was not overlooked. We do find many of the same "pick and shovel" men in many of the camps; the reason for covering such a wide area of mining history.

The list of references used may have unintentionally overlooked someone, but the hope is that they knew how much help they gave me. The book is indebted to marvelous help from the Spokesman-Review with permission to quote from it's news stories, and with the great amount of photos they permitted me to use. (Rather than use the full name of the credit line it will say S-R.) The same thanks go to the Spokane Chronicle (with S-C on the credit line). Other special thanks go to Richard G. Magnuson for permission to verify facts from his "Coeur d'Alene Diary;" to Dorothy R. Powers and her "Powers to the People" articles, and to Al Nugent, Elmo Thomas, and Ray Horsman for their early help in my learning about mines. My thanks also to the last 100 years of Stevens County newspapers to include the Statesman-Examiner.

Possibly two of the better statements made about this type of life were; Harry Magnuson from Wallace, Idaho who said after hard times hit the area – "Hope springs eternal from the breast of a hard-rock miner." And it is said of the rush to Rossland, British Columbia – "As dividends follow dividend and poor men were suddenly lifted to the plane of millionaires, the world gasped, rubbed it's gold-famished eyes, and made a wild rush to the snow-capped peaks where such delightful things came true." – "Oh! for that one big strike!" – but be aware that prospecting and mining are two of the greatest gambles on earth – even in "Colville Country."

The Gold Seekers

Giants of industry are awakening to certain mineral shortages, caused by nearly total mine closures due to depressed prices in recent years. This gives cause for re-telling stories of some of our great mines; of some that are little more than prospects, and brief mention of many "heart-break diggins" that did not prove out – but may if need is shown.

This is a history of the mines; of those who participated, and events of the time, within the radius of our hub and mining capitol, Spokane, Washington. It is the history of hardships and struggles, greed, bloodshed and misery. It is of the terrible loneliness of men pressing forward in search of "that one big strike" that they may again be with their loved ones. Or – the finding of that one great nugget that would lift man's soul above the dirt and weariness and make possible the development of a prospect to a great mine.

It is the story of large companies more able to cope with loss or gain; and the story of capital – without which, there could not be the dreamed of success, and more probably failure.

It will recall many locations long forgotten, made by old-timers who never received recognition for searching and scratching out minerals we take so for granted. It will tell of fortunes made or lost in building the great mining industry we have known – and will assuredly know again. This story is dedicated to the unsung heroes who have had the intestinal fortitude to buck the elements and prospect our moutainous terrain.

We will follow many, but in particular there will be two – the late Joseph Moris, better known as Spokane's own Mr. X, who died just short of his 100th birthday; and Alfred Stiles, prospector and assayer who died at Addy, Wash., just over 101 years of age. But – we have many miles and locations to cover before we meet those two ---.

Let Us Start

It was just after the turn of the 18th century. Lewis and Clark had started their canoe trip down river into Lewiston and Clarkston (as they are known today). The first explorers were invading these Native held lands in what would be known as Oregon and Washington. It was 1805 and not for twenty-two more years would anything of mineral value be discovered.

The first was coal and it was found near the junction of Cowlitz and Toutle Rivers in northwest Washington. It is said to have been discovered in 1833 by Dr. Tolmie, an Englishman employed by the Hudson's Bay Company at Fort Vancouver. It would take many settlers with needs for the use of coal to bring about it's mining – which has been credited with being Washington's first industry. (Remember this is long before statehood for either Oregon or Washington.).

Indians called Captain William Prattle's attention to the black outcroppings on the shore of Bellingham Bay. Along with Morrison and Thomas, claims were laid to 160 acres of ground and a "wheel-barrow" mine was opened. A small amount of coal was taken but the location was soon abandoned. In 1853 the MA-MOOSIE mine was developed on Morrison's claim with about 150 tons shipped to San Francisco. This mine was soon abandoned.

That year a large tree blew over two miles north of the original Prattle claim and is said to have exposed a coal bed which was located by Brown, and Hewitt. Bellingham Bay Company was organized but large scale production did not come until 1897. Most of that was from Kittitas, King, Pierce and Skagit Counties. Peak production in coal in 1918 was four million tons valued at more than $14 million. By 1953, output had dropped to a million tons.

By briefing the coal industry we can return to 1834 when immigrant wagon trains ventured into this great Oregon Country. They stopped for rest and to replenish supplies at Fort Hall which had been established by Nathaniel Wyeth in that year. Not until 1836 would Fort Boise become another welcome sight for weary travelers over the Oregon Trail.

In 1836 Marcus Whitman and Henry H. Spalding arrived with their wives, the first white women in the area to become known as Spalding, Idaho. Spalding Memorial Park, the first school, church

Cataldo, Sacred Heart Mission is Idaho's oldest standing building. Photo taken Sept. 23, 1946, prior to recent renovation. (S-R)

and printing press and a mill was built by Reverend Spalding that year. They would all be sad remains after the Whitmans were massacred by the Cayuse Indians in 1847.

It is difficult to actually pinpoint Whitman's Mission on maps today because it was at Waiilatpu, not to become part of Oregon Territory until 1848. The town of Spalding is east of the present day line of Washington State and to go to Fort Walla Walla one had to go west. Walla Walla City and Spalding are nearly 100 miles apart. It was to the Fort Walla Walla of the Hudson's Bay people that a Mr. Osborne tried to take his family when they escaped the horrible massacre of Marcus and Narcisse Whitman, and some 11 to 13 more who made up Whitman's little mission. Reverend Spalding had escaped and continued to bring the "White Man's Book" to the natives.

A second and important site in what would later become Idaho is the Cataldo Mission. The location credited to Father DeSmet in 1842, and by differing historians to Reverend Anthony Ravalli in 1848-49, brought about Christianity in the area and was later named after another Jesuit Priest, The Reverend Joseph Cataldo. Under the good Father's direction, Coeur d'Alene tribal Chief Vincent and his people are said to have constructed the mission all without nails. An important effect was the quieting of Indian attacks on settlers moving into the west.

It was to Chief Vincent's Indian people that Father DeSmet brought potatoes to plant their gardens. While visiting other missions, he had found the precious potatoes in the Colville Valley of what is now Northeast Washington.

With a population of some 9,000 at that time and feeling the need of more governing, Oregon Territory was created in 1848 but would not become a state until February 14, 1859. Washington, severed from the Oregon Territory, was credited with only 3,000 population when it became a territory March 2, 1853, though it included all of Idaho, parts of Montana, and Wyoming. Major Issac Stevens was appointed the new Washington Territorial Governor and this foundling area was to see a great deal happen in the next 36 years before it became a state November 11,1889.

Gold in

Colville Country!!

Hudson's Bay Company had established Fort Colvile (spelled with only one l – named for Lord Colvile), in 1827. Angus McDonald, the chief trader at the Fort, (north of the falls on the Columbia River) tells the story. His teamster Joseph Morel drank from the cool clear river and saw scales of gold in the black sand. He was claimed to be the first finder in 1854 (historians disagree but more likely 1845). That sighting was soon spread to the prospectors to the east around the mouth of the Pend Oreille River near Idaho's present western border. The panic and excitement grew! News items described the scene as a "flood of picks and pans." The romance and history of the search was on.

Placering

Placer mining was king! As we travel through the next wild years we learn proper equipment was needed – a strong back, a good shovel and pan, a little luck – and a grubstake. It all depended on which side of the purse strings one stood, but demand to give, or to receive, was everywhere. Speculation was rampant. Absolutely necessary was the pan – three to four inches deep and 12 or more inches in diameter; used for panning gold, mixing bread and even for bathing. Flour, bacon, beans, tea or coffee, sugar, frying pans and a coffee pot made up the larder and utensils.

If one was lucky, one bought a cayuse for $50-$75 that should have cost no more than $25, without which one started out with burro and pack – and more often on "shank's mare." Gleanings were lean, but the pouch of gold dust became the monetary exchange for supplies usually obtained at Fort Colvile.

The urge to travel each way across the Cascades and the need to investigate possible passes over those mountains was met by Governor Stevens. He assigned George B. McClellan (later to be known as General McClellan during the Civil War) the task. The two met at Fort Colvile October 1853. McClellan noted that he had found traces of gold as he ascended the Yakima River in September. Word spread and from that time through 1860 prospecting became pretty general in central and northern counties of Washington Territory.

Placer mining had extended into the Similkameen mining district (Okanogan County now). It was often many months before an old-timer could return to the county seat at Pinkney City (origin of Colville), where he filed his claim for recording. (Many of these old records are still in the files in the Stevens County courthouse).

The earliest gold find in Okanogan in 1859 was credited to a soldier surveying the Canadian boundary just to the north. Some years later this claim was profitably developed bringing about a boom town of 3,000 people, around Ruby. By 1862 placer miners had spread out locating deposits at Ruby Creek (Whatcom County), Sultan Basin (Snohomish County), and the Peshastin Creek district near Blewett.

First quartz ledge

THE CULVER

The rush to stake placer claims would decrease by the early 1900's but statistics from the State of Washington show 125,000 claims were filed up to

1936, and possibly 1000 more to 1953.

The search for riches saw many miners returning from the Cariboo and Fraser River area of Canada and looking towards the Wenatchee Mountains. John Shafer is credited with the first quartz location in 1874 in Culver Gulch. (Quartz is a very hard mineral deposit composed of silica). Hardrock claims would soon be located - the GOLDEN CHARIOT, BLACKJACK, BOBTAIL and others. The town of Blewett, west of present day Wenatchee, counted several hundred people with a store, schoolhouse, assay office and saloon, and around 1879 a wagon train was built from the west side. (All are Chelan County locations.)

The arrastra (taken from a Spanish word) had come into use, it's original construction never traced, but proved to be the best method of grinding ore found to date. A pit-like formation with drag stones, the crude mill was powered by horizontal or overshot water wheels which were geared to a center post. Heavy rocks dangled from the four outstretched arms which were dragged around the shallow basin to grind egg sized pieces fed into the pit.

As one travels today along Highway 97 north of Swauk Pass you will find the old townsite just off the road - and will see an old stone arrastra. The first stamp mills were used along the Peshastin Creek from 1870 to one of the larger 20-stamp mills built in 1892. One can still view parts of that old mill and the remains of a tramway which carried the crushed gold ore some 4,000 feet to be dumped into the mill for crushing. With information limited, the production in gold from this area has been set at $1.7 million in bullion from the 1870's to 1910.

The Arrastra (or arrastre) as it appeared in 1924; found to be the earliest means of grinding ore. (Division of Mines and Geology)

The Columbia River and areas to the central and north of Washington were being populated by gold crazed prospectors. That left the coast communities not directly in the path of the mining advance, feeling their lesser importance. Those people began to fear they would be outvoted in legislature by representatives from the eastern side. Also, benefitting population growth for central Washington, was that peace was being established with the Indians.

There were no laws to cover this advancing gold-hungry mass except rough and ready judgment; the gun frequently first, and often the last authority. Possession was ten points of the law. California miners spreading through the area brought order into things by naming a judge, sheriff and recorder.

This authority managed well until British officials came to maintain law. Powerful authority was vested in their gold commissioner. The first to be named November 19, 1858 was Chartres Brew, who served as judge and jury until a board was appointed to lighten his load. His law was effective on both sides of the border that was just being defined as the 49th parallel or International Boundary between Canada and the United States.

To the east, the Mullan Wagon Road (official name "Military Road") from Walla Walla, to Fort Benton, Mont., was being built by Captain John Mullan and his crew, who completed the 624 miles in 1863. Taking just under five years, the road cost $230,000 or $369 per mile. A number of years later, a branch came through Spokane just west of Cataldo - traces of which can still be seen along Highway #10.

John J. Lemon writes that General McClellan was once commander of a camel corps. Years later Frank Laumeister had tried using camel mule trains in the British Columbia mining fields, but found their feet did not hold up in that terrain. Other camel trains were said tried on the Kootenai Trail in 1866 and 1867; and camels were said used as pack beasts on the Mullen Road carrying supplies from Walla Walla to mining camps in north Idaho. Lemon did not want their bones taken for prehistoric beasts.

Pierce, and North Fork of Clearwater, Idaho

To preserve some semblance of continuity we must return to the placers of Idaho. A fur trader and Army explorer both told of seeing gold, but it was Captain E.D. Pierce who made the first great hearing the tales told by Captain Mullan's crew encouraged him to venture into the Nez Perce country.

Pierce prevailed upon Chief Timothy's daughter, Jane, to lead them by secret route, and that winter the party camped at Canal Gulch. Following the Pierce finding of gold in the spring, Jane married a Virginian in the party, John Silcott. Many years later she was to be proclaimed one of three red heroins of Idaho.

By July, 1861, 2,000 people were swarming the gulches – some claiming finds of $5-$8 in gold dust per day. That year the ORO GRANDE was discovered by William F. Bassett, one of the Pierce party. The city of Pierce was established at the location.

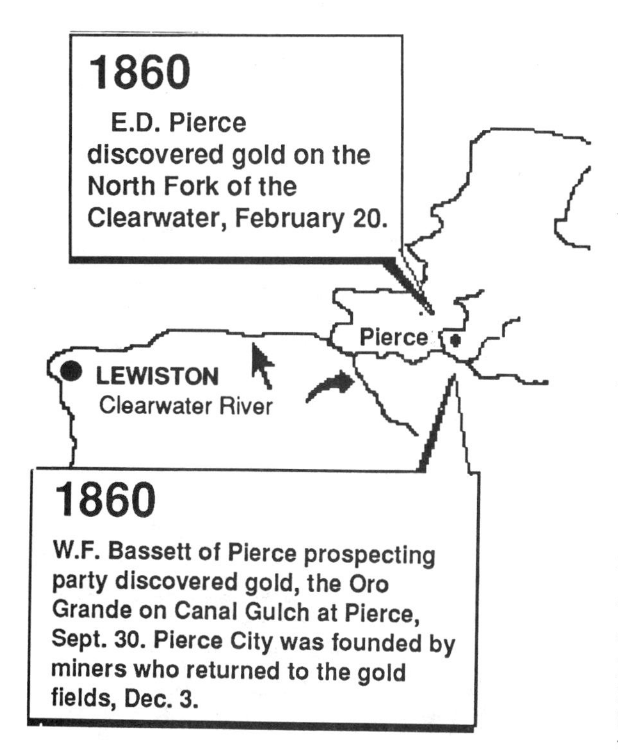

Still in Washington territory, Shoshone County was named – with Pierce City becoming the County seat which it retained until 1885. Pierce was the only seat of government at that time for what is now Idaho and Montana. Walla Walla, Wash. was becoming the supply center. It was said that after Pierce's visit there for supplies, that 25,000 men rode through the city heading for Idaho's first big gold camp! Surprisingly, Seattle would not surpass Walla Walla in population until 1880.

With the advent of the Oregon Steam Navigation Co., developed by R.R. Thompson, president; J.C. Ainsworth, and D.F. Bradford, both miners, began traveling up the Snake River on such famous old steamships as the Colonel Wright, and many others to follow. With gold in their pockets, many miners were able to return down the river and retire for the winter in the cosmopolitan city of Portland, which counted 6,000 population in 1866.

Lewiston, Ida., – often called Ragtown, came into being around 1861. Said to have 3,000 population, that count consisted of miners, teamsters, soldiers, gamblers, and camp-followers. As Lewiston grew, the Army was worried about Indian uprisings but that fear did not keep gold-mad men from coming. Time was when Lewiston vied with Walla Walla and Olympia for permanent capitol of the Washington Territory. It grew to become Idaho's first city when Idaho became a state in July 1890 – nearly 30 years away.

An old-timer told that at one time the Forest Spencer family, then of Spirit Lake, Ida., are said to have turned down an option to buy Lewiston's townsite for $500 – one dollar an acre. If only for hindsight – or is it foresight??

As we move into Idaho, you will note the pace picking up to a feverish pitch to be in a new field of discovery – and you will note more names with a familiar ring, and be able to follow them into each new area.

South Fork of Clearwater – Elk and Florence

Gold was discovered on the south fork of the Clearwater on June 14, at Elk; and at Florence on Aug. 12, both 1861. Both developed into growing towns with Elk being a point for supplies. Five thousand miners and tradesmen were said to be in those golden gulches by June, 1862 and 1319 claims were recorded on the books at Florence.

The Elk City rush was made up of many men from San Francisco, The Dalles, Portland, and Utah. Too many miners were ill fitted for that bitter winter; stock died and men died, and the men developed scurvy, which some fought off with raw potatos soaked in vinegar. Many were charging from one "diggins" to the next in the frantic rush to be first. From Baboon Gulch came word that Mr. Weiser had taken $6,600 in one day while placer mining . . . a new field to be reckoned with.

D.H. Fogus and Moses Splawn had joined up with the Grimes party. As they moved into this new Idaho mining area they had searched for the famous **"Blue Bucket"** site along eastern Oregon. That legend came about when a wagon train of people had camped along a creek. The youngsters were sent for drinking water but soon returned with their little pails full of tiny gold rocks. Overweary parents paid no heed and it would be many years before that mine would gain recognition through Oregon history.

Florence, Idaho

Janice Ruark writes that Main Street in Florence was the first recorded public road in Idaho. Said to be the first town settled in Idaho County after the

1861 discoveries, production ceased around 1900, after nearly $30 million was said taken out in gold.

James Ayers, Lemuel Grigsby, Hull Rice and John Healy have been credited with first claims. As winter progressed, prospectors followed over the treacherous mountain trail and by November ,after fighting deep snow, there were said to be 1,000 men. Living in tents and fighting starvation, those who exited were met by men and families heading into the area – and within a year or so there were said to be 5,000 people, some of whom had wooden buildings.

The town boasted a Masonic Hall, library, school, courthouse, laundry, newspaper office, a stage and livery barn – and a Wells-Fargo office to name a few businesses. Placer mining gave way to quartz mining in the late 1895 period and even saw a second townsite. The Chinese who followed the other miners counted many among the remaining people who mined on a limited scale into the middle 1900's.

The criminal element said to be thieves and gamblers from the Pacific Coast, and the East, caused a killing nearly every night, according to Alonzo Brown, who owned a store. H.J. Talbotte, better known as Cherokee Bob, lies among those in Boot Hill. His major offense was defending the rights of his "painted lady," said won in a card game – when she was not allowed to go to the public dance and be treated with respect.

Ruark explained that graves were laid out west to east, and some north to south; those who died with their boots on were one way, those of natural death faced the other way. Boot Hill is about all that remains of the ghost town Florence. Many of the wooden markers have been replaced by the U.S. Forest Service.

Orofino, west of Pierce, became a gold rush town with richest claims located on Rhodes Creek. It was reported in the Portland Oregonian, August 1861 that 2,500 practical miners were working that creek, Orofino Creek, Canal Gulch, and French Creek districts. It noted that four or five thousand more were making their living in other ways.

Methods of mining were advancing – first the pan, and now the "cradle", or rocker, to be followed by the sluice. Five times more pay dirt a day could be trapped behind wooden strips over which water washed to separate the gold from the tailings (dirt). Timber and water were a necessity; ditches had to be dug, and sluices built. Whipsawers would make good money for sometime yet, before the arrival of sawmills. In the meantime the patient Chinese were following the footsteps of the placer, the ground often yielding a good living for them.

The north central Idaho country was developing into one of the nation's top gold producing districts and has been credited "that the steady flow of gold from Idaho during the Civil War 1861-65 was an important factor in turning the tide in favor of the Union." Estimated gold taken from Idaho placer mines to 1870 is said to be $70 million.

Idaho becomes a Territory

With this phase of Idaho's first boom over, towns dying, placer miners moving on, business had so deteriorated in 1870 that farmers had no sale for their produce. Earlier the mining advances into Nez Perce and Salmon River had resulted in Idaho's becoming a territory from what had then included much of Montana and Wyoming. There were ten gold camps and 20,000 people March 3, 1863 when W.R. Wallace became Idaho's first Territorial Governor. BUT – with the decline of placer mining it has been said that Congress theorized that Idaho had only bears, Indians, and abandoned gold mines. Farmers and sawmill operators tried to hold the interest but Idaho's first census showed only 14,999.

Boise Basin

"Like rats from a sinking ship," the miners were fleeing the old territory – as early as 1862 when Boise Basin discoveries were made by George Grimes, and Moses Splawn on Grimes Creek. Grimes was killed by the Indians and his friends buried him in a prospect hole, – and went on to mining.

In 1864 there were some 16,000 people in Boise Basin (25) miles N.E. of the present city of Boise; half of these were miners. An estimated $17 million was taken the first four years and Idaho City became the largest town at the time. Boise, on the Oregon Trail, was said to be populated largely by a migration from Missouri coming to escape Civil War conditions.

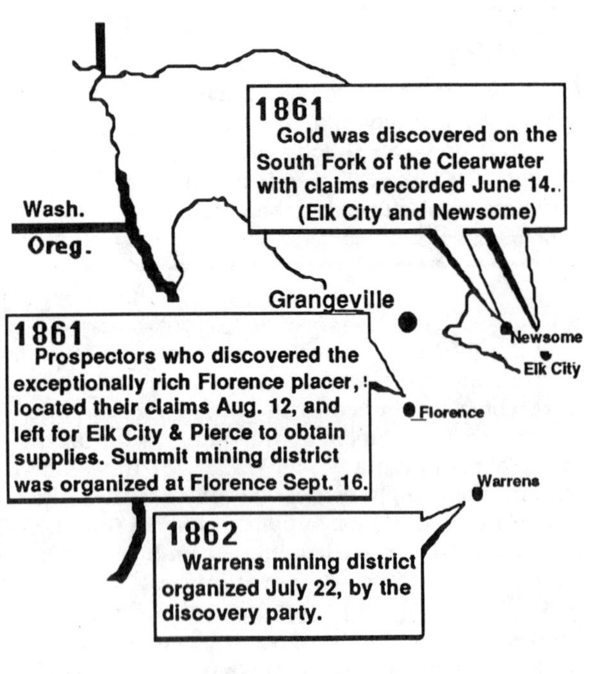

Placerville, Pioneerville, and Centerville, were also leading mining centers. Quartz was discovered at Rocky Bar, May 7 and at Banner August 23, 1863 when a search was being made for a better road between the two. Warrens mines to the north, first discovered July 22, 1862 were doing well. The entire south-central mining districts were enlarging. 1940 figures give uncounted millions taken from those mines, some still in operation at that time.

Silver City grew into a camp of importance. J.H. Chittenden, Superintendent of public schools in the Idaho Territory, and Silver City Assayer, was instrumental in bringing about the first schools. One report stated three school houses where 12 classes were held.

Possibly we do injustice by passing so lightly over south and central Idaho's mining history. Today's decreasing silver supply and the price fluctuating, is causing another look at regions of which we should know more. For that reason – how about tying a lead halter on that "Rocky Mountain Canary," grab your sourdough fixin's, and try a second go-around – prospectors did!

Placers around Idaho City were said to yield some $2,000 the first week, and at Placerville, a prospect showed return of $300 a day. The word was out and more hardy souls must rush many hundred miles south from the Pierce, Orofino, Elk and Florence camps to these newer camps to the south.

Bitter winter and four foot snow did not deter those who hardly found standing room by spring, 1863 as they scratched for claims along Granite Creek. Claims of "richest mines that have been struck since the '49 gold rush in California" were many. Miners were forced by water shortage to gather dirt in flour sacks and carry it several hundred feet where when washed was paying off at $50-$60 a day. Due to water shortage, things slowed down at Placerville in a few short weeks, but those on More's Creek fared better. One company installed a hydraulic giant and with a five man operation was averaging around 100 ounces a week.

A stampede of supplies started moving south to the gold placers by September 1863. Idaho City, fostered by J. Marion More, would count 6,267 people, 360 of whom were women, and 224 children. Pierce, Florence, Warrens, Elk City, and Lewiston had dropped to a few hundred people apiece. Miners were working around the clock at Idaho City – even in the very streets where they were causing houses to collapse. Very close to city limits, four partners were said to have netted $7,771 for one week's placering in May, 1864.

The **Gabrinus-Landon** lode was said to have yielded $2,000 a ton returns by use of the arrastra. Improper development coupled with lack of operational knowledge was blamed for troubles that arose. The **Elk Horn** ran into unfortunate circumstances shortly after the Boise Basin treasurer was alleged to have invested County funds in that property.

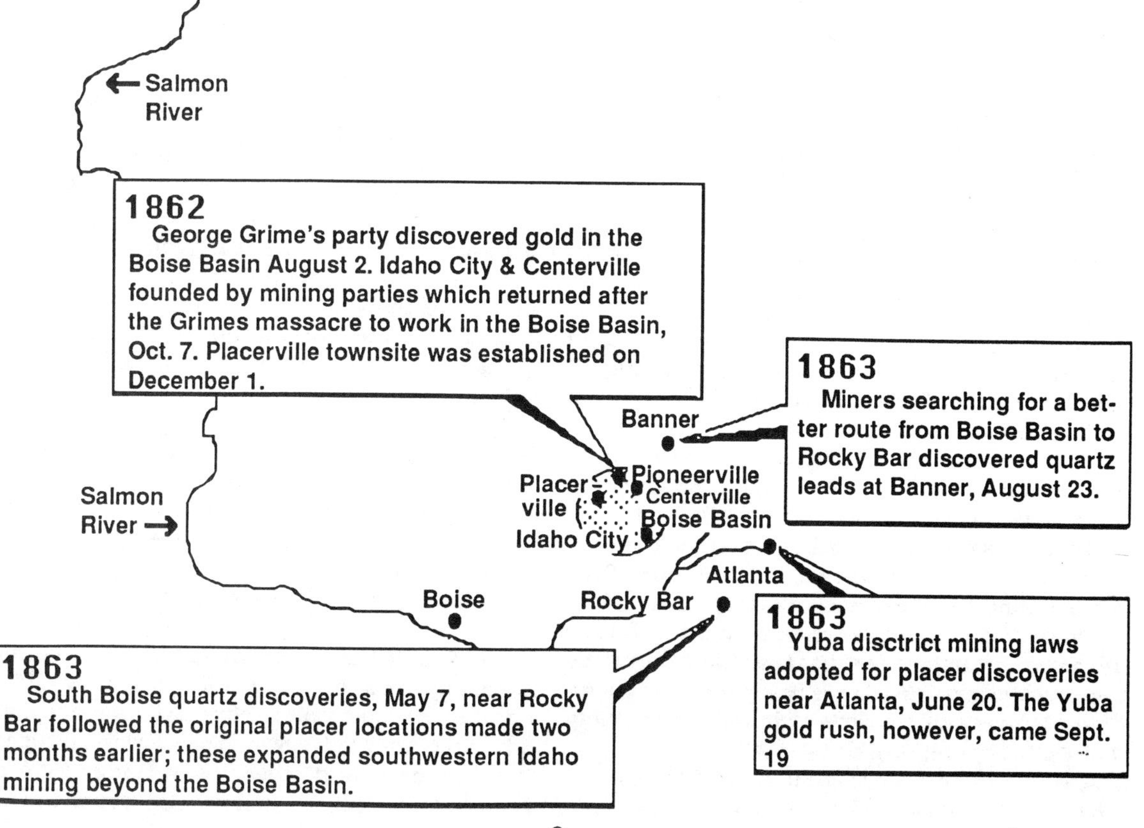

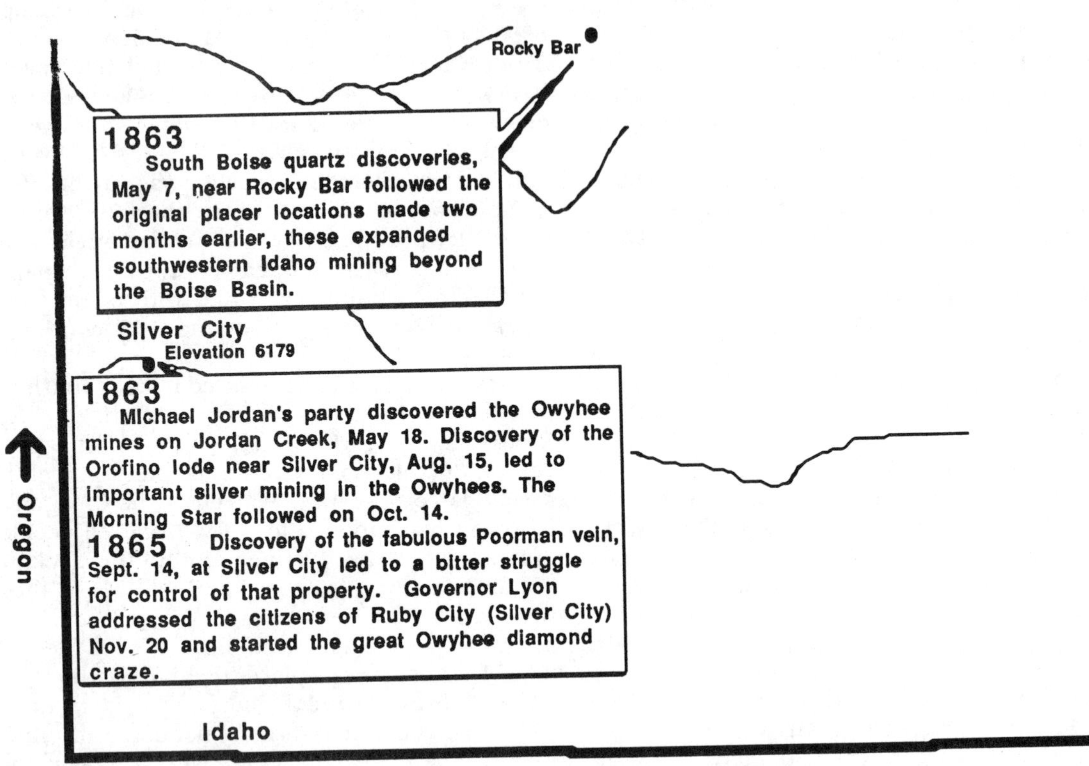

William Lent, and James M. Classen re-opened the **Pioneer** on Gold Hill, attempting to use more business-like development. Historians tell that this company engaged the experienced quartz miner George Hearst, already a noted California lode miner. (Part of the Hearst fortune is said to have come from this region of Idaho, and that this laid the foundation for his election to the U.S. Senate – and to his son William Randolph Hearst's eventual newspaper empire).

South Boise Basin

The word lode (vein of metal ore) will be used more often as we see South Boise Basin being promoted after discovery of the **Elmore** in 1863. Reports say that H.T.P. Comstock promoted it as one of the richest mines, "as rich as the lode in Nevada named for him." The **Ophir**, the **Idaho**, and the **Bonaparte**, were big news at the time and all within a radius of Rocky Bar. Values ranged from $75 to $300 a ton in 1864 and $130,000 to $160,000 was taken from South Boise Basin quartz mines that season. From ten to 80 arrastras were in use at the time.

Those assays told from the **Comstock** ran as high as $7,112; from the **Confederate Star** $5,589, and the **Elmore** $7,434. By 1864, stamp mills were being shipped into the area with little regard to exorbitant freight costs, and very little thought given to having ore ready for their operation. Recovery losses were high and many stamp mills proved utterly useless.

Because this era was actually ahead of their use on Peshastin Creek in 1870-92, in upper Washington, a stamp mill might be explained. It could be described as a boxlike structure into which ore is fed to be pounded by iron weights weighing from 300-800 pounds apiece. The ore is first broken by sledge hammers into apple sized bits before feeders bring the ore under the weights, the crushed rock finally coming out as a wet powder.

Hard labor has developed claims to this point but now the future of a mine is dependent upon capital. An ordinary prospector-miner might discover and have his quartz lode tested. He might even get it recorded. Even a small mine is usually out of "pocket" reach for the claim owner. One must also keep in mind that in these early mining years a few thousand dollars represented a great deal of money. As million dollar investments were made, or sales for their ore output materialized, such sums earlier were only in the "chicken feed" class with today's costs and/or gains.

Accusations of mismanaged funds, swindles, poor operation, and over promotion of South Boise Basin mines became common by 1866. It was said that there was more stamp mill capacity than in any other Idaho district. New York investors questioning high costs and failures, often refused to back the mines at the needed time.

Wilson Waddingham was reported to have sold his interest in the **Elmore** at $50,000. A Pittsburgh Company took over operation but it would be another 20 some years before British money, and advanced knowledge, would see large scale operation. Gouging, and other unsavory mining practices, helped bring South Boise Basin rush to a halt in the fall of 1866.

Owyhee Mines

Placer mines of the 29 member Jordan party at Happy Camp were soon stripped. When the promising surface gold of the **Orofino** on one side led to the eventual discovery of **Morning Star**, (not to be confused with **Star-Morning** at Mullan), silver property on the other side of Eagle Mountain, a new town and new situations were bound to be the outcome. Silver was becoming king!

Then came the Owyhee movement – reported as the "special 48 hour insanity for Owyhee." Michael Jordan's party discovered the **Owyhee** mines on Jordan Creek May 18, 1863. The **Orofino** lode was located near Silver City August 15, and the **Morning Star** find was made October 14, that year. Then in Sept. 14, 1865 the **Hays** and **Ray** vein was discovered, along with the cross vein the **Poorman.** That gold and silver find was reported to run 60% bullion and it was said the **Poorman** people took out $250,000 in two weeks.

The district was costly to investing capitalists and control of property led to bitter struggles. By 1866, ten stamp mills had been transported into the area via river – and the 300 miles by wagon trail which cost 25¢ per pound. Full production of the district was not possible for another 20 years due to lack of transportation.

J. Marion More and D.H. Fogus were doing the most development on Owyhee mines, and were buying up property as rapidly as possible at $100 afoot. The Oregon Navigation people, already holding a monopoly on transportation, are said to have profited greatly from this district. Among those investors were Simon G. Reed (who will become owner of **Bunker Hill** and **Sullivan** some years later – and even later, the endower of Reed College in Portland); and J.C. Ainsworth (one of the miners to first travel north on that Company's Colonel Wright steamship).

In 1864, Ainsworth erected a ten-stamp mill at a cost of $70,000 with mill yield said to be $90,000 from it's first month and a half operation. Problems did develop when ore turned largely to silver. On the other hand – Thomas Donovan was well pleased with a three quarter ton load which was said to yield $5,000 – the assay at $6,706 in silver, and $489 in gold to the ton. Another ton milled in Wales (while only less valuable Gould and Curry ore was being processed in Virginia City), ran $6,500. It was the practice of shipping rich **Comstock** ore to the Swansea Mill in Wales, so Donovan had followed suit.

Silver City

The construction of more and more stamp mills kept 40 men apiece busy with their building. As these advanced toward the southeast, the first county seat and short-lived town of Ruby City gave way to Silver City becoming the leading center. This was not accomplished until 1866 at which time ox teams moved buildings over the road built by Colonel W.H. Dewey. Old timers report that move was not without it's lighter side as a half-way point had been established to assuage the thirst of the burdened mover, and relieve him of having to carry his own "jug of juice."

Charles S. Peck had discovered the **Poorman** but outsmarted himself by covering his find as he held out for a better deal. In the meantime the claim was rediscovered by D.C. Bryan. The reported 18 inch vein assaying 80% gold and silver – mainly silver, was the center of litigation between the **Poorman** people backed by the Oregon Steam Navigation Company's Owyhee interest; and the **Hays** and **Ray** people who combined with G.C. Robbins' New York and Owyhee Mill Company. Judge Milton Kelley granted an injunction to the **Hays** and **Ray** people who claimed the **Poorman** vein was on their property, and the outcome was a compromise. The **Poorman** was temporarily closed down while development went ahead on the **Hays** and **Ray** vein.

As this mining district settled into more systematic operation, the More and Fogus mines were supplying a number of mills. **Orofino** mine was said to employ some 60 men, and when **Poorman** resumed operation in July, 1866 it was mined by 30 some. Miners were from many places but the Cornish "Cousin Jacks" were said to be the hardest working". Their wives, the "Cousin Jennys," settled in comfortably wherever they found themselves.

Transportation problems were being met by such as John Mullan (earlier of the Mullan Road from Walla Walla to Fort Benton), who opened a competitive freight route from Chico, Calif., to Silver City, Ida. John Hailey, Ben Holladay, and Hill Beachy were providing stage service to the Owyhee.

Recreation in Silver City ranged from horse racing, to prize fighting, church goin', and "just sittin' a spell." More often than not, the hard working miner was found in the saloons gambling away his sweat-earned pay. The hurdy-gurdy dancer; the painted lady from the houses of ill-repute, and plenty of "Rocky Mountain dew" vied for any shekels a miner might still be blessed with after hitting town.

The More and Fogus combine finally failed financially and the Mine Workers' Cooperative stepped in to settle accounts which they did by sinking shafts which revealed richer ore on the **Orofino** (more often spelled Oro Fino). The **Poorman** had resumed work and was reported yielding in excess of a million dollars in less than six months operation. The Ida Elmore and the Golden Chariot Companies came to blow as being on the same vein.

In that same period 1867, the Owyhee County Courthouse was the point of meeting for the first labor union organized in western mining. California wages at the time ranged from two dollars to three dollars and fifty cents a day but miners in the out-of-the-way Owyhee were asking five dollars out-of-the-way Owyhee were asking five dollars and fifty cents a day.

War broke out on March 2, 1868, when gun-play took place between the **Golden Chariot** and **Idaho Elmore** factions. John Holgate and Meyer Frank were said to have fallen victims in this underground battle. The **Golden Chariot** forces gained control of D.H. Fogus' **Ida Elmore** shaft and by this time word of the war had reached Governor D.W. Ballard, at Boise. He dispatched Orlando Robbins, the noted marshal and Indian fighter. Proclamations were issued and new agreements reached when J. Marion More became the final casualty. At the end of that first week of April, the Owyhee war was over!

By the last of 1869 five companies; **Orofino** and **Morning Star; Poorman, Ida Elmore, Ainsworth,** and the **Golden Chariot** were reported to have shipped the largest percentage of $4,900,000 in bullion from Owyhee. Total production figures from the Silver City mines and War Eagle Mountain have been said as high as $30 million.

The important development of dynamite had decreased expense and increased production but new mining methods and transportation were sorely needed. Then came the failure of the Bank of California in 1875, creating a sad shortage of capital. It would be nearly 1890 before lower grade ore, mainly on Florida Mountain, would revive the Owyhees to bring about major production.

For those who enjoy "word-of-mouth" versions of yarns about a district's people, "Historic Silver City," copyrighted by Mildretta Adams, Homedale, Ida., is the best "derned" book you will find.

Banner, Stanley Basin

By 1867, Nathan Smith of earlier Florence mining fame, had drifted down to Deadwood and was reviving old placer locations. Those who were using the "giant" were said to be cleaning up as high as $5,000 in two weeks. T.A. Patterson had reported earlier assays of $2,100 a ton in the Volcano district. Little Smokey quartz discoveries of 1879 were said to have reached a production of $1,200,000 in 1884. Where snow was measured in eight foot depths, and there were no roads, it would be 1874 before G.W. Craft got a mill into his Banner properties. By 1882 **Banner** was said to have produced three million in silver.

Following a July 4th celebration in 1864, Civil War officer John Stanley and his party headed out to prospect the Bear Valley and went on into the Stanley Basin (later named for him). Yuba River findings were to be the beginning of the **Atlanta** mines. William R. DeFrees located the **Greenback** silver lode some two miles from the **Altanta** lode, in 1865. The **Lenora** gold lode was located July 1866, but attempts to get stamp mills into this isolated district; and problems of operation, caused many set-backs before any degree of success could be counted. During this period, British investors were developing the **Lucy Phillips** under guidance of Matthew Graham. The Monarch Company soon followed with important capital from Indiana. Cyrus Jacobs was reporting good assays from **Minerva.**

Test milling had shown good yield in gold but much silver was left in the tailing piles due to problems of recovery. Entry of the Union Pacific Railroad in 1869 made possible the shipment of some high grade ore but the country was held back for lack of even a wagon road until 1878. It would be nearly 60 years before **Atlanta** district would prove itself "up to expectations," and produce some $16 million in mineral wealth.

Leesburg and Lemhi Valley

Still engrossed in Idaho mining history we forget that miners and prospectors were covering other areas to include Washington, and British Columbia. The "Dewdney Trail" over the summit towards Trail Creek, B.C., was completed now and would be in use for some 20 years. It was built in part by a namesake who would later become Lt. Governor of the British Columbia Province. Miners were covering the Fort Steele district, but were suddenly drawn back down over old ground to central Idaho where placers were discovered in Leesburg. A new gold rush was on as miners stampeded into Lemhi Valley.

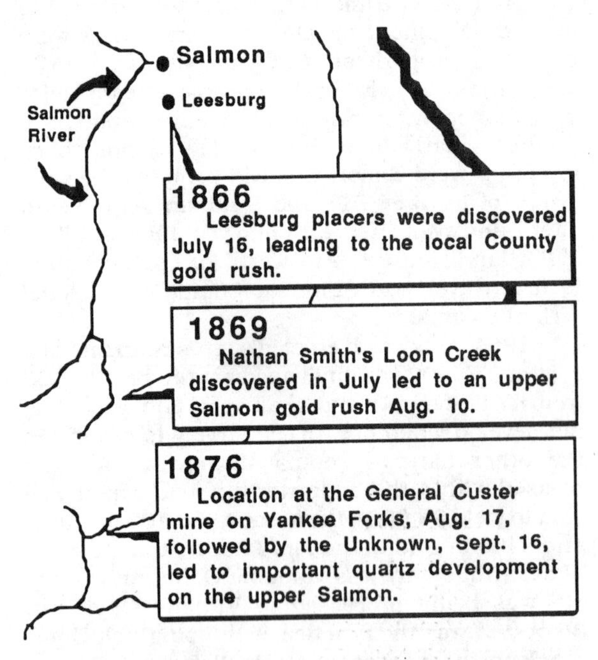

F.B. Sharkey and a five man party had found pay dirt on Napias Creek, in Leesburg Basin June 10, 1866 (sometimes said to be July 16). The Confederates had started Leesburg, and as northerners invaded Lemhi Valley, Grantsville was founded. Eventually grown together as one town, Leesburg retained the name.

That first bitter winter took heavy toll on both pocket and physical condition. For those who braved the Salmon Country, there was a shortage of supplies, food, and particularly medical help – all were badly needed. A toll trail 18 miles long was dug through five feet of snow, and was just wide enough to allow pack animals to pass. Rates of 75¢ each way was charged by the enterprising group who had dug the trail, and those in isolation were glad to pay the fee. By spring some 2,000 people were on hand and in search of gold. Sad as it was, the snow continued to fall and by May there was reported hundreds of broken men waiting mining activity in Leesburg.

"Kilroy," J. Marion More had been in this camp (sometime before his demise at Silver City), and had dug a ditch. For once, with all the snow, there was no shortage of water.

Leesburg was reported to be a quiet and stable mining town minus much of the gambling and brawling. There were no exciting discoveries but the eventual mass of prospectors did succeed in finding "colors" enough to raise 1867's Idaho's production of $250,000 up to $750,000 in 1868. Miners were moving to new ground and small mining did continue at Leesburg but the population dropped to around 180 in 1870.

Still in search of that big strike, Nathan Smith was the cause of changing the course of some rapidly departing miners from Idaho, when he made a discovery on Loon Creek, May 1869. The rush came from all points of central Idaho, and in August it was estimated that 2,500-3,000 people had arrived at a town called Orogrande. D.B. Varney and Company were first to get into production with sluicing returns said at $40-$50 a man-day. This was said "as handsome as the Kootenay gold," and prospecting began. By fall, nothing was evident but prospect holes and shattered hopes!

Some small findings by Jamieson, (an associate of Captain E.D. Pierce – of earlier fame on North Fork of the Clearwater), led prospectors to Clark's Ferry. The most encouraging news for the region in June 1869, was the advent of the Central Pacific Railroad; in addition to John Hailey's new stage line from Boise to Kelton.

A Mine–Or Not?–Term Explanation

Possibly we have hurried with the history and have overlooked making some explanations. Also, with apologies to learned historians researched who may have used other terms, we explain: The term "mine" has been used when perhaps "prospect" would have been more nearly correct, and with so many locations, we are bound to do the same, but few will quarrel with facts long forgotten. Geologist Alfred Nugent, of the Coeur d'Alene district, pointed out for our enlightenment: "A claim may be made up of many prospects; claims are developed, but not every development makes a mine – and not until investment capital is returned does a mine cease to be a prospect."

Chronology, and trying to be too many places at once, may find us retracing a step or two. Many years of traveling are ahead – our penalty will be missing the history of some new creek bed and gulches, but we will attempt to keep up with the weary prospectors.

Montana, Virginia City, Butte

First slowly, then a mad rush, these men were venturing into the present part of Montana which was still Idaho Territory. In 1863, 10,000 miners gathered within weeks to dredge gold from creek bottoms around Alder Gulch. Virginia City became the capitol of Madison County, which it held until Helena took over in 1876. 190 murders were said committed within the first six months, the bodies fast filling "boot hill" facing the city. The men asked for, and got, the vigilantes who based their laws on Lewiston's Protective Organization which had been so effective in 1862.

Miners rushed from the Boise Basin to Jefferson Basin (Montana), and established camps in The Big Hole; The Beaver Head, Stinking Water – or Passamari, and on Grasshopper Creek where **Bannack** Mines were located. To the south was Alder Gulch where highwaymen, freebooters, gamblers, and many good sturdy men and women came – and became neighbors to the Indians. In 1864, between 75,000 to 100,000 people were estimated in the gorges and valleys of Montana – principally at Virginia City. The city population was reported then at 11,493. Many came from Missouri River towns, by way of Kansas and Nebraska. Alder Gulch is said to have produced $30 million in three years.

Prospectors on their way to Virginia City struck gold along the banks of Silver Bow Creek. By 1867 Virginia City had turned to a ghost town and Butte was reported to have less than 60 people!

In this late 20th century, tourist's will thrill with reliving the history of Virginia City. Much of the mining heyday can be felt as we saunter through the saloons, and spend our "hard money" on the tinkle of the old time music machines. Or take an exciting trip up Boot Hill behind six lively horses and spend a few minutes to contemplate the past as the trail winds down the now silent gulches . . . how long silent, remains to be seen.

Montana-Helena

Below Three Forks, where the Jefferson, Madison, and Gallatin unite to form the Missouri River, is a large creek from the west called Prickley Pear. A few miles from it's mouth is Last Chance Gulch where Helena lies; and just above is Oro Fino, and Grizzley Gulches. Nelson's Gulch, a branch of Ten Mile Creek, gains access from a low divide. **Whitlatch Union Vein** with a reported yield of three million up to 1876, brought Helena into becoming the leading city.

Miners moving into Deer Lodge, Bannack, Alder Gulch, Last Chance Gulch, and Grasshopper Creek are said to have hastened the creation of Montana Territory, May 26, 1864, but which would wait 25 years before statehood, Nov. 8, 1889. Judge Sidney Edgerton became the first Montana Territorial Governor.

During the activity 1863-64, expressman David D. Chamberlain carried letters for one dollar apiece from Walla Walla to East Bannack. He said his worst fear was snow-blindness. Joaquin Miller (later became a famous poet) rode express from Walla Walla to the Salmon River country. Following the trail of miners who either walked, rode saddle horse, or traveled with saddle trains, came the pack animals wearing a collar of tinkling bells. This pleasant sound was soon replaced by the "gee-haw of bullwhackers" and the crack of teamster's whips.

J. Ross Browne, U.S. Commissioner for mining regions west of the Rockies, gave total yields from the beginning of known mining fields up to 1867; Washington at $10 million, Idaho $45 million, and Montana at $65 million. George Dawson estimated from 1858-1867 a total in British Columbia at $26,110,000.

While this prospector-miner frenzy was going on in Idaho, Washington, and Montana territories, H. Bauerman was proceeding with his geological survey 1859-1861 for British Columbia. According to a reference library of L.K. Armstrong's, no such survey was made in the United States until years later and not published until 1883.

Kootenay, British Columbia

News of coarse gold findings on Wild Horse Creek, running into Kootenay River (Fort Steele district), led Robert L. Dore and party to prospect that region. "Honest Old Bob's" findings, April 1863, was cause for an immense rush from all mining camps. A wagon road from Colville, Wash., to Pend Oreille gave access to a mule trail to Wild Horse Creek. The Hudson's Bay Company opened a trail from Hope, B.C., by way of Similkameen, Rock Creek and Pend Oreille. By August 1864, 5,000 miners traveled into Kootenay, or **Fisherville** camp, many coming thru Colville. $20 million in gold dust and nuggets is reported to have been taken before the mines petered out, and the gold hunters scattered by 1865.

By 1899, that area would be known as Chinatown. It was being worked by improved hydraulic systems with a good yield in silver, gold, lead, copper, and coal. Between 1866-69, Chinese were playing a large part in the economic mining

Chinese followed into Wild Horse Creek and took out their daily needs. There were never any records to say how much was removed. (1988-S.R.)

advances. There was estimated to be 800 Chinese in Montana; and in 1866, of the population of 13,800 in British Columbia, 1,800 were Chinese. Communication was possible due to eight post-offices in British Columbia by December 31, 1863, and telegraph services extended north and south of the border by 1870.

Joseph Moris

So far our prospectors have been, to most of us, just names in a story. We have read briefly the names of some who became famous – and learned a bit about some who became famous – and very

Joe Moris, the little prospector-miner that plays a leading part in this book, especially in Canadian mining history of the great Le Roi Mine. Pictured at age 95, in 1959 (Beaudry)

rich. Finally we learn a bit about one prospector who became one reason for this history.

Joe Moris (Maurice) was the fifth of a family of ten children, and was born in Quebec, Canada, in 1864. For reasons known only to himself he later chose Christmas, instead of a few days earlier, as his birthday. Joe left his home when only 11-years-old and went aboard a boat as a galley boy. His job was peeling potatoes. Even at that early age he showed great consideration by sending his first pay check home to his mother. (We are indebted to his nephew and late niece-in-law, Mr. and Mrs. Albert Beaudry of Spokane, Wash., for what limited early history we have about Moris.)

Joe left the boat on one of the Great Lakes and approached a hotel asking for work of any kind for room and board. For the next year he worked in the kitchen. Choosing a partner, the two headed west with the Canadian Pacific Railroad, whipsawing their way as construction proceeded, but Joe was horrified when three people were killed by Indians and he struck off all alone. He was only 20 years old but he was headed for new adventure. It is not known just how he came about his little donkey,but he headed for Colville, where he had heard of mining at the **Old Dominion.** We learn little more about him until 1885 when we will meet him in Colville.

Alfred Henry Stiles

Another prospector and second reason for this history was first interviewed at Addy, Wash., Alfred Henry Stiles was born at Dallas, Ore., May 17, 1868. His father, George B. Stiles, fresh from the Union Army, had come to that town 14 miles west of the present city of Salem, in 1864. With George's partner Charles McDonald, they opened and ran the town of Dallas's first business – the Dallas Tannery. That business celebrated it's 100th year of existence in 1964. Ground for the town of Dallas was Al Stile's uncle's 600 acre claim.

Al claimed he'd been on his own since he was 12; his dad had died and his mother was running a general store and millinery at Corvallis. After some rambling around, Al went to Pomeroy, Wash., in 1883, and in 1884 bought land. He was staying at the St. George Hotel then, and said he inked hand presses for F.W.D. Mays, publisher of the Washington Independent.

He also attended school as time permitted. At 96 years of age, he still recalled early day teachers Tim Driscoll, Mrs. Frank Morrison, and a Mrs. Rush; and he mentioned a friend, Oren Burt at Pomeroy. He also dug wells around Pomeroy, but nearing 19 years old in 1887, he was itching to try this mining game. Joining up with the Marsden Brothers, Henry, and Jupe; and with Charles St. Clair, they left for Philipsburg, Mont. We will meet him there in 1889.

Butte, Montana's First Comeback

Butte, Montana was founded in 1865 and met near death by 1867. By 1882 it had come alive and was said to be "the richest hill on earth." By 1924 it would be one of the world's greatest mining cities with 9,500 miles of tunnels revealing rich copper, zinc, and silver ore.

In later years, the **Kelley** mine would become one of the "musts" on any tour; as also would **Washoe Smelter,** at Anaconda, where copper was refined, and where the largest smokestack in the world, at 585 feet high, would be pointed out. (That stack, also stated to be 506 feet high, was not built until 1908. It was for Anaconda Company's poisonous fume dispersal from the metal refinery. The plant closed in 1980 and the stack was blasted to the ground in September 1982). The companies would employ over 20,000 but see their mines closed in 1983. Butte has seen great turn-abouts, and still does.

Alfred Henry Stiles at age 96. Stiles lived to be 101 1/2 years old and could still recall when he mined, and hauled ore from the Le Roi in Canada; and worked as a chemist at GERMANIA, and TURK in Stevens County.

An aerial photo taken Sept. 16, 1956 of mining in the Butte region from Berkeley Pit looking northwest; Leonard Mine in right center, Rarus Mine in center, Kelley Mine in left center, High ore compressor plant and central heating plant in upper center. (S-R)

As Butte proceeded to become the great mining center in Montana we will learn how it fits into the overall picture. Silver Bow was the early center of mining bringing about Butte's being the county seat. 45 miles west of Helena, Butte is the local of the Montana School of Mines established in 1893.

The fourth largest State in the Union, Montana was not being arrogant when it became a state in 1889 and chose the motto Oro y Plata, meaning gold and silver. Gold was discovered in what is now Deer Lodge County in 1852 by Francis Finlay, but by 1882 we will find copper taking leadership, especially in the Butte-Anaconda region.

We will learn more about the **Berkeley Pit** that devoured several communities; and about many ethnic groups that make up the population of Butte – a town that refuses to die – when we return to the history of Montana.

Brief return to Swauk Creek

Prospectors were moving about but with less panic when placer gold was found on Swauk Creek in what would later be Kittitas County. It was in the Liberty area which would soon boast 3,000 population. It later became a ghost town. In 1921 the creeks were dredged and gravel and streams turned upside down with little result. In the late 1900's this area would be great for the gold panner out on vacation – with occasional colors being found.

On to Okanogan

It was geographically natural for many to go north into Okanogan's mountain area. The prospectors had followed the Idaho discoveries and many were aiming back into northern Washington when the first lode gold was found at Conconully in what would become Okanogan County. Much of the work carried on in development was by Hiram

F. Smith who later became a legislator. This lode was on the reservation set aside for Chief Moses and his tribe. Pressure was put on and it was not long before miners were moved out by soldiers and were kept out until 1896. Before leaving the area, placer miners along the Boundary Creek in lower British Columbia had taken out $50,000 in dust – but had paid no attention to quartz deposits and iron – except "Tenas George."

Camp Keller – Iconoclast

Whether it be truth, or legend, here is the story of George W. Runnels, better known as "Tenas George," of Camp Keller; once referred to as the J. Pierpoint Morgan, of the Pacific Coast. Tenas George landed on the Pacific Coast in 1860. A sailor from Maine, he took to following the miners, often finding good properties. He was a typical miner in his generosity to those less fortunate, and a great frontiersman. He operated pack trains and trading posts through Indian country and was affectionately respected by the people. When he made his "find" the **Iconoclast,** on Mt. Tolman in 1873, he claimed it was "the greatest mine I have ever staked." Dr. Day, a Walla Walla assayer, checked the ore and found it was good copper.

Some of that respect no doubt came for Tenas George when he is supposed to have fought a duel with knives, and from which a young buck of the San Poil tribe was to carry long scars for life. Runnels had told his wife, a maiden of the San Poil people, that if he lost his life there he wanted to be buried on the claim. He stood his ground with a Winchester for some time when the Chief Moses reservation was thrown open to miners in 1896.

In 1904, a report stated that Runnels had taken out $300 daily from placer claims, and that he staked such noted quartz mines as **Triune** in Okanogan County; **Golden Eagle** at Fairview, B.C.; **Trailor, Tenderfoot, Mountain Lion,** and **Last Chance** in Republic camps, **Iron Mask** in Kootenay, besides Tolman Mountain strikes – all quite enough to make him a man of wealth.

Return to Salmon River Country

Repeat returns to the upper Salmon River County of Idaho were paying off with the discovery of the **General Custer** mine, and the **Unknown** located in 1876. **The Vienna & Sawtooth City** mines at the head of the Salmon in 1878, brought about the town of Vienna. Galena City was founded, and the **Galena** mine was located by Daniel W. Scribner. That would become one of the most famous and long-lasting mines of Wood River. One of the first railroads to reach anywhere was the Utah-Northern narrow gauge line across the southeast border of Idaho with it's first terminal at Franklin in 1874. But – it was not so peaceful for miners crossing Nez Perce land.

In 1877 the Indians of the Nez Perce went on the warpath. Many versions have been written of this war, but a recent report about Mrs. Sarah Pugh, of Grangeville, Ida., recalls when as a four-year-old child she and some 250 others fled to a fort on Mount Idaho to take refuge. Hardship and suffering followed and eventually the war passed into other regions. Sarah Jarrett (Pugh) at one time worked at the way station at Adams Camp southeast of Grangeville, a stop-over for many miners going to and from Florence, and Buffalo Hump gold fields.

Major Indian wars came to an end in the area when Young Chief Joseph is claimed to have pulled quite a coup by attempting to escape and protect the women and children of the tribe. The war ended there when he and some 150 of his people were eventually transferred to the Colville reservation where he died Sept. 21, 1904. He is buried at Nespelem, Washington.

Prichard-North Fork Coeur d'Alene River

It was a long weary hike but Andrew J. Prichard is said to have "hoofed" it over Mullan Pass, arriving on the west side in November, 1878. Moving northward with his party over Evolution Trail, he arrived at what would later be named for him – Prichard Creek. In 1882, he is credited with the first lode location in what would become the fabulous Coeur d'Alene mining district in Idaho.

Prichard is said to have taken his first gold from Prospect Gulch, and to have discovered other placer locations along Prichard and Eagle Creeks. Beaver Creek discoveries soon followed. The town of Murray was founded on Prichard Creek and became the Shoshone County seat in 1885, which it lost to Wallace in 1898, due to the fickle nature of the people.

It was a member of the Prichard party named Gillett, who found placer gold and located a claim on Prichard Creek. The story goes that Prichard left his district, and later returned in 1883, with documents giving him power of attorney to act for Gillett. Prichard then proceeded to stake out several claims to include the **Widow,** at Murray (some 20 miles north of Wallace). His arrogance is said to have led to long litigation between the two friends.

Possibly it was when Prichard was away from the claims that he is said to have sauntered into the office of the Spokane Falls Review, in Spokane. The editor wrote, "This bewhiskered fella by the name of Andrew Prichard tossed a pouch on my desk and said, 'I've discovered gold'." The story was written – The Review sold out fast! Word spread through the west and back as far as St. Paul – and "The

Idaho gold rush was on!"

Patrick Flynn located the first quartz claim on Prichard Creek, the **Paymaster,** near Littlefield, Sept. 21, 1883. All of these early discoveries led to the greatest mining district in the west – the "Silver Belt" and the Coeur d'Alene district of Idaho.

We are told that by 1964, the Coeur d'Alene mining district consisted of an east-west belt running about 30 miles long, and ten miles wide. It became one of the most productive districts in the world in lead, zinc, and silver.

The Union Pacific Railroad came into Idaho in 1880, a real asset to transportation. (It would be this same company that would be instrumental in building Idaho's most widely known recreational area, Sun Valley near Ketchum, in the middle 1900's).

Gold was $20 an ounce and news items stated that placer diggings were yielding $40 to the pan. Suddenly a swarm of tinhorns, gamblers, riff-raff, adventurers, and saloon keepers filled the camps by the thousands. Prichard had done his best to keep his secret and to encourage only followers of his religious clan, but some of his own party revolted against the colony development. Many of those members were farmers from the east and middle west. Wholesale claim jumping was responsible for congress stepping in to establish law and order.

Trails into Idaho Camps

Prospectors and miners rushed back to Idaho from all of the Northwest and Canada. They came over the Mullan Road and over the Thompson Falls route following it's construction and completion in 1884. That trail ran from Thompson Falls, Mont., past Prichard Creek, Ida., thru Murrayville to Eagle. Some came the Belknap Trail from Belknap, Mont., to Eagle. This became the mail route when Eagle had a post-office, and later a telegraph line was built. Other miners traveled the Trout Creek Trail, called "great snow trail" between Belknap and Thompson Falls – and on 35 miles to Eagle, Ida.

Phil Lynch carried the first mail over this trail, on snow shoes. He charged one dollar a letter and claimed he lost money. Letters cost 50 cents from any point later on but took as long as two weeks to arrive, even from Spokane Falls.

The Jackass Trail was popular from the west. Prospectors arrived at Rathdrum, Ida., by rail, then took a stage, a steamer, and finally a saddle

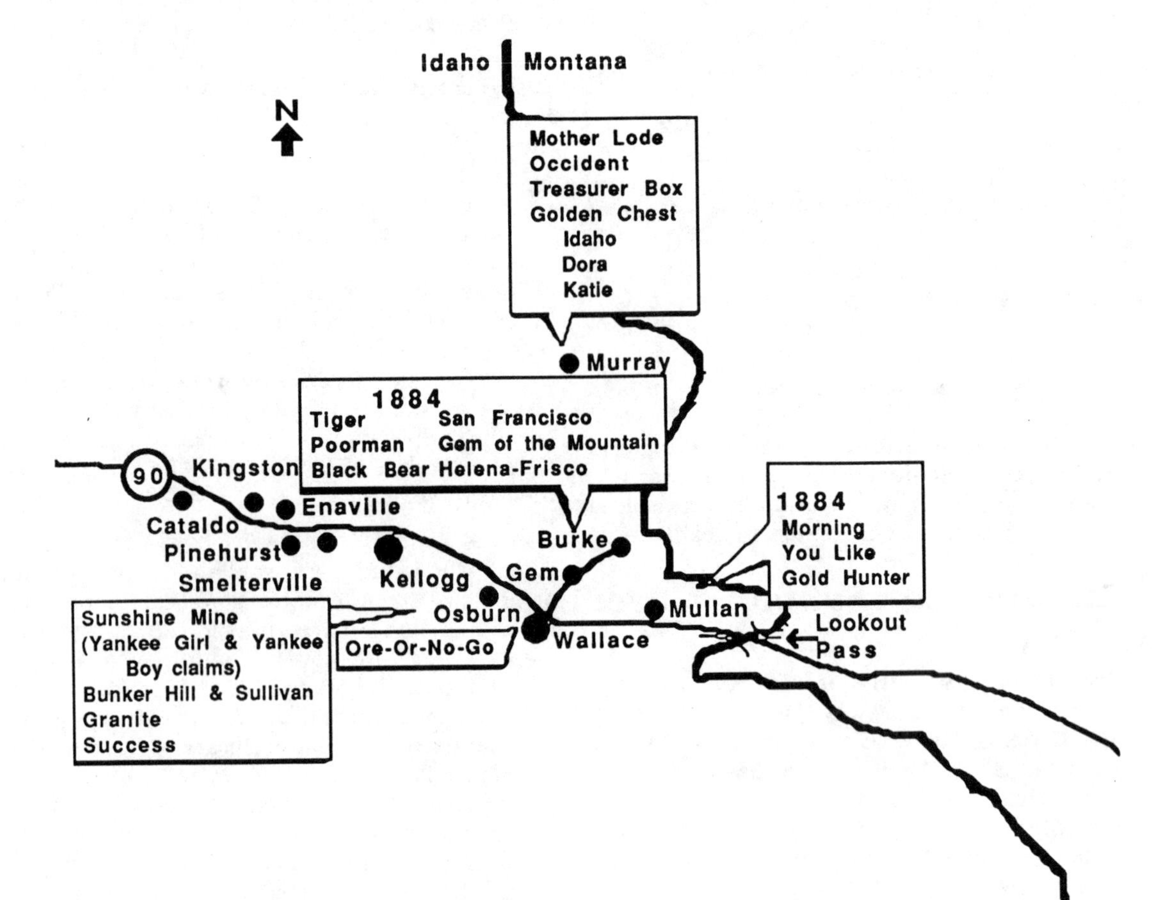

horse to Eagle City, by way of Beaver – arriving at the end of the second day. All were heading to the boom town at Eagle City, and other creeks and gulches in Idaho. And they came – and came – for another chance at "that one big strike."

Freight was said running $40 a ton for hay, for a 38 mile haul. Loads "poled" up the North Fork of the Coeur d'Alene River ran 25 cents per pound, and five cents per pound additional by dog team. The three mile haul from Hummel's Landing to Eagle City cost as much as the rate on a 2,000 mile run from Chicago to Kellogg would cost by 1940. Pack trains soon followed and brought much needed supplies.

Murray, Eagle, Mullan, and Wallace

Following the boom of 1883-84, Eagle City was on a rapid decline. Up until then, 29 new homes had been built with lots selling from $200 to $2,000. Stoves were the highest priced commodity, made of sheet iron – they sold for $30 to $40 apiece. The "jingle of money, and the jangle of music" was heard everywhere.

It was practically the only city in Idaho until Murrayville (later Murray) outstripped Eagle in size. 2,000 locations were recorded in the Coeur d'Alene district; of which only five withstood the test of time – Canyon and Nine Mile Creek, Summit, Beaver, Evolution, and Eagle. The rich placer mines in the panhandle country produced more gold than those of California and Alaska (according to a Spokane Daily Chronicle release Aug. 9, 1976). It did not last longer than two years and the gold boom was over.

South Fork of Coeur d'Alene River

IT FINALLY HAPPENED – after nearly 75 years of following prospectors, the big break came for many – for many, the hardship was over, for others there was work. The first lead-silver discovery was made in 1884 on the south fork of the Coeur d'Alene River. The **Tiger** on Canyon Creek (now known as Burke Canyon) was located by John Carton and Almeda Seymour. It was bonded to John M. Burke, and later sold to S.S. Glidden in 1885 for a reported $35,000. The **Poorman** claim was staked on the same lode and the two became the famous **Tiger-Poorman Mines.** Serious litigation held up progress, but was eventually resolved. The **Poorman** was sold to Marcus Daly (later of Montana fame), Kingsbury, and William A. Clark (from Butte) for a figure of $136,000 in 1887. The two mines were consolidated in 1895.

The **Polaris** was located nearby by Weldon B. Heyburn, August 30, 1884. (Heyburn will be remembered as later becoming a Senator in Idaho –

MORNING Mill, north of Mullan. Photo taken June, 1935 (S-R).

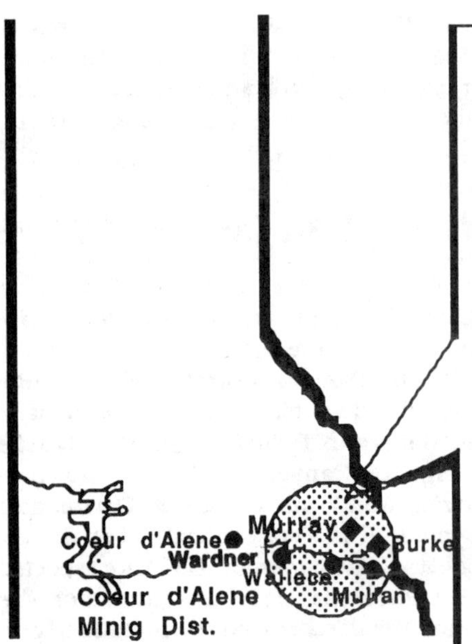

1865

Reports of Coeur d'Alene mineral discoveries, May 27, led to a gold rush there, but no gold was found until after 1880.

1882

A.J. Pritchard located a Coeur d'Alene quartz mine, April 25, but kept his discovery quiet for over a year.

1884

Murray was founded Jan. 22 and soon replaced Eagle City as the center for the Coeur d'Alene gold rush then underway

Discovery of the Tiger Mine near Burke, May 2, followed immediately by the Polaris at Mullan, marks the beginning of lead-silver operations in the Coeur d'Alenes.

G.S. Good & C.C.A. Earle located the Morning Mine - eventually the deepest lead-silver operation anywhere at Mullan, July 2.

Weldon B. Hayburn and his associates located the Polaris mine near Wallace, Aug. 30. The Yankee lode (unprofitable for 45 years until deep workings - the best below sea level - made it the biggest U.S. silver producer) followed Sept. 25.

Noah S. Kellogg discovered the Bunker Hill & Sullivan Mine at Wardner, Sept. 10. 1885

The old Town of Murray, Idaho, the Shoshone County seat in 1885 (Nugent)

and having a town named for him). The **Morning,** two miles north of Mullan, was located by G.S. Good, and C.C.A. Earle. This mine will later be said to be the deepest lead-silver mine to be operated anyplace. It was reported sold to Charles Hussey, for $10,000 in 1887. He also acquired the **You Like,** at Mullan in 1889 and sold in 1891. That same year the **Morning** was said sold again for $400,000. In 1895 Larsen and Greenough reportedly bought the **Morning Mine** and secured a lease on **You Like.** Production in tonnage was reported in 1904 as greater than that at what we will know as the **Bunker Hill** and **Sullivan.** The **Morning Mine** was sold again in 1905.

Colonel W.R. Wallace had a cabin and small store at what he called Placer Center, when he discovered the **Ore-Or-No-Go** (has also been called **Orinoco**). With his partner D.C. McKissick, they built on the site of what would later become the Federal Mining and Smelting Co. The partners offered half of the townsite (later to be known as Wallace); and half interest in the claim if Wallace's friends from Portland would build a lead mill at the location.

The San Francisco; Gem of the Mountains (known as the **Milwaukee** by 1901), and the **Black Bear** were located northeast of Gem, in 1884, on Canyon Creek. By consolidation in 1908, these would comprise the **Helena-Frisco.** The **Gold Hunter,** northeast of Mullan, discovered by Hunter and Moore, was producing under ownership of Gold Hunter Mining and Smelting Co., in 1908. **California Consolidated,** or **Tamarach and Chesapeake,** north of Wallace at the junction of Ninemile Creek, was located in 1884. Production up to 1908 was said to be $200,000. Later new and better ore bodies were found.

Some activity was still going on the south side of Prichard Creek. The **Mother Lode, Occident,** and **Treasure Box** quartz mines on Ophir Mountain, near Murray, worked gold ores in arrastras at a profit for several years following their location in 1885. **Buckey Boy,** in Dream Gulch, was a profitable pocket of free gold for some years producing around $25,000. **Treasure Box** was cleaning up as much as $10,000 in a week in 1887 and continued use of arrestras until 1892. By this method, 30 tons of **Mother Lode** ore in 1890 averaged $92 a ton. A five-stamp mill was erected along with one at the nearby **Dandy Mine.** The **Yosemite** first gained notice in 1893, and had produced some $500,000 by 1908. This was the only mine in the vicinity with development still in progress according to a 1904 report.

Miners, prospectors, claim jumpers and even gunfighters poured into these new camps. Among them was; Wyatt Earp, and two younger brothers who are said to have been "well healed" when they came into the Murray-Eagle district in 1884. Within six months their substantial stake was gone and they left for greener pastures (via a dastardly plan we will learn as they reach Colville in April 1885).

Sunshine Mining Co.

THE GREAT SLEEPER! – in Idaho mining is the **Yankee Lode** situated in the west end of the silver belt, or so-called dry belt of the Coeur d'Alene region. This claim was not to become the great profit maker until 1929 (45 years after it's location), and when it was developed as the **Sunshine Mine.** Dry smelting ore is low in base metal content.

Ranked since 1930 as the largest single lode producer of silver in the world (this may be disputed in later years), well over $25 million had been paid in dividends.

On Sept. 25, 1884, True and Dennis Blake had filed claims called the **Yankee Load,** then added additional claims and patented them in 1909 as the **Yankee Group.** The Blakes had worked secretly for over 30 years as regards to output. Rumor has between several thousand, and a half a million dollars in ore was shipped to East Helena Smelter. Valuation and development of claims was reported in 1909 as $18,140. Lessors in 1919, Sidney L. Shonts, and later – Dan Price, developed the claims. Said "not up to expectations!" they were abandoned.

The Sunshine Mining Co. was organized in Spokane in 1918, but was inactive until 1920. In 1922 and 1923, by purchase of **Yankee Girl** and **Yankee Boy** claims, these two groups compose the original part of Sunshine Mining Co. properties. After incorporation, a mill was constructed by Charley Lynch. An assay in the early 1920's showed a concentrating product of 150 ounces of silver per ton, based on 80 percent recovery – as compared to a 1,000 ounce showing with better than 98 percent recovery in 1940. Heavy indebtedness, reportedly about $200,000 was said paid off from "the Chinatown level," and after the summer of 1927 the company looked to a good future. C.S. Samuels had been retained as manager and continued to enlarge the plant until 1935 when a 50 ton mill replaced it.

Wardner

But we move ahead too fast – When Murray captured the County Seat from Pierce in 1885, the population was said to be 1,500. Prichard and Beaver Creek was still the center of activity, but there was much moving about. The **Golden Chest,** in Reeder Gulch east of Murray, was the largest gold mine in the district and one of the first successful operations. 25 men were working and ore was treated in a ten-stamp mill. The mine was reported sold in 1900, for $25,000 and the **Idaho, Dora,** and **Katie** claims were acquired by the The Golden Chest Company.

The new town of Wardner was earlier called Kentucky, in honor of the Kentucky owners of the **Golden Chest.** The town was growing and the

SUNSHINE MINE, photo taken Oct. 31, 1945 (S-R).

name was changed to Wardner before a post office was permitted, around 1885.

Still another great producer, the **Granite**, or **Success Mine** was located on the southeast side of East Fork Ninemile Creek. Staked in July 1885, it was sold about two years later for $18,500. By 1908, it had produced some $800,000 gross and paid around $200,000 dividends. First known ore bodies faded about 1890, but new finds around 1904 saw important zinc production. 1906 reported production at the **Granite**, then operated by Success Mining Co., was 10,289 short tons carrying 22,840 fine ounces of silver; 685,800 pounds lead, and 1,929,000 pounds of zinc. And prospectors still poured in as they heard of these great finds.

Bunker Hill and Sullivan

The really "great one" was the **Bunker Hill and Sullivan!** The lead-silver lodes in Milo Gulch have been called "the biggest of 'em all." That mine played the largest part of the economy of North Idaho and the Inland Empire. It was of the longest duration – and its demise was the saddest!

There are so many legends and amplified "truths" as to the discovery of **Bunker Hill and Sullivan Mines** on Sept. 10, 1885, (even the date) that a "middle of the road" report will have to suffice. (New plaque says Sept. 4, 1885.)

The story is that three men around 60 years old, who had observed the mining activity and were desirous of improving their financial status, formed a partnership. Retired medical officer Dr. John T. Cooper; building contractor Origen O. Peck, and their aspiring prospector whom they grubstaked – Noah S. Kellogg, formed the first business arrangement. The grubstake of bacon, beans, sugar, coffee, and flour are said to have cost $22.85, of which only two dollars and forty cents was claimed ever paid. For three dollars extra they obtained a burro for Kellogg. A long time returning from his first prospectors trip, it was said that Kellogg showed his samples to some very disgruntled partners who (in his opinion) called the whole thing off. He had even lost his long-eared friend!

Those samples were shown to Jim Wardner, and a second grubstake was arranged with Jacob "Dutch Jake" Goetz; Harry Baer, and Phil O'Rourke participating. O'Rourke had recognized it as galena, (the most important source of lead). The more romantic story goes that they found the burro braying it's heart out, it's feet squarely planted and it's vocal cords pointed at what was staked as **Bunker Hill** (named for the Revolutionary Battle).

It is assumed the other men followed them because by the following night Goetz, O'Rourke,

GOLDEN CHEST, the largest gold mine in the district. Photo taken about 1895 – (Al Nugent).

Capitalists who owned GOLDEN CHEST MINE, and built the Louisville Hotel at Murray, watch their men at the favorite sport. (probably around 1895) – (Al Nugent.)

and Cornelius "Con" Sullivan are claimed to have found the vein – and called it **Sullivan.** (Some historians say it was for the prospector – others say in honor of the fighter).

From these great finds – credited to a "jackass", **Bunker Hill and Sullivan** has claimed to be the largest single lode producer in lead and zinc in the United States.

Word got back to Cooper and Peck, the first grubstakers, and claims on the find were taken to court. Held in Murray, the jury found in favor of the new partnership, but Judge Norman Buck reversed the decision to hold that the original owners of the "jackass" were entitled to half interest in **Bunker Hill,** and quarter interest in **Sullivan.**

Settlement to locaters and grubstakers has been reported at; $200,000 to Goetz and Baer; $75,000 to Cooper and Peck; over $200,000 to O'Rourke; $300,000 to Kellogg; $75,000 to Alex Monk (a partner of O'Rourke's), and $75,000 to Sullivan. There were also substantial attorney fees. These were large payments indeed – but very small in contrast to a reported $60 million paid in dividends up to 1946.

In the immediate days following the great **Bunker Hill and Sullivan** discovery, little peace was enjoyed by residents in and around Murray where the forgotten burro brayed day and night. Stories vary as to it's actual demise. Some say dynamite was tied to his legs and he was driven from town, thus guaranteeing the first good nights sleep for the citizenry. Others say Kellogg brought the burro to Forest Grove, Ore., where he had it tenderly pastured and cared for – for $50. Still others claim that following the burro's death south of Kellogg, it was mounted and kept for some time at Spokane's famous Coeur d'Alene Hotel.

The town of Milo took on the name Kellogg about 1893. A news release May 1, 1964 states the Kellogg Senior Chamber of Commerce recently ornamented their city entrance honoring the "jackass" (instead of the "burro" that has been credited with finding these ore bodies). Cards expanding on the theme, and said to be in use in the area, read;

Wallace Mayor Frank Morbeck (front) inspects new plaque honoring discovery of BUNKER HILL MINE above Kellogg. Wallace McGregor (rear), Al Nugent,(center) (S-R).

Sign outside Kellogg honoring a specie of animal that supposedly helped locate the BUNKER HILL & SULLIVAN MINES.

"You are now near Kellogg, the town which was discovered by a jackass – and which is inhabited by it's descendants." The designation of specie might be challenged by animals lovers, but there is nothing "mulish" about a 96 years production since that first plaintive bray resounded through the canyon.

Taking over business management of the properties in 1886, James F. Wardner eventually located a Portland buyer for the prospects; Simeon G. Reed (who earlier had mining interest in the Owyhee district.) Historians disagree on the purchase price said to be $1,500,000 cash and shares in the company, for five claims. Others have reported it to be an offer of $65,000 "take it or leave it." On July 29, 1887, Reed paid $731,765 for mine, concentrates and ore on hand and equipment. The Bunker Hill and Sullivan Mining and Concentrating Co., was incorporated under the law of Oregon, July 1887.

Governor Hauser (Montana) in a long drawn out transaction, is credited with financial assistance towards the building of the first mill with a 100 ton daily capacity. It began operation July 20, 1886. It was built near the portal of the Reed Tunnel, later located in Milo Gulch at Wardner. The first lead-silver ore shipment from the Coeur d'Alene district was made from there to Cataldo landing by wagon. Rail service was established March 1887, the **Bunker Hill** enjoying continuous yearly production until it's closure in 1981. To July 1963, there had been extracted 26 and a half million tons of ore; over two million tons of lead, 87 million ounces of silver, an 321 tons of zinc according to one news report.

Spokane Falls was fast becoming the "treasure house of a vast mineral empire." It was to this budding metropolis that Goetz and Baer turned to invest some of their money. It went into a tent covered beer hall which extended from the present vicinity of Post and Riverside, towards the river. Later on, they built the Coeur d'Alene Hotel and the

BUNKER HILL & SULLIVAN Smelter – (photo taken 5/18/1925) (S-R).

Lead by the carload from BUNKER HILL. (Photo taken 1/26/1936) (S-R).

old Frankfurt Block.

The Bunker Hill Company began with a capital of three million and increased assets brought about through expansion showed $30 million in 1963. It was said the nation's number two lead mining company; a refiner of nearly 24 percent of United States produced special high grade zinc, and 20 percent of the primary refined lead.

Many great men of America aided in pulling the company through its usual ups and downs to include: Simeon Reed, who became the first president of **Bunker Hill**, and later became founder of Reed College in Portland; John Hays Hammond and General N.H. Harris, both early presidents of the Company; San Francisco banking men W.H. and George Crocker; the American Harvester Company's Silas H. McCormick; James H. Houghteling of Peabody, Houghteling and Company, Chicago; and Frederick W. Bradley, and Stanly A. Easton, both presidents of the company. Daniel Guggenheim of American Smelting and Refining, was another who aided materially.

BUNKER HILL & SULLIVAN MILL (photo Jan. 28, 1936) (S-R).

Ready to roll - and mine - BUNKER HILL & SULLIVAN (Photo Jan. 21, 1938) (S-R).

Victor M. Clement became manager of the new corporation and in 1891 a new concentrator with daily capacity of over 400 tons, was constructed. Open cut mining commenced. By 1940, there were 60 miles of underground workings, and in 1964, there were over 100 miles of tunnels, crosscuts and drifts. The uppermost of these is 3,600 feet above sea level, the lowest is 1,350 feet below sea level. At 1964 prices, the value of refined metals would show well over three quarters of a billion dollars.

The Coeur d'Alene mining district is one of only eight districts in the world where more than one billion dollars worth of ore has been removed from the earth. ($740,880,186 reported by 1932)

Into Colville Country's Embry Camp

Wait a minute!! - You may have to set your time clock to 1885, and holding, because things were happening in other mining districts. While the great rush was into the Coeur d'Alene district - and why not?, great mines were developing - but some prospectors were not overlooking other districts. Slowly, but steadily, they were spreading into the Colville Country (in Washington). First discoveries

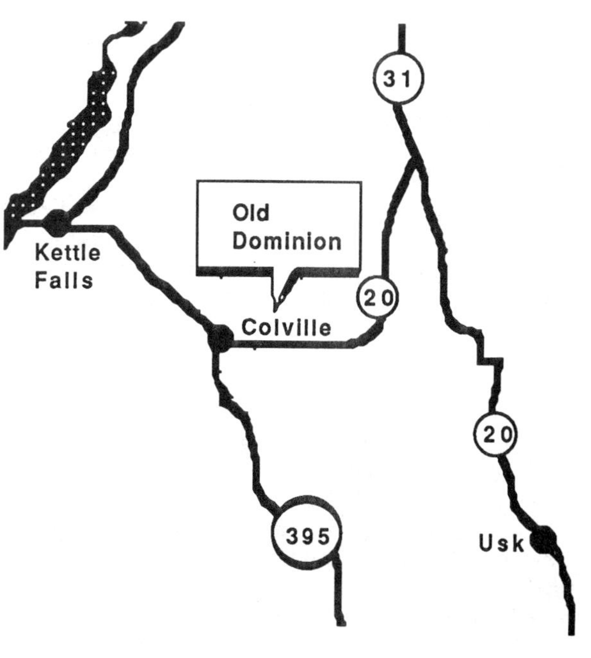

were made in Embry Camp, the first lode mining in northeastern Washington (two miles east of the present town of Chewelah, Washington). Prospectors had been sent out by John Squire of Spokane. Then suddenly – the BIG strike at Colville.

Old Dominion Mine

The most factual story of the location of the **Old Dominion** mine was told in the Spokesman-Review on April 8, and April 15, 1928 by E.E. Alexander, and added to by our little prospector Joseph Moris. (Mr. Alexander was one of the locators).

Pat Kearney was clerk for the Great Eastern Dry Goods Emporium, in Spokane. He knew nothing of mining but grubstaked Albert Benoist, a clerk at the Sprague Hotel (then near the corner of Post Street), and sent along his brother William Kearney to watch out for his interests. Benoist's wife was said to have some money so she supplied horses for the prospect trip. George B. McAuley was added as a third member.

Alexander, also of Spokane and only 23 years old, had earlier struck out by himself and was at Embry Camp. In the camp were some reputedly desperate characters suspected of being claim jumpers. He was said to have named them as Tom Fuller, a fat gambler; Theodore Erickson, Bob Stokesbury, Charles Sands, and Wyatt Earp. (We recall that Earp and his brother were known to have lost a considerable stake in the Murray-Eagle, Idaho camps and apparently they joined the group Alexander named).

Alexander was running out of food and, not owning a rifle, he borrowed one from a member of that notorious gang. Being a good shot, and with plenty of game available, he downed three deer. One he kept for himself; one went to the owner of the rifle, and the third was given to a hungry prospect party just arriving – Kearney, Benoist, and McAuley.

Benoist and McAuley had a falling out and McAuley left. Thankful for the meat, Benoist asked Alexander to join his group. The next day the three men headed north towards Addy, an area where many Indians lived at the time. On the way they talked to a half-breed who showed them samples of ore supposedly taken from a mountain east of Colville. With great anticipation the men hurried on towards Colville.

Locating the mountain, their search was to no avail and they became discouraged. Two reasonable stories have related to what would become a great find. One states that after several days of what Kearney thought was a wild-goose chase, he was ready to leave. He is said to have angrily swung his pick against a rock – and to their amazement the men stared at broken fragments of stone full of silver ore.

Alexander recalls it differently. He said that Benoist leaned wearily from his saddle and half-heartedly flicked a boulder with his pick, revealing what looked like rich ore. At almost the same moment Alexander said he discovered a lime dike. A location notice was posted April 12, 1885, signed by William Kearney and Albert Benoist, and added Pat Kearney as grubstaker. Alexander complained, and his name was added for one-eighth share.

Colville, still in it's infancy and unincorporated, was the center of business. Kearney and Benoist practically raced the seven miles to file their claim. They must have assays made and buy supplies so Alexander was left to guard their claim.

In the night, Alexander claimed, the same hoodlums he had seen at Embry Camp, jumped the claim and were firmly entrenched with rifles by morning. When Kearney and Benoist returned and saw the situation they returned to Colville for help. It is believed the "greasing of certain palms" had the desired effect and the claim jumpers soon departed.

Wyatt Earp's participation may never be confirmed but became a possibility. A news report from Tucson, Ariz., Feb. 11, 1984, states the finding of a handwritten Earp letter dated Nov. 9, 1880, addressed to Pima County Sheriff Charles A. Shibell. It reads, "I have the honor herewith to resign the office of deputy sheriff of Pima County. Respectfully, Wyatt Earp." It goes on to say that Earp moved to Tombstone where his brother Virgil was town marshal — and where on Oct. 26, 1881 with his brother Morgan, and friend John "Doc" Holiday participated in the gunfight at the OK Corral (actually in an alley). The Earps and Holliday were charged, but acquitted of murder.

In 1881, Virgil was said shot in the back and crippled, and three months later Morgan was killed. Wyatt Earp is said to have shot Frank Stilwell, the man believed to have killed his brother. Whether there were other brothers is not known, but the "Earp brothers" are said to have been in the Murray, Ida., camp in 1884 where they lost their funds. Maybe broke, and still on the move, Wyatt Earp may have joined the claim jumpers at Embry Camp, and moved on to the **Old Dominion** claim. Wyatt Earp, said to have been a buffalo hunter, deputy marshal, stagecoach driver, saloonkeeper, and rancher in various areas of the country, died in Los Angeles, California in 1929. (He registered to vote at Eagle, 1884)

The name **Old Dominion** was given the mine partly because Benoist was from that part of the country to include Virginia; and partly in honor of the Dominion of Canada where the Kearney's were born. From naming to immediate development, this was one of the first lode silver mines in eastern Washington; and without question, became one of the state's most famous and productive.

A certain amount of prospecting had preceded this great find, and satirically, Colville's own great booster G.B. Ide, had thought samples he had seen were coal. Even as the locators tried to work the claims on short funds, they were told by a Major O'Neal that a Mr. Sherwood of Colville, had earlier made the find and had "intended" to file his claim.

Concentrator at OLD DOMINION MINE (S-R book 1899).

O'Neal that a Mr. Sherwood of Colville, had earlier made the find and had "intended" to file his claim.

Minor problems soon settled, John "Big Jack" Hanley was hired to help. Along with Alexander, Kearney, Benoist and John McIntosh, they went to work with pick and shovel. In this crude manner they are said to have taken 80 sacks of ore in a day. Reports say the ore was taken down the mountain by horseback, and others say by stone boat to where it would be loaded on wagon and trailer. It was sent to Spokane for shipment to a smelter in San Francisco.

An act of 1878 had authorized coinage of silver dollars effective Feb. 28. Silver was in great demand for that, and for other uses and was at one dollar and one cent an ounce by 1885. An early day Spokane news item tells of eleven large covered freight wagons arriving from the **Old Dominion Mine.** Resacked and shipped to San Francisco, the ore was reported to assay all the way from $300 to $9,000 a ton. The 31,996 pounds of ore was said to yield a net return to the owners of $3,569.

When last heard of heading for Colville, our prospector Joe Moris tells in the 1928 Spokesman-Review story of the part he played in some of that early history. He was a rank amateur at prospecting and mining but a hardy little fellow. Along with his little burro, he arrived on the mountain one day and asked about work. Ore had been extracted to make quite a large chamber of high value material, and a narrow vein twelve to eighteen inches of rich silver ore had been revealed. As "Big Jack" talked to the staunch little man, he saw the answer to one of his problems. Joe was to wriggle, and crawl through into the chamber of ore careful not to break off a lot of waste material, and retrieve the good. In this manner, Hanley was very pleased when Moris got out one or two sacks a day.

This was termed a "high-grader" (which by the way – under different circumstances than these, is not otherwise complimentary). Joe would continue to work until he made enough for a grubstake for himself, then he would take out over the hills on a prospect trip of his own. He continued this for several years until 1890 when he made his own big strike in Rossland, B.C.

In the meantime he made a trip to Ruby and LeClerc Creeks where he prospected Pend Oreille County; and when he needed to renew his resources he returned to the **Old Dominion.** Mrs. Beaudry recalls "Uncle Joe" saying that if he ever needed work all he had to do was go back, and he did again in 1888. Early in 1889 he joined Oliver Bordeaux, a Colville miner. The two headed up Trail Creek to do assessment work. We will meet him again the the Rossland district.

Colville Came into Being

Colville was platted in 1883, but not incorporated until 1889. Sometime in 1880 a soldier, John U. Hofstetter (often called the Father of Colville), had put up a building said to be the first one – at the corner of what would be Birch and Main Street. It became a brewery and in 1882 Daniel Pohle was a brewery worker there. John Layson, first superintendent at **Old Dominion Mine** also had some connection.

Records are very limited, but an apartment upstairs must have housed John B. Slater who was publisher of the Stevens County Miner in 1886 and had been asleep in the building in 1885 when residents felt an earthquake, something new to

Exact location has always been questioned; used by the Pohle Brothers operating the Colville Brewery. (Burned Jan. 18, 1969 the building came about mid-1890's as a three-story malt house. (St. City Hist. Society).

those early residents.

When streets were being laid out, a large brick building was constructed on the west side corner, slightly adjacent to a wooden structure (then owned by Pasquale S. Paradis). The brick building housed the Stevens County Land Bank; a commercial bank, and the Colville Valley National Bank which was sold in 1938. Herbert I. Minzel was a director of the third bank. The first bank was just completed when the railroad of D.C. Corbin's Spokane Falls and Northern arrived in Colville in 1889. That bank may have been the one started by J.H. Young. The door to one of the early bank vaults now serves as entrance to a recreation room belonging to the writer's son, Scott Battien.

The brick building cost $4,500, and a schoolhouse just going up cost $3,800 – both made with bricks from Colville's own brickyard. Some 800 people resided in this new city and were willing to pay $500-$2,500 for business lots. Erin Chandler of North Dakota was just designing the Hotel Colville across from the two-story bank, the cost would be nearly $30,000. Main street had board sidewalks and soon had electric lights, a water system, and telephones.

Lumberman "Peggy" Robbins bought the first brick building and the old wooden structure, added to it and the building housed many local businesses. This writer's husband, Roy Battien, owned the building from 1967 to 1977 for his Insurance business. He sold it on contract and from then on the building was only briefly occupied, ending in a dilapidated condition. Sadly, Colville's first brick building was totally torn down in 1986 to make room for – yet another bank use; this time a drive-in and parking for the new bank across the street where the old Hotel Colville had burned in 1982.

Jimmie Durkin

Due in part to the **Old Dominion,** and other mines following closely, Colville was growing. In a story in the Spokesman-Review, Nov. 22, 1969, Jay J. Kalez relates Jimmie Durkin's arrival in Colville in 1886. This English born was then only 27 years old.

In Durkin's words, "When I got off the daily horse stage running between Spokane and Colville that spring day in 1886, I immediately started parading the streets looking for a job. It didn't take long for me, an Easterner stranger, and a midget compared to most of the brawny street crowd, to see that a job was going to be hard to find."

"I counted nine saloons in Colville and while all of them seemed to be doing plenty of business, competition was obviously plentiful in that field."

He recalled that the **Old Dominion Mine** was going full blast, and that supplies were going north into the Canadian mining fields. He pictured Colville as having the makings of a frontier boom town and in any mining town liquor was treated as a necessary commodity.

"That first day in Colville, I learned that freight moving between Spokane and Colville carried an average wagon train rate of two cents a pound. From my past experience I quickly appraised the fact that if a person purchased liquor wholesale and shipped in barrel lots, he could reduce the freight rate between Spokane and Colville by one-half. He could also undersell his nine competitors. I had $2,500 in my pockets when I hit Colville and I decided to risk it all on my transportation idea."

"During the next few years, my idea netted me some $65,000. For a young fellow that was something."

So too, was the panic of 1893 – the year the silver coin act was repealed. Hard hit but having saved enough cash, he went to Spokane in 1897 to make a further name for himself. No one would ever forget his admonition – "If your children need shoes, don't buy booze."

Within the 1887 period, things were going well at the **Old Dominion.** Assays were showing something like 300 ounces silver, and 35 percent lead, with some ore running as high as 1,000 ounces of silver. The Proutys had built three miles of road to connect with one from Colville, and shipments were being made to Spokane at a cost of $25 to $40 a ton. (Earlier rates by horse, and even by wagon, had been quoted as high as $100 a ton.)

Developing Old Dominion

Silver price was still dropping and was down to 90 cents an ounce in 1887. Alexander chose to sell his one-eighth interest to Major Sidney D. Walters for a reported $12,000 cash. He had also received $5,960 in dividends and considered in all, this was pretty fair returns for sharing one of the three deer he had killed at Embry Camp.

Benoist was said to have sold out to the Kearneys for $25,000. Under the Kearney management, the properties were reported to have paid between $400,000 and $600,000. The mine was only down 75 feet at that time.

Benoist must have become interested in the **Young America Mine,** also discovered in 1885, because his estate was filed after his death Jan. 8, 1887 to be; 30,000 shares of **Young America Mine** stock worth $15,000, four gallons Port wine nine dollars; five barrels Schlitz beer $90, and other belongings.

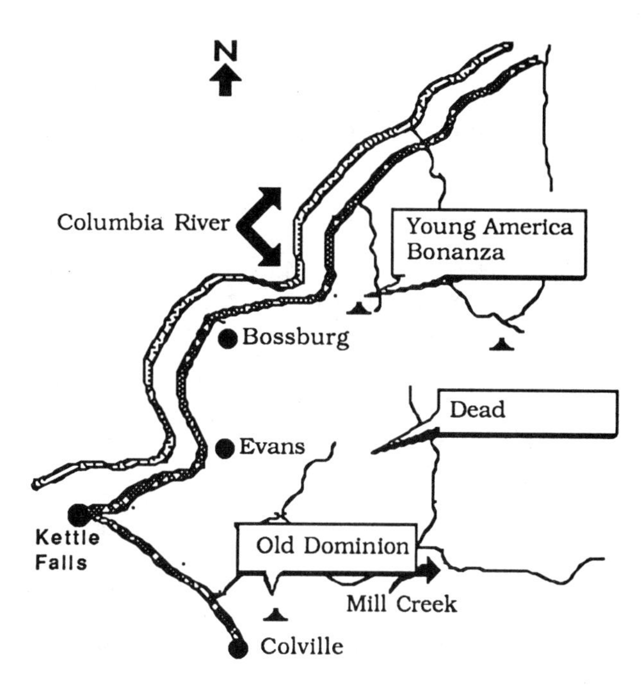

Miners had spilled into these new camps near Colville, Wash., from the Coeur d'Alene camps. Four paying mines were within 15 miles of the town; the **Old Dominion, Young America, Bonanza,** and **Dead Medicine,** reported the news Dec. 3, 1892.

Major B.P. Moore, with backing from New York capitalists had built a smelter on the north hill of the town, in 1887. The Mutual Smelting and Mining Co., is said to have cost $25,000 with $100,000 in shares, out at one dollar each. Ore at first was being hauled to the smelter from the **Old Dominion,** and from Chewelah's Embry Camp, and when there was ore there were 40-50 men working at the smelter.

The obituary for this smelter was written in the Colville Chamber of Commerce notes, Jan. 29, 1892. It said, "the smelter is certain to become the property of some individual or company who will run it for all it's worth after making a few changes in construction." The entry goes on to say, "the cause of the smelter lying idle was it's unfortunate ownership by men lacking in capital, and devoid of business ability."

Unfortunately, at the time the need was indicated, the smelter passed into the hands of receiver John B. Slater and was disposed of by him, Dec. 10, 1892. The Colville Republican reported the smelter sold at sheriff's sale for $9,777 and interest, to satisfy the judgement, and cost of the suit $543.85.

At the **Old Dominion,** things were happening. E.J. Brickel, then president of the Traders National Bank, in Spokane, came into the picture. With Simon Oppenheimer, they produced enough capital to build a 50 ton concentrator at the Kearney held mine. This was in 1889 when the average grade of ore was showing 350 ounces of silver to the ton.

In 1890, the Sherman Silver Act was approved on July 14, restoring silver as legal tender. The price of silver went back to one dollar and twenty-one cents an ounce.

The Spokane mining association – The Northwest Mining Association, was formed in 1885. One time president of that group, Graham B. Dennis, is said to have purchased the **Old Dominion** from the Brickel estate in 1891, for a stated $160,000. It consisted of the **Old Dominion** and **Ella claims** and 51 other claims. Ferdinand and Clarence I. Peck, of Chicago, and A.H. Revell, were financially interested. Dennis is said to have operated the mine until 1899.

Some of the earliest claims to be kept in mind include the **Ella, Phoenix, Paris Bell, Ophir,** and **Reeves.**

News articles of 1892 tell that G.B. Ide filed a complaint against Joseph Kaufman asking one fifth interest in the **Paris Bell** and **Phoenix** claims for Kaufman's failure to pay his share of assessment work for 1891, and cost of court. (That article said some of the best ore in the district was found at **Paris Bell.**) F.H. Bell and Superintendent Morrison, Chicago capitalists, were also mentioned in connection with those two claims.

On Dec. 3, 1892, the Colville Republican announced the **Old Dominion** is about to be sold to Chicago capitalists G.W. and F.H. Peck. (Is this a misprint and G.W. and the earlier mentioned Clarence I, the same man?) A deed was filed with a consideration of $500,000 stated. The article went on to say that amount had been taken from the mine to date "though with very bungling management."

On Dec. 24, another article stated G.B. Dennis and Cyrus Bradley recently engineered a sale for the mine for $265,000. (Probably closer to actual price.)

Some of those reported to be working at the mine were: Frank Hammond, George Lodge, John Morrison, Tim Keefe, and E.F. Larios. Other "Old Dominion boys" were: Teddie Farrow, and J.H. Fratey.

In that same issue of the Stevens County Standard, "Big Jack Hanley", Superintendent of the Old Dominion Mining and Concentrator Co., (as it was then called), received from Fred W. Peck, (who was then vice-president of the World's Fair Association), a 92nd Columbia half-dollar. Peck said, "it's out of the first bag; out of the first keg of the first consignment and very valuable." The connection between the silver-lead mine, and the new coinage, was inferred.

Claims were being transferred to E. Dempsie consisting of the **Ophir, Ella, Alexander,** and **Reeves,** all to benefit the purchase of the mine by the new company.

In 1893 Hanley was reported turning out a carload of concentrates per day with production of about five times that amount in crude ore, with net profit running about $20,000 a month. Working 60 men, his payroll was said to be $7,000 per month and he intended to double the force within the next 60 days. The company will soon spend $100,000 for new machinery (according to the report), and the **Old Dominion** is declared the most lively camp in the state.

It was stated to be the biggest and richest mine in the State, and said to have produced over two million dollars worth of ore in the eight years since discovery in 1885, and a new strike in the upper workings was reported. Said to be a six foot vein of pure lead concentrates, it ran as high as $500 in silver a ton.

Professor Clarence King was a mining engineer for the **Old Dominion** for several years and in the days of inaccurate reporting, his1909 report said the production from 1892-1909 was 3,061 tons, valued at $602,536. His estimate of ore ran 450 ounces silver and 33 percent lead to the ton.

A still unsolved mystery of the **Old Dominion** is reported to have happened in 1893. A huge boulder of solid ore was found northeast of the **Ella** claim. Hanley said the great boulder was blasted into fragments aggregating 130 tons of material – all of which was sold and realized over $100 to the ton. Documents show the search for the origin has continued down through 1926, with varied opinions written. Others since have searched – where did the boulder come from? To date the source has never been pinpointed. The find was made when silver was below 85 1/2 cents an ounce.

In 1895, a notice to contractors stated a bid wanting ten men to work and wanted an extension of 1,500 feet driven in tunnel one. The order was placed by Dennis, president of the mine.

Having lost track of William and Patrick Kearney in the several transfers of ownership, we find grubstaker Patrick took the wealth he had amassed and went into other fields. His last days were reported spent in comparative poverty while he attempted to develop claims at Chesaw, and where he died Nov. 6, 1921.

The late Earl Wheeler, of Colville, recalled living at the Old Dominion camp, and of his father, John L. Wheeler working there. Earl was about six years old in 1896, and he remembered "Big Jack" Hanley, and a little Jack Hanley whom he played with. He recalled the camp being quite a settlement then of 50-60 people. He said his father later went on to mine at **Bonanza,** and still later at Fernie, B.C. As a man, Earl worked for the **Magnesite** at Chewelah.

The mine made a slow-down due to insufficient high grade ore, and fluctuating prices. Early 1897, bar silver was 59 cents an ounce; copper was $11.12 1/2 and lead at three dollars and fifty cents. When prices slowly returned, Hanley prepared to resume work.

One of the strangest events reported April 30, 1897, was another huge boulder found by Jack Rounds while plowing the Vivian place northeast of the St. Regis Mission, some ten miles west of the **Old Dominion.** This one weighed 150 pounds but was of like material to the one found near the **Ella** claim in 1893. It assayed 70 ounces silver, and 65 percent lead. Where did it come from – was it thrown from a distance by volcanic action – or what?

And – even stranger, back in 1892, a huge boulder was discovered at Kaslo, B.C., way to the north of Colville. When that $20,000 lead ore boulder was discovered, and short on communication, little was thought of what would become a sequence.

By 1897, the Old Dominion Mining and Concentrating Co., had incorporated and capitalized for one million dollars (some reports said five million). Dennis had spent around $500,000 including the purchase price. 5,000 feet of development (some reports say 12,000 feet), had been done with the best production within 75 feet of the surface – when the Rockefeller Syndicate offered one million for the mine.

The offer was made through A.L. Dickinson, who was then chief engineer for **Bunker Hill** and **Sullivan,** acting for Rockefeller. A statement later attributed to Dennis was – "I was anxious to accept the offer but my Chicago associates thought if it was such a good thing for Mr. Rockefeller, it ought to be good for them, and the possible sale fell through."

Tragically, the ore body died out within 15 more feet of work! Three thousand more feet of work has since been done but the "main" ore body has never been found again. The mine was shut down in late 1897-early 1898.

A later report said the mine had a mill with daily capacity of 60 tons, a power house said to have contained an 80 horse-power boiler and 80 horse-power engine, a six drill air compressor, and a blower. A Spokane paper reported in 1898, that regular shipments were grossing $16,000 monthly and employing 75 men. (This had to have been an earlier production report.)

President McKinley was inaugurated into office and possibly the change launched more interest in mining. The auditor was kept busy with filing a jump from 43 locations in March, to 179 notices and articles of incorporation in May.

Prices were going up by late 1896. A.A. Barnett & Company, advertised pack saddles rigged – at three dollars; gold pans all steel 16 inch – 50 cents, poll and drifting picks – 75 cents, and giant caps-triplex, per hundred – 75 cents. You could also buy miners reflectors, dutch ovens, folding stoves, tents and miners clothing.

During the time, or shortly after the renowned Professor Clarence King was geologist-engineer for Dennis, at the **Old Dominion,** King was credited with breaking up a California swindle evolving around a said to be "salted" diamond mine promotion. How wide-spread this scam has worked is not known, but diamonds were placed in old workings – the mines then sold on the basis of even greater value than they earlier – or actually contained.

In 1915, W.H. Linney bought the **Old Dominion** properties at a tax sale and organized the Dominion Silver-Lead Mining Co., then started to reopen the mine. Local news reports of Aug. 20, 1915, and in October, stated that 14 men were working and that $40,000 worth of ore was taken under this new management.

Reports told of the boarding house being re-opened, of men clearing the lower levels, and re-timbering the mine. One report said the main tunnel was opened to 1,200 feet.

A post office was established at the foot of the mountain on the east side in 1915. Mrs. Willie Ginder was named postmistress. Many prospects were being opened on the mountain to include the Exley brothers opening a ledge of carbonates, two feet in thickness in a 50-foot tunnel.

The Linney operation was not to be smooth. The Peck brothers brought suit in Nov., 1915, alleging they had not been properly notified of a tax sale. Litigation continued until sometime in 1918, when a stock settlement was made.

Even with silver down to 50 cents in 1920, activity was still great and the Chamber of Commerce had requested a mineral survey be made. News was often distorted. One report was that R.E. Lee, as receiver for the **Old Dominion,** had deeded the property to Ferdinand Peck and that development work was going on. (This is difficult to follow because the Dominion Silver-Lead Mining Co., continued to own the property until 1952.) One report did say two carloads of ore had been shipped to the Northport Smelter.

Frequent articles tell of old-timers connected with the mine. June 12, 1920, H.D. Kay, of Spokane visited Colville and told of hauling ore from the mine 35 years prior. He said he operated a freight line of eight mule teams from Spokane to the Little Dalles. Another article as late as May 10, 1929, recalls J.T. and C.R. McDonald having leased the mine; and it stated over one and a half million dollars had been taken over the years. One, on July 5, tells how those men narrowly escaped death when flood waters backed up, hit the main tunnel and took out timbers as rocks rushed out the entrance.

Death was too often the result of this reach for glory. On June 20, 1920, the news was asking who Ed Sheldon and George Shannon were, where they had come from and were there any living relatives to be notified; they had been killed in the mine June 11.

A news story July 24, 1920, may have solved that mystery. A letter received by the Stevens County sheriff from a Mrs. Mildred Denny in St. Louis, Mo.,

The crew at the OLD DOMINION MINE about 1923, identity not possible, but known to work then; James Slentz, Paddy Rion (or Ryan), John Wheeler, George Shannon, Ed Sheldon, H.D. Kay, J.T. and C.R. McDonald, Jack Hanley, John Layson, and Clarence Reed, besides ones earlier employed in the late 1890's. (photo Stevens County Historical Society)

said that George Shannon is probably her father. She had enclosed a picture of her father which bore a great likeness. No information had been received concerning Ed Sheldon.

In 1923, news tells of a body of black sulphide ore seven feet wide having been found. Work was doubled and sometime between 1924 and 1930, a rich ore strike was made. An ore body of about 40 tons was knocked down, according to Archie Wilson, foreman for W.H. Linney at that period. (Archie Wilson, a prominent Colville mining man died July 25, 1946.)

In 1924, an option was given the Hercules Group, including August Paulsen, (Paulsen Building, Spokane) Jerome and Harry Day (Idaho mining men), F.M. Rothrock, and Louis Hutton, of Spokane. The option was later dropped, but a carload of handpicked rock was said to bring $17,000.

In 1925 a new strike was recorded and it was said $59,000 ore value had been shipped since September. Some work continued and June 10, 1926, a ten and a half foot vein of ore was reported by drilling. This was the biggest strike in years.

Prominent mining men, I.M. Hunley, and Earl Gibbs acquired the **Old Dominion** property in 1952 under their Bonanza Lead Corporation. Money was spent on exploration and opening a tunnel. Some strikes were made and a machine shed built when Hunley sold out to Gibbs. Production was said very good but the figures not revealed were often included in production from that company's **Bonanza** Mine.

A Spokane paper did estimate that between two million and seven million dollars had been taken over the life of the **Old Dominion** (a very safe bet). In the early 60's **Bunker Hill** had a lease on the mine, along with other leases in Stevens County, which they later dropped. Following the death of her husband, Mrs. Gibbs did retain ownership.

Bonanza

Two other great mines were discovered in "Colville Country" in 1885. The **Bonanza** about five miles southeast of a new camp called Bossburg, was to become one of the state's heaviest producers of lead. The **Young America,** just outside that camp, was a second good producer.

Norman D. Lindsey, chief engineer, said in 1949 that early development of the **Bonanza** was erratic

Northport Smelter about 1915.

due to failure to understand the ore deposits. He described them as a blanket, or "manto" formation – roughly dome shaped and varying from paper thin to 12 feet or more. He also said that transportation difficulties, and "pessimistic advice" along with low lead prices, all tended to thwart early-day miners.

A news report Dec. 31, 1892 names a Mr. Gross, of Spokane as one of the several owners of **Bonanza.** Ed Patburg was said to be a general foreman, and Fred Burbridge was general superintendent at that time. Twenty-five men were said working.

Development did eventually include a 700 foot inclined shaft, several levels, and a total of 15,000 feet of underground work. In the same month, 1892, E.W. Kirkpatrick was said to have told the assay from a fissure ran 75 percent lead and $14 silver. Prices were not good, but lead was running $4.45 in 1890, and down to $2.98 in 1896. Silver was $1.01 in 1885, and bottomed out in 1893.

On Nov. 2, 1893, the local news reported the **Bonanza** Company was in a bad plight. 40 men had been thrown out of work and when it closed that week – all of them were behind one to two months in wages . . . and the Company had recently placed a $30,000 mortgage on the mine.

In 1897-98 Le Roi Mining and Smelting Company of British Columbia, began operating a copper smelter at Northport, some 30 miles north of Colville and near Bossburg. It later became known as the Northport Smelting and Refining Co., Ltd. In 1916 Day Brothers (Coeur d'Alene mining men) purchased the smelter to process ore from their mines in Idaho. That smelter was beneficial to some of the new mines in Colville Country, especially the Bossburg region.

"Big Jack" Hanley, recently with the **Old Dominion Mine,** took over **Bonanza** on Aug. 30, 1897. The mine had fallen into the hands of the miners who held liens on it and they leased and bonded it to Hanley in 1898. Production was said to be 1000 tons a month.

F.H. Davey was said to have a lease and bond on the mine in 1914. Later, Oscar Nordquist, from the Coeur d'Alene area, was reported having put up a mill but having financial difficulties. Two other Stevens County men operated the mine in the next number of years to include G. Vervaeke of Evans (reported both in 1924 and 1943, according to two different manuals). The other was Russell Parker and his son-in-law William "Bill" Parrot, formerly of Colville, who operated the mine around 1942. With continued development, 23 percent lead and five ounces of silver per ton was stated produced as per a 1942 report, and $40,000 in lead-silver ore was said shipped prior to a 1944 report.

Suddenly the **Bonanza** boomed –

In late 1944-45 miners I.M. Hunley and Earl Gibbs purchased the holdings and incorporated under Bonanza Lead Corporation, which included other operations. Under them, the **Bonanza** did live up to it's reputation as the state's heaviest producer. They were said to have grossed around three million dollars. Development included the $200,000 mill at Palmer Siding.

Plans included a 300 ton mill and a camp for 150 men according to a July 1, 1949 report. Housing facilities for mine and mill workers will be built on the flat near the mine, said Earl Gibbs.

At least a 75-ton mill, 14 miles from the mine and on the highway north of Colville, did come about. Employment at the mine and mill were said to number 65. The mine itself is in a treeless hill partly surrounded by farmland and pastureland. Gibbs and Hunley were said to have purchased the farms and increased their holdings to 1900 acres – the **Bonanza Mine** sitting in the center. A June 17, 1949 report said 50 men are employed, and a fleet of trucks haul ore from the **Bonanza** to the mill at Palmer Siding.

"It remained for a new operator, Earl Gibbs, who resolutely followed his own ideas despite opinions of ultra-conservative technicians, to put the mine on a paying basis," said one engineer.

On Nov. 13, 1950, Anaconda Copper Co., from Montana, purchased **Bonanza** from Gibbs and Hunley for a two million dollar option, paying to them $700,000 before turning it back to them in March, 1953. They reported a similar figure of three million dollar production before returning the property. Hunley sold out to Gibbs and this property also remained in Mrs. Gibbs ownership at her husband's death. (**Bonanza** closed about 1952-53.).

It is of interest to note that Mr. Gibbs had operated, under lease, during 1939 and to 1943, two famous mines of the 1885 discovery period. These were the gold producers, **Mother Lode,** and the **Golden Chest** near Murray, Ida. Mrs. Gibbs recalls the renewed gold excitement in that district during the four years until the "war effort" closed down this type of non-essential operation in late 1942. That was the Government War Production Board Limitation Order L-208. He had also operated the **Monarch** and the **Terrible Edith** in Idaho, during this period – and he had done exploration at what would one day become important, the **United Copper,** out of Chewelah.

Young America

The **Young America Mine** is at an elevation just under 2000 feet and can be seen from Bossburg. Another of the 1885 discoveries, it was a good producer with values in lead, zinc, silver, gold and copper. It did contain zinc which until a means of handling was perfected, zinc was more frequently termed "black jack", "rosin jack", or just "poison."

The property was developed including over 1200 feet of work, six tunnels, and two buildings. Whether worker, or owner, an 1890 news story tells of J.R. Hall and Vic Adams resuming work and will be shipping ore.

A tragedy occurred on Dec. 11, 1892 when Olivia Anderson disappeared. She had operated the rooming house and her brother had worked at the **Young America.** Her body was found March 25, 1893 on the river bank below Bossburg. According to a jury, she was declared a suicide.

Sometime in 1897 litigation between J. Sullivan et al, and Thomas Burgoyne was being settled. None-the-less, the mine was producing and in 1898 they reported having shipped $80,000 worth of high grade lead-silver ore.

That same year, our prospector-miner Al Stiles was in the area and was contracted to drive a tunnel at a cost of $1,000. Colonel Haymer was his boss, and Ed Neiholm and Eric Wickland worked with Stiles.

A 1914 State mining bulletin tells that William Vessey, of Spokane, was listed as owner. An overall figure of output had been over $100,000 shipped. Values were said 25-30 percent silver and 25-30 percent lead to the ton.

Perry Leighton, of Colville, was once listed as owner. From 1897 production reports, The Young America and Cliff Consolidated Mining Company were said owners until 1941 when the property was owned by Cuprite Mining Company, and leased to a Mr. Riggles, of Bossburg, at a figure of $25,000.

In 1951, Mr. Hunley, and Mr. Gibbs (of the **Old Dominion** and the **Bonanza**) were said owners, and operators of the **Young America.** They were credited with putting in a flotation mill in 1952, and having 1500 feet of tunnels.

Young America closed around 1954.

Colvile, Fort Colvile and Colville

Prospectors were working the area north into Canada; east in the Pend Oreille area of Washington, and west into the Okanogans, but many claims had to be filed in Colville. To again clarify – the first Fort Colvile (named for Andrew Colvile) was built by the Hudson's Bay Company at Kettle Falls in 1825. In 1859 the military required a fort built at Harney's Depot (actually Fort Colville) three miles northeast of the present city of Colville – thus causing some confusion. Across the creek, the pioneer town of Pinkney City (named for Major Pinkney Lugenbeel) became the actual county seat of what is now eastern Washington, north of the Snake River in Idaho, western Montana, and even for awhile a small part of Wyoming. (A note of interest – Major Lugenbeel had earlier been responsible for the building of Fort Boise.)

The military fort, and the town of Pinkney City existed from 1859-1882 when residents abandoned both and followed John U. Hofstetter to his new location called Colville. Among the first buildings moved was when he absconded with the jail and the records and moved them to Colville. B.J. Yantis was the first postmaster back in 1858, but

YOUNG AMERICA MINE one-fourth mile north of Bossburg had hopes of a revival in 1950. (S-C–Oct. 21, 1950)

that post office was also moved to Main Street Colville in 1882. Colville became the county seat of Territorial Stevens County in 1863, and was county seat in 1889 for Stevens County at Statehood – having fought off a number of attempted moves.

Because of very early placer scrapings in the Columbia and Pend Oreille Rivers, the Chinese had followed and worked the gravel bars from as early as 1860-1877. Mrs. Wade Bailey of Colville had pictures of the Chinese at Marcus flats, and Colby Gleason recalled them being active at what became known as China Bend around 1889.

Tye and Nim were well-known in the Daisy area and a news article in July, 1937 told of the death of Wong Fook Tye, pioneer of the Daisy vicinity, and Stevens County's only surviving Chinese resident. He had been one of the few survivors of the hundreds of Chinese engaged in placer mining along these rivers in early days. He had come to Daisy from China in 1877 at the age of 16.

Edith Bauer – (nee Pelissier) was born at the Columbia River town of Marcus in 1889. She recalled her family telling that placer mining was in full swing along the Columbia, with Marcus as the outfitting point. The steamer "Forty-nine" with Captain Pingston at the helm, was going from Marcus to the head of Kootenai Lake in Canada. It took three weeks to make a round trip and placer gold and bear pelts were the medium of exchange.

Pelissiers had come to Colville in 1881 and located on what was later Our Lady of the Valley Convent. (This St. Regis Mission farm had been started in 1870.) Walla Walla was the nearest railroad point and supplies had to be hauled by teams. Edd and his brother were in the freight business and Edith recalls it took six weeks round trip with a weight cost of eight to fifteen cents a pound.

Edd Pelissier helped build the first road from Marcus to the Canadian border in 1888. He also ran a ferry at Marcus where it crosses the Columbia. All of these means of transportation were important to the development of the **Young America Mine,** and others of the area. With gold being the major interest then, he invested heavily at one time in mines at Republic which failed to produce and left him badly off financially. He died in 1917.

Eliza Gendron, Mrs. Bauer's grandmother, was the daughter of Patrick Morrow, a full-blooded Irishman who was murdered while placer mining on the Pend Oreille River. His empty gold poke found beside his body, was mute evidence of what happened.

Other Stevens County pioneers remember steamboat and ferry use. Joe Class was once a steamboat captain. Henry L. Lillianthal helped organize a company to build a landing at Northport for a steam ferry boat. According to 1895-1907 reports, F.M. Davis was conducting a ferry on the Columbia at Northport. Lew Davenport told of his dad freighting from Spokane to the China Bend area and knowing about the ferry that ran across to deliver material to Bossburg. He also remembered the steamboat up the Columbia.

The grubstaker was equally important to the discoveries of the mines. R.E. Lee had come to Colville in 1890. (He was a fourth cousin to General R.E. Lee.) From 1900-1925 Mr. Lee operated a hardware store in Colville and grubstaked many a prospector. With John Hope, they also prospected near Williams Lake with little return. R.E. Lee also invested heavily in marble properties.

Mrs. Lee (Belle Lindley Anderson) was born in Indiana. Her mother's people came over on the Mayflower. Possibly that heredity encouraged her to back her husband's philanthropy. Their son Clare Lee was a later day postmaster who died in 1963, the last living member of the Lee family.

As settlers moved in, many brought by mining fever, Colville town developed. Streets were going from muddy, to seeing the board sidewalks being built when the new Hotel Colville was built. For some time, blankets covered the doorways for privacy until shipments could be made of proper doors. By 1899, Colville was said to have 800 people.

Northport

The Northport Smelter was a factor in the development of mines in that area. Most employees of the smelter were Finn or Austrian. Kelly and Wisley Jackson were named locaters for Stevens County – they met the train when it came with men to work, and wanting to settle here. Kelly located the Finns on Deep Creek; Wisley located the Austrians on Onion Creek. (Wisley was killed in 1913 in a western gunfight – one of the last known in this area. Kelly died in 1941 when he was Stevens County Sheriff.)

Dr. Wells and his nurse, Mrs.Travis, ran the Smelter hospital – needed mainly because there were few, if any safety features then when several men were killed on the job. He would ride horseback to deliver babies on Onion Creek. Those mothers fortunate enough to get to the hospital paid only $40 for room expenses and delivery of their baby.

Dead Medicine or Silver Trail

Even with so many interesting people recalling those early days, we must return to the fourth of those earliest Colville area mines, the **Dead Medicine.**

About 18 miles north of Colville, at the head of Bruce Creek, was a mountain from which the Indians took ore to make their lead bullets. They showed this to John Keough – little knowing they were actually making silver bullets. In late 1886, Keough staked his claim calling it **Dead Medicine,** the term used by the Indians. (Verifying past history is also a letter from Keough's granddaughter Alta – Mrs. J.H. Knapp Sr., of Colville, and written Oct. 26, 1967.)

By 1887, over $100,000 in ore had been taken from the silver-lead deposits. By 1891 a five-stamp mill produced 129 oz. silver and 20 percent lead. By late 1891-92, the Keough partners John, Albert and Harry reported shipping 100 tons with a product of 70 percent silver and 15 percent lead. A later report showed 5000 tons of milling ore had been shipped with seven percent silver and one and one-half percent lead. A Colville paper reported March 4, 1893 that 30 men were working and 30 tons a day were going through the concentrator with a value of about $70 a ton.

On September 30, 1892, an extension called **Prince Mine** was made on the east side of **Dead Medicine.** That claim listed Pat Kearney (of the **Old Dominion Mine**), Richard Downs, H. Hoskins and a Mr. Raulett as Company men. Little more has been heard of that claim.

Dates are too distorted from the many reports researched, but the Keough's must have sold the mine to a combine of Spokane investors around late 1892 or early 1893, consisting of I.N. Peyton, Col. W.M. Ridpath, George Forester and George Turner. (All men who returned their wealth earned in mines, to invest in Spokane City in later years.) One report said that I.N. Peyton was in the area in 1891, had married Victoria Ide from Colville, and built a cabin on the near grounds to the mine where they stayed a couple of winters. That cabin still stands.

The Keough's are said to have received $15,000 for the mine at a time silver had bottomed out and depression was setting in. The mine had two shafts, and a news report said that a 25 ton concentrator was on it's way to **Dead Medicine** from Chicago. It added that the ore on the dump would run it, and it would meet the requirements of a 50 ton concentrator. S.D. Allen was said working at the mine.

From news reports, the four Spokane men mined until they earned $50,000 and each took their share and invested it in the **Le Roi Mine** in Rossland (one of the greatest British Columbia discoveries of which we will soon read about). A 1918 report said $62,000 in silver-lead ore had been shipped from **Dead Medicine.** Peyton was said to then have shipped the mill to Rossland. Ownership of the mine went to Sig Dilsheimer (at that time part owner of Colville's Barman's Store); and to Charles Cox. The mine stood idle until 1910.

That year, J. Richard Brown, and associates, of Spokane, bought the mine and reorganized it in 1913 under the name Silver Trail Mining Co. Mary M. Markham was Secretary-Treasurer and John Goetsche was a director. 1912 reports were of 42,560 pounds shipped, 54.5 oz. was silver, 14.1 percent lead, nine percent zinc (they were not paid for the zinc in earlier mining). That company worked the mine intermittently until 1930 when they leased it to a Mr. Wiltermood. A September 25 report had said a 1,000 foot tunnel had given the Silver Trail new life. A wood burning steam boiler operated the plant. Wiltermood made some shipments of ore before the mine was closed down.

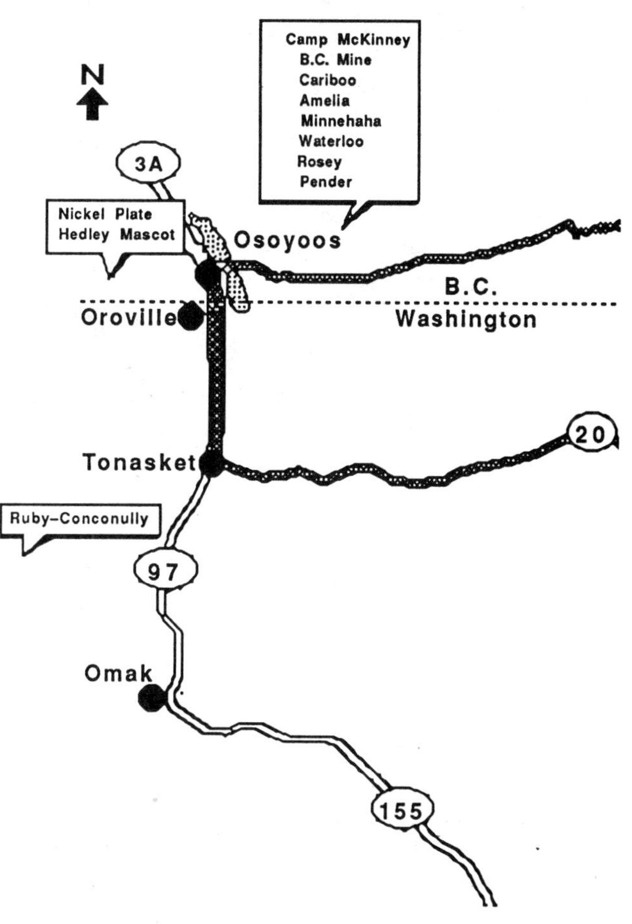

McKinney, Nickel Plate, in British Columbia

The search had spread into the Osoyoos district of the Yale mining division of British Columbia and one of the first locations was made at Camp McKinney in April 1887. The **Cariboo,** and **Amelia** claims were located by McKinney, Rice, Burnham, and LaFevre. These formed a valuable portion of the Cariboo Mining and Milling Co. Ltd., of Toronto. The camp was some 44 miles west from Greenwood, and 56 miles east of Penticton. Other claims of note were the **Minnehaha, Waterloo, Rosey,** and **Pender.** Values ran from a small amount to several thousands of dollars to the ton. Following the boom, and the purchase by the Toronto firm, the **Cariboo** settled to be the leading mine. By 1898 it was said to have paid $236,965 in

Hole in the mountain on the Blenz farm west of Colville – drilled by gunshot by young members of the Hall family, where some SILVER KING money went.

dividends, much to Spokane and Colville investors.

Several Keough homes in Colville were built from Camp McKinney profits, and the Monaghan side of the Graham family "were the only ones who ever did make any money in mining," according to the late Colville newspaper publisher, Charles Graham – and that was at McKinney.

One of the Monaghan family was James Monaghan, whose statute graces a corner street in Spokane.

As early as 1860, placer miners went through Similkameen and on into Hedley, B.C. (northwest of Oroville), but it was not until 1894 that the first claim was staked. In 1897, large ore bodies were found on Nickel Plate Mountain but lay dormant until 1904. From that year until 1955, some four million tons of ore were mined from the **Nickel Plate.** A sign at Hedley states that $47,000,000 in gold was taken. Dividends were reported to amount to more than $7,500,000.

The **Hedley Mascot,** in the same vicinity, operated for 13 years from 1936 to 1949. High on the bluffs up Twenty-Mile Creek, this mine was connected with the mill in Hedley by an aerial tram. Companies have since worked in the district with an eye towards retrieving gold deposits that earlier appeared unprofitable to mine.

Silver King

The **Silver King,** on Toad Mountain, was the first major discovery south of Nelson, B.C. and was in 1886. The town itself was first known as Salisbury, then changed to Stanley. In 1897 it was officially called Nelson after the Lt. Governor of British Columbia, and had some 3000 population. Mining and steamboating has increased that figure to some 10,000 or more people today.

Years of re-telling the finding of a great mine adds much fancy, and loses some fact. Based on one report, Marcus merchant and ferryman Wil-

The CARIBOO MILL at Camp Mckinney, British Columbia. (S-R-1899).

Osner Hall, one of the locators of SILVER KING MINE (photo taken later, when he was near 50 years old - courtesy of family members).

Winslow "Bill" Hall in older years. One of the major locators of the SILVER KING MINE near Nelson, B.C. - (S-R 1899).

liam V. Brown is said to have grubstaked a party to prospect in West Kootenay. Those in the party are remembered by some to have been Winslow "Bill" Hall, and five sons; his brother Osner Hall and son Oscar; three men of the Oakes family; Charles Brown; William White; and two Indians Dauney Williams, and Narcisse Downing.

After many days of discouragement the men were short on rations; the horses, short on pasture, took out to forage for themselves. Bill Hall's adopted son William White, along with others, were sent to the north side of what would be called Toad Mountain to retrieve the mounts. Stretched out for a brief rest, William White idly tossed a rock at a Pine squirrel and hit a rock jut that fell off to reveal peacock copper ore. Filling his pockets with samples, White and his group returned to camp but after "potshotting" at grouse on the way back he barely had samples enough to show. What was left caused excitement enough to hasten the prospect party back to Colville for assays and supplies.

Arriving in Colville, one assayer named Lindsley discouraged the group, but another man, Jake Cobaugh, found high silver and copper in the samples. Cobaugh joined the venture and in the spring of 1887 the party took off up-river to return to their "bonanza". Word had leaked and noting they were followed, they disembarked at China Bend. Loading for pack they backtracked across to what would be the general direction towards the Little Pend Oreille Lakes; then on north through Idaho and into the Kootenay district. One story says they camped when a member of the party gave out, and the creek "Give Out Creek" still carries that name. Halls called the location Mineral Mountain, but a group that followed changed it to Toad Mountain when startled by a wart toad.

Other locations were soon staked near the Hall claims; **Dandy Group,** by J.R. Cooke, and a man named Fox. This one was grubstaked by A.H. Kelly, of Colville, later of Nelson. Charlie Townsend and Ben Thompson located **Iroquois,** and **Jim Crow.** Still other Toad Mountain mines of importance by 1899 were the **Goldendale Group, Starlight Group, Princess,** and **Great Eastern Group.**

The first ore from **Silver King** was packed out by Joe Wilson, and netted some $8,000. By 1889-90 shipments some 700 miles to Montana were grossing $75,000, market cost said to be $57. Hand picked and cobbing, (breaking out the less valuable rock) continued as the dump grew waiting for concentration machinery. The company is said to

First shipment out of SILVER KING was made by Joe Wilson and netted over $8,000. They refused $300,000 for the claim and lost it by being "jumped". A settlement was made. (S-R 1899)

Shipments were made from Hall Mines Smelter (S-R-1899).

have refused $300,000 for their holdings and about that time the claims were jumped. An Englishman named Atkins offered $25,000 to fight the Hall party case in return for half interest leaving the other parties each holding 1/26th interest apiece. A nephew named Daly reportedly received Atkins stock shortly after that man's death.

By 1890 word had spread of the fabulous **Silver King** ore; prospectors and promoters congregated around Kootenay Lake and that area. Demands for transportation and services increased. John Stevens, (late of Panama Canal fame) pushed through the Great Northern railroad, and D.C. Corbin hurried the Spokane Falls and Northern. Steamboats replaced canoes as the Kootenay Steam Navigation Co. connected with Canadian Pacific subsidiary using the "Columbia" and the Kootenay." Shipments were then made to San Francisco and other U.S. smelters. In the meantime a 13-mile wagon road was built in switchback fashion down the mountain from **Hall Mines** to Nelson.

Much capital to bolster the fluctuation at **Silver King** came from Colville. Heavy investors included Harry Young, Jimmie Durkin; Sig Dilsheimer, and David Barman (two owners of Barman's Store. That store closed in September 1987 after 100 years in business.) Some went in deeper than others as John McDonald convinced them that pure silver flakes as big as half dollars was being yielded. Stock went to ten dollars with many hoping to sell at $30. Much of what was profit did return to the economy of Colville and the upper Columbia.

One writer of Colville history claims that Jimmie Durkin gained, in respect to stock owned in the **Silver King,** by his knowing that Jake Cobaugh liked his drinks – and he inveigled Cobaugh into helping him obtain that stock. It was that stock that was added to the amount of money Durkin had when he left his liquor business in Colville and moved to Spokane.

Variations in transactions have been reported but one tells of business agent McDonald obtaining three million dollars capital and forming a Scotch company **Kootenay Bonanza,** under management of J.E. Crossdale. Depressed business and lowered silver price was occurring at that time due to monetary debate over gold or silver standard. In 1893 Crossdale went to England for additional capital. The holdings were reportedly sold for over a million dollars and the company was well known as Hall Mines Co. Ltd.

Under the new management, 640 tons of ore was sent to various smelters in 1894. Suspension of shipment by local transportation came as a smelter was built at Nelson in 1895-96; and a bucket line was erected from Toad Mountain to the ore bins at the smelter. The stamp mill and one-stack water jacket smelter with 100 ton capacity (later 300 ton); the four and a half mile tramway dropping 5000 feet between the mine and smelter, all were activated Jan. 2, 1896.

Business was great by Christmas 1896 and the company made the gift of a 300 man boarding house, and saloon. Just by coincidence, within a week following the issuance of the saloon license the smelter had to close temporarily for lack of ore, liquor had caused a holiday. Rough cabins were replaced by single and multiple homes; a store, blacksmith shop, mine buildings, and offices were built – but there was no plumbing. There was an electric light line for use at the mine and in the homes, and food was packed by team, or sent up via the tram bucket. Families moved there trying to make a home for their men.

In 1897, production fell and smelter operation was interrupted. In 1900, the company reorganized the Hall Mining and Smelting Co., Ltd. Official records of 1902 give total copper production for the entire province of British Columbia

18s/- PAID. See Endorsement. 19s/- PAID. See Endorsement. FULLY PAID. See Endorsement.

CERTIFICATE No. 01069

ORDINARY SHARES.

REGISTER No. 381

THE HALL MINING & SMELTING COMPANY, LIMITED.

INCORPORATED UNDER THE COMPANIES ACTS, 1862 TO 1898.

BRITISH COLUMBIA.

CAPITAL £325,000,

DIVIDED INTO 325,000 ORDINARY SHARES OF £1 EACH.

No. OF SHARES	NUMBERED FROM	TO
100	221,361	221,460

This is to Certify that Sigmund Dilsheimer, Esq of Colville, Stevens County, Washington U.S.A is the registered holder of One hundred Ordinary Shares, numbered as per margin, in the above-named Company, subject to the Memorandum and Articles of Association thereof, and that upon each of the said Shares there has been paid up a sum of Seventeen Shillings. GIVEN under the Common Seal of the Company this 11th day of October One thousand nine hundred

Entd. J.W.W.

SECRETARY pro tem.

DIRECTORS.

NOTE.—NO TRANSFER OF THE SHARES COMPRISED IN THIS CERTIFICATE WILL BE REGISTERED UNTIL THIS CERTIFICATE HAS BEEN DELIVERED AT THE COMPANY'S OFFICE.

Stock certificate from The Hall Mining & Smelting Company, the SILVER KING at Nelson, B.C. representing 100 shares purchased October 11, 1900 and fully paid for by June 28, 1901, issued to Sigmund Dilsheimer (then of Colville) and belonging today to, and courtesy of, nephew Robert Strauss of Colville.

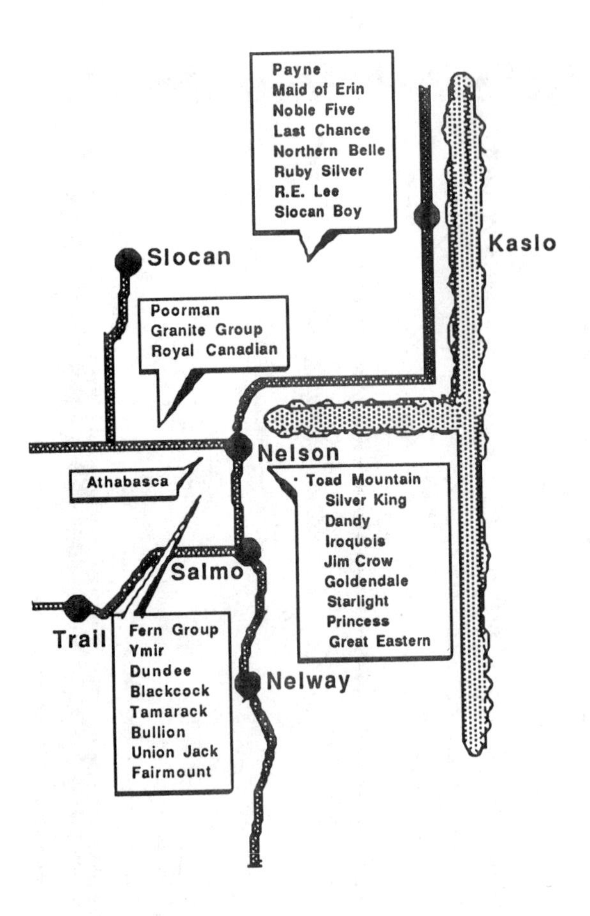

as 29,636,050 pounds. Of this, 491,144 pounds was from the Nelson district, the remainder from the Rossland mines. **Silver King** had contributed a bare 22,661 pounds. By 1907, tonnage had fallen some 90 percent and silver content had dropped over 50 percent in a ten-year period.

In 1905 the Canadian Consolidated Mines, Limited (later The Consolidated Mining and Smelting Company of Canada, Limited), was formed. Walter Aldridge was made managing director. In the company was S.G. Blaylock who had risen from the assay office to become chairman of "Cominco." At that time the Bank of Montreal appealed for a man to wind up the Hall Mines smelter, and Blaylock was the ideal choice. The Hall smelter had drawn custom work from 125 mines but with the "blowing in" of the Boundary smelters much of this was lost. Blaylock went about cutting costs and it was evident that operation might soon have been profitable, but the Hall Company was in receivership. In 1910 operations were closed and the Hall Company sold to a group of Pacific Coast capitalists. In 1912 Cominco took over **Silver King** on an exploratory basis.

Profits from **Silver King** were spent in many and varied ways, many legendary. Jake Cobaugh had sold his 1/26th share for $25,000. The White family, and some of the Halls spent to improve living conditions on the Colville reservation. One remark was that **Silver King's** greatest claim to fame was through social, economic, and industrial

Not the rope ferry of earlier day but a cable ferry of the 1925's carrying many a car across the Columbia River to the old town of Marcus. (photo from Jim Gendron of Spokane)

development of it's respective border areas. In 1895 Charlie Hall bought the old rope ferryboat at Marcus and converted it to a steel cable ferry. In 1898 Will Miller bought this ferry and enlarged it to handle the traffic heading for Republic. That same year Albert Hall built and installed a ferryboat across the Kettle River south of Laurier. This was later replaced by a toll bridge long since taken over by the County but still known as the Hall bridge. Charlie Hall built a hotel at Marcus, and another member of the family set up a saloon. William Oakes established telephone connections between Spokane and Rossland; and later farmed extensively along with bringing in new fruit growing methods to the upper Columbia River country. He became a renowned horticulturist and had a beautiful farm home overlooking Colville valley.

On the lighter and more hear-say side is that some members of the original party stood on the Columbia River bank and skipped $20 gold pieces across – or lit their cigars with various denominations of paper money. One was said to have tossed a good gold watch in the river because it's time did not coincide with the time of a passing train. A red Wilton velvet carpet unrolled from the house to the edge of the river was said to cushion the steps of his daughter's feet as she left from one man's home following her wedding party – that had come and gone by boat. Gambling and high living is said to have used some Hall profits. A portion of the money was spent wisely to develop good citizens but others wasted theirs and ended in dire need.

Two things remain to carry on the Hall name – a creek near Toad Mountain, and another on the Colville Indian reservation. A third and lesser known is at the present B.J. Blenz ranch west of Colville, on Highway 395. A rock juts from the mountain face where one can see a hole which was cut by some of the younger Halls as they fired round after round of ammunition in an attempt to drill through the rock.

Coeur d'Alene, Idaho Mines

Many of the prospectors and miners would remain to strike again in the British Columbia fields but in Idaho the mines of the CDA region were ready to ship ore. In 1887 the narrow-gauge railway reached Burke and that year something over 50,000 ton of lead-silver ore was mined. Principal producers were the **Tiger, Bunker Hill** and **Sullivan, Tyler** and **Stemwinder, Last Chance, Sierra Nevada, Poorman,** and **Granite.** The **Mammoth** and **Standard** veins were only good prospects at the time. Some hydraulic operations were around Murray and Dream Gulch. A hydraulic elevator operated for some time in the bed of Prichard Creek, just below Murray; and some dredging was done near Delta until 1904. The trend was shifting as early as 1888 to the lead-silver mines of the south Fork of Coeur d'Alene.

The steamship GEORGIA OAKS operated on the Coeur d'Alene Lake as far as the Mission up to the late 1890's. (Courtesy Al Nugent, via Bill Zimmerman).

Transportation in Coeur d'Alene region

Transportation rapidly became a major factor in mine development in Idaho for movement of ore in the vast Coeur d'Alene region. In 1886, the Coeur d'Alene Navigation and Transportation Co. operated the 45 miles from Coeur d'Alene to the Mission. This was sold to the Coeur d'Alene Railway and Navigation Co., March 1887, who operated until 1888 when it was leased to Northern Pacific, and who operated until January 1897. The company ran steamers including the famous Georgia Oaks. This boat was eventually sold to the Coeur d'Alene St. Joe Transportation Co., and used for a long period as an excursion boat on the St. Joe River.

The boat line connected the city of Coeur d'Alene and the Mission, from there the old narrow gauge ran to Wallace. The Spokane Falls and Idaho Railway was built by D.C. Corbin of Spokane, and completed to Wallace by 1888. That portion was abandoned in 1897 with much of it's right-a-way used by Highway #10, according to a 1940 report. The Oregon Washington Railroad and Navigation

Co., built from Harrison to Wallace, in 1899; and the Northern Pacific Railway was extended from Missoula to Wallace. Part of this was abandoned later.

Idaho Northern Railway was built from Enaville to Murray in 1910 by E.P. Spaulding. The passenger train was called the "Merry Widow." During the flood of 1917, waters of the St. Maries, St. Joe, and the Coeur d'Alene River, and Lake, over-ran the country; the line was washed out and never replaced. The Union Pacific had just completed a branch from Delta to the **Red Monarch** mine in 1917 (Sunset Peak region) when a flood washed it out. Sunset Peak branch of the Northern Pacific was built in 1895.

After the **Success** mine was presumed "worked out", and the **Tamarack** and **Custer** had been opened at Dorn on the Burke branch, the sunset branch was abandoned (about 1934) just above Bunn. A 1940 report states the Northern Pacific took up their tracks from Wallace to Burke and operated on the Union Pacific tracks to avoid congestion of traffic up the canyon. (In September 1964 the Union Pacific asked permission to discontinue its agency at Burke.) In 1912 the Chicago, Milwaukee and Puget Sound established Olympian passenger service from Chicago to the coast via Avery, Ida. Serious floods did persist in doing their damage in 1953, and again in 1964 over some of the old lines.

Chief Moses Reservation

As one hurries from one field of excitement to another thru the pages of this story one can only marvel at the quality of man, and the hardships he endured, as he traveled from one camp to another by what ever means offered him at the time. Back in Washington, Chief Moses reservation (in what is now parts of Ferry and Okanogan Counties) was being thrown open in 1896. To the north, placer miners had taken some $50,000 over a 30-year period but were turning their attention to quartz finds. For a brief few years locations were made and mining camps opened so fast it is impossible to do justice to chronology so the best one can offer is try to keep in the wake of the moving masses.

Up in the Boundary district, along lower British Columbia, W.T. Smith is credited with first discoveries at **Smith's Camp.** This portion of the country was already being settled by a tough breed of men known as miners, "tough" considering they had to withstand long cold winters and find their existence in heavy mountainous regions.

Granby and Phoenix

Henry White is credited with staking the famous **Knob Hill** and **Ironsides** claims in 1891 at what would become known as Phoenix. In 1898 the Granby Consolidated Mining and Smelting Co., purchased Phoenix Camp consisting of **Knob Hill, Ironsides, Gold Drop, Snowshoe** and **Curlew**

The OLD IRONSIDES and KNOB HILL Hotel in Phoenix Camp. (S-R 1899).

mines and that year started construction of a copper smelter at Grand Forks. J.P. Graves has been credited with starting this Montreal firm of **Granby.** Furnaces were blown at the smelter in August 1900, at what was called the largest copper smelter in the British Empire. A.M. Campbell was the managing director of the company, and Jay P. Graves, both of Spokane, was in charge of construction. This venture is said to have raised Graves' fortune from zero to half a million dollars.

In 1923 the city of Grand Forks purchased all lands owned by the Granby Co. in Grand Forks, to include the abandoned slag dumps from the famous smelter. In the middle 1960's the Sunshine Mineral Inc., with Marion Bumgarner of Wenatchee, and the late Delbert Scoles of Colville, as principal stockholders, had a purchase contract with the City of Grand Forks, for those slag dumps.

The Dominion Copper Co. owned and operated the **Brooklyn, Stemwinder** and **Idaho** mines at Phoenix, and constructed a smelter at Grand Forks. The B.C. Copper Co., owned and operated the **Rawhide** and **Athelstan** mines and constructed a smelter at Greenwood, B.C. about the same time. All these properties were located at a town called Phoenix.

In this period of build-up, placer mining had been the reason for early existence of Danville, Wash., (known as Nelson until around April 1903).

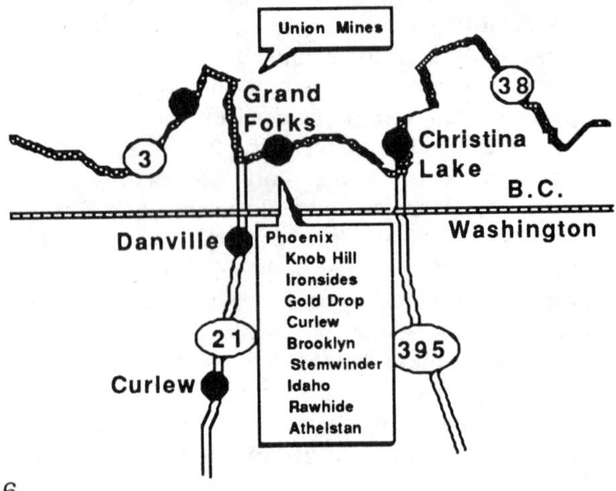

The BROOKLYN MINE at Phoenix (S-R-1899) **The city of Greenwood, B.C. where The Dominion Copper Co., operated a smelter. (S-R 1899)**

Phoenix, just to the north, was said to have had some 3000 residents but following closure of the mines the buildings deteriorated and have not been rebuilt with later re-openings of the mines. Danville boasted around 2000 population and dropped to less than 100 people. Gone with the people were such glorys as 13 saloons, and three large hotels. Remaining is a small grocery store and a two-man sawmill operation. Ten miles to the south is Curlew, once a town of some 3000, later less than 200. This area's main claim to recognition is recreation, scenery, and past history. It was on the tide of mining activity that Ferry County was established in 1899.

The mines and smelter at Greenwood and Grand Forks operated continuously until 1919 when they were forced to close due to low copper prices, and by reason of a coal strike in the Fernie, British Columbia district. The Granby Co. acquired properties at Anyox and moved it's operation to that area. The B.C. Copper Co. (formerly Keough property) acquired prospects at the **Copper Mountain** workings. **Copper Mountain** was eventually taken over by the Granby Company who operated until the late 1950's when they once again established at Phoenix. Miners later drove to work from surrounding areas, but the town of Phoenix is gone now. Mining and milling was reported carried on with concentrates shipped to the Tacoma Smelter.

Much high grade mining was done north of Grand Forks in those days and during the 1930's the Hecla Mining Company, from Idaho, operated the **Union Mines.** Development work is said carried on by the Heustis Mining Company.

A news article Nov. 19, 1976 tells how Grand Forks would benefit from more than a century of accumulated slag laying in twin mountains – and from the **Granby** copper mine. Mayor James Robertson told that the British Columbia City of some 3000 would agree to sell the slag to Pacific Abrasive and Supply Incorporated for two dollars and fifty cents a ton. Much of this was sent to shipyards on the coast to use as abrasive material in the ship-building industry. With millions of tons in the slag heap, the Mayor said the contract called for the extraction of 100,000 tons a year for 30 years, with renewal option for another 30 years.

Tacoma Smelter

For many reasons the year 1889 should be remembered. On Nov. 8, Montana became a state. Three days later Washington attained statehood. Colville was established that year. Important in mining and continued operation until recently, was the opening of the then 45,000 ton Tacoma Copper Smelter at Tacoma, Wash. In 1898 Bunker Hill and the Alaska Treadwell group purchased controlling interest in which it held until 1950 and sold to American Smelting and Refining. The smelter employed some 800 in 1964.

Before the Spokane Fire

Spokane was fast becoming the mining center of the Inland Empire. Chas. M. Fasset, (Spokane's first mayor under commission form of government) opened an assay and chemical business. Population of the city was estimated near 5000, and nearly that many were traveling to and from the city a month, by train. The Spokane Hotel and the Silver Grill were the gathering places of mining people. Such things as; school, cable cars, carriages, buggies, the electric train, the Natatorium Park, the Davenport Hotel, the Auditorium with the Harry Hayward stage shows, the Crescent and the Whitehouse stores, the Northwest Exposition, F.O. Berg's car, and the fabulous residences of Frank Rockwood Moore, A. B. Campbell, Patrick Clark, John A. Finch, and Austin Corbin were rapidly becoming part of this growing eastern Washington city.

Spokane Fire

On a Sunday, Aug. 4, 1889, the young city of Spokane burst into flames. (Population has been given all the way from 5,000 to 20,000 including transients.) It is reasonable to say the devastation nearly leveled the city and had serious effect on every soul who lived there. Out of the tragedy came many stories as to how the fire started but this humane and romantic story was told by "Uncle Joe Moris" to his niece-in-law, the late Mrs. Albert Beaudry, of Spokane. Whether myth, or truth – it could have happened.

It goes that Rebecca Woodruff Trego, a school teacher in Spokane (who would become Moris's bride five years later) was being squired by R.A. Jones, the man in charge of the pumps. They had been to Coeur d'Alene to a picnic. Upon returning by train they were stopped at the trestle at Olive and Trent streets. "Aunt" Rebecca is said to have told many times how as they crossed over on foot they saw chickens with their feathers afire trying to escape the spreading flames, and at that moment they first knew an awful tragedy had befallen. Her version was that Mr. Jones had the keys to the pump in his pocket, water could not be pumped on the devastation. Who among you readers can truthfully say you have never unintentionally carried away keys with you that had best been left somewhere else? If this is what happened, it is far too long ago to make any difference now and Spokane has come "thru it's trial by fire" to become a greater city by far.

Monte Cristo

But, we must hasten to the west side of Washington where some activity was going on in the Cascade range and on July 4, 1889 Joseph Pearsall discovered the mineral ledge of **Monte Cristo** (in Snohomish County). He named the mountain after that famous personage, then continued to climb to the spot where he made his first location **Independence of 1776,** and called it "Seventy-six" for short. While in search for galena (the most important source of lead) Pearsall had gone up the east bank of Silver Creek and from the steep sides of Hubbart's Peak is said to have been able to look down the divide to where the two creeks unite to form Swauk. From that vantage point he viewed a glistening streak on the side of Williams Peak.

J.M. Williams became interested with him in a number of claims. Seattle men also associated were H.G. Bond, L.S.J. Hunt, H.C. Henry, Edward Blewett, and three of the Wilmans family, J.M., F.W. and S.C. Wilmans. By 1890, claims at **Mystery Hill** were said to be worked by 250 men, and **Pride of Mountains** mines were reported operating a concentrator. In the fall of 1891, the Rockefeller Syndicate was said to have purchased controlling interest in **Monte Cristo, Pride of the Mountains,** and **Rainy.** The Wilman brothers retained control of the **Wilmans** and the **Golden Cord.**

The Rockefeller Syndicate built the Everett and Monte Cristo railway in 1892, and in 1893 extended from Everett to Snohomish. A smelter in Everett was built around 1893. The **Monte Cristo** mines are reported to have produce some seven million dollars from gold ore up to 1903. Disappointment followed some development. Low grade ore was said proven at great depth indicating the Cascades was not a "poor mans" mining field generally speaking, but required large capital to put prospects into production. The **Sunset** mine discovered in 1897 with development through 1932 was reported one of the largest copper producers of the state, with an annual output valued between $100,000 and $200,000.

Slocan

Many of the prospectors that had stayed over the Nelson, B.C., strikes began to fan out in all directions. John Seaton and Eli Carpenter headed up the north fork of Kalso Creek, the nearest town being Ainsworth on Kootenay Lake. While eating lunch one day on Payne Mountain, Seaton picked up an ordinary looking piece of rock and was puzzled by the weight. Staking 600 feet north and south they called the location **Payne** and took samples for assay. According to a lead pencil report by Carpenter the assay showed 25 ounces silver which was so disappointing to Seaton he all but abandoned the Slocan at that point.

William H. Hennessy, "Old Bill," procurred some ore samples, had them assayed and came up with 174-75 ounces silver and 75 percent lead to the ton. He kept his assay report secret. In the meantime Carpenter took G.B. Wright into his confidence, and who outfitted Carpenter and E.A. Bielenberg. On July 22, 1891, they left for the location by way of Slocan River and lake.

"Old Bill" guessed what was occurring and hastened to the Seaton's claim to tell him and in so doing was followed by Frank Flint to whom Old Bill had told the correct assay. Seaton agreed to return to Slocan for wages and an interest in all new locations. Bill Hennessy's brother John had been refused permission to join the expedition so he and Jack McGuigan gathered their prospect outfits and left for Kaslo Bay.

The outcome was that McGuigan, John and Old Bill Hennessy, Flint, and Seaton agreed to go on together. John discovered the **Maid of Erin,** which is said to have been the correct trend of the **Payne** ledge – and this ensued into another argument. Final agreement was that each should have equal interest in all locations except the original **Payne** claim which was left in Carpenter and Seaton's names.

The group of men located 23 claims to include: **Payne Group, Noble Five, Last Chance, Northern Belle, Ruby Silver, R.E. Lee,** and **Slocan Boy,** some of which turned into bonanzas and started a stampede into the Slocan. In 1898 Slocan mines were said to have shipped crude ore and concentrates aggregating 30,057 tons valued at $100 a ton, with an annual yield of three million dollars. The **Payne, Idaho, Ruth,** and **Slocan Star** became leading producers with around 900 men employed.

Sullivan

Still other prospectors were advancing to the east and north, where in 1892 the world's largest lead-silver mine the **Sullivan,** was discovered at Kimberley, B.C., by Walter C. Burchett (from Spokane Falls); Pat Sullivan, John Cleaver, and Ed Smith. All except Burchett died before learning the extent of this amazing discovery. The location is 47 air miles north of the boundary and is claimed the deposits are so immense the ore is measured in acres. George Turner, Spokane, was president of the company formed. For some time the mine was operated by American Smelting and Refining Co. In 1910 it was acquired by Consolidated Mining and Smelting Co. (name changed to Cominco in late 1960's), who were then operating copper and gold mines at Eugene (East Kootenay), and a smelter and refinery at Trail. Cominco is the largest producer of nonferrous (not containing iron) metals in British Columbia, with production to the end of 1950 reported to be more than $1,276,872,000.

Nelson, B.C.

Between 1892 and '99 many prominent mines came into being around Nelson. The **Poorman** is said to be the oldest free-milling gold mine in the district. Some five and one-half miles west of Nelson, on Eagle Creek, the owners with the use of a ten-stamp mill recovered $100,000 in gold within a seven year period. The **Granite Group** was on a parallel to **Poorman; Royal Canadian** a mile away, both owned by Duncan Mines, Ltd., who erected a 20-stamp mill for operation.

The **Athabasca** is said the most famous on Morning Mountain, some three miles southwest of Nelson. Development began in 1896. Under a London company ownership 200,000 shares on the London exchange were offered at 65 cents in 1899. The mine shipped over 300 tons to Hall Smelter with average ore value reported between $60 and $70 a ton, almost all gold. This was said to be first class ore from the surface with a second running about $20 to the ton. The company operated a ten-stamp mill.

Hall Creek siding was ten miles from Nelson on the Nelson and Fort Shepherd road, and three miles from the siding was the **Fern Group** of claims – said to be one of the best known gold mines in the province. It was owned by a Nelson, B.C. company, and under development since the spring of 1897. The company mill capacity was 30 tons per day. They operated a tram to transport ore, and a cyaniding plant. One dividend of $10,000 had been paid by 1899.

The district closely rivaling Rossland in 1899 with a summer output of 1,600 tons per week also included the **Ymir Mine,** 18 miles south of Nelson. The **Ymir** was said to have sufficient ore for two years without further development by the London and British Columbia Goldfields Corp. Their stamp mill was said the largest in British Columbia.

At the **Dundee** mines a concentrator was treating 50 tons of ore a day in 1899 with 100,000 tons ready for the concentrator. **Porto Rico,** under development since 1897 had a ten-stamp mill with first month output 590 ounces gold from 542 tons crushed ore. $30,000 gold was reported extracted from 1,400 tons ore crushed in six months. **Blackcock** mine was reported shipping 50 tons of ore a week for six months, to the Northport smelter, averaging $32 a ton. Other recognized locations at the time were the **Tamarack, Bullion, Union Jack,** and **Fairmount.**

Joe Moris, Le Roi and Rossland

There was some movement up from the Columbia River and towards Trail Creek when two prospectors George Bohman and George Leyson first located the **Lily May** claim in 1887. Oliver Bordeaux, of Colville, and Newlin Hoover, of Nelson, relocated the ground in 1889. Joe Moris had just returned to Colville from one of his frequent treks when Bourdeaux asked him to go along to do assessment work at the **Lily May.** They left Colville in March of 1890.

For the re-telling of where that fantastic journey led we are indebted to the Rossland Miner publication, and to the Beaudry's of Spokane, who speak for their "Uncle Joe" – for this story has been told, and re-told many times.

According to Moris, the Bordeaux party left Colville by sleigh on St. Patrick's Day, 1890, traveled to the Little Dalles, and then boated up the Columbia. Upon arriving at the mouth of Trail Creek the men found snow too deep to use pack horses. You'll recall Moris being a slight built man and he tells of the hard trip over five feet of snow as the two carried everything on their backs. By early April, bare patches were appearing and Moris was able to look across the south slope of the mountain where he noted the red nature of the earth. The coloring was considered some clue to possible ore and Mrs. Beaudry recalls Moris saying "they wanted red showings – and by golly! There it was."

It was not until the assessment work was complete at **Lilly May** that Bordeaux told Moris he had no money to pay him but had some in Nelson, and that Moris would have to go after it. On April 18, Moris started across country but on the way was stopped by an interesting outcrop which he said he located as the **Home Stake** mineral claim. He returned to camp and the next day both men started for Nelson. After waiting around for several days Bordeaux admitted he had no money to pay Moris so Moris went to work for 17 1/2 shifts at the

Silver King mine on Toad Mt. to get a new stake.

Starting down the river with a load of new supplies, Moris found the weather too bad to prospect so went to work on the **Home Stake** until a Joseph Bourgeois and his partner Pat (no last name given) drifted by. Pat gave up prospecting in disgust on the third day out so the two Joes headed out to find out about that red soil (on what would later became famous as Red Mountain). On the second of July, 1890, the two located the following claims: **Centre Star, War Eagle, Idaho,** and **Virginia,** and Moris said he put two stakes on the extension of **Centre Star** and called it **Le Wise.**

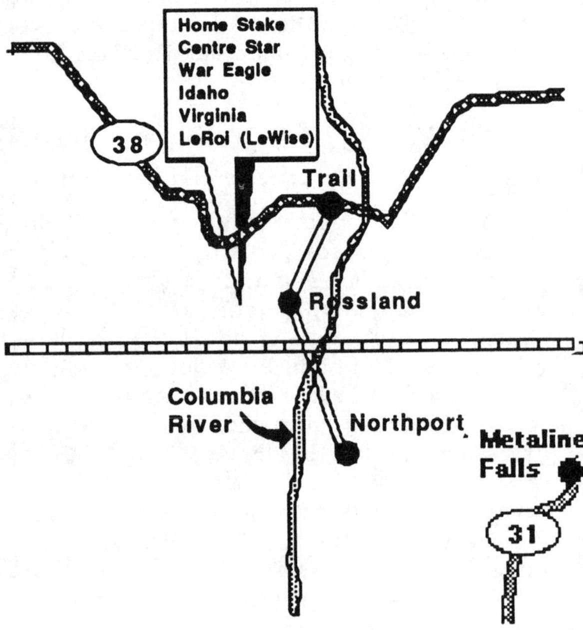

Hurrying off to Nelson for assays and to record their finds, the two Joes arrived on July 4 (after some 60 miles of travel). Out of ten samples the best went $3.25 and showed no trace, according to Moris. Bourgeois was all for giving up but Moris encouraged recording these "traces". By law then, they could only hold two claims apiece so they must dispose of the additional staked ground in some way. Bourgeois offered a choice of the claims to Col. Eugene Sayre Topping the deputy recorder, if he would pay for recording the group. For the sum of only $12.50 Topping chose the **Le Wise** which he later called **Le Roi** (meaning the king), and thus became the owner of one of the richest mines this country has ever seen – and a mine destined to make world history.

Many stories have been written that the offer was made because the two prospectors were broke. Moris said this was not true – that Bourgeois had over $700 at the time; but as for himself, he was as "good as broke" with only $18 to his name.

Col. Topping rushed off to Spokane, armed with samples and aiming to interest capital. Aboard the Spokane Falls and Northern train he met Col. W.H. Ridpath, and George Forester, formerly mining the **Dead Medicine** north off Colville. In Spokane they approached George Turner, Col. Isaac N. Peyton, Col. W.W. Turner, Oliver Durant, Alex Tarbet, and Frank Graves with a proposition (it has been suspected most of these titles were of the "Kentucky" variety). The group was to receive 16/30ths of the **LeRoi** claims for $16,000 and would perform assessment work to the value of $3,000 before June 1, 1891. W.J. Harris was an added member of this group, and sometime later Col. Peyton purchased the other 14/30ths of the claim from Topping and disposed of it to friends in Spokane.

A company was registered in Washington to be known as the Le Roi Mining and Smelting Co., of Spokane, with share issue of 500,000 at a par value of five dollars. 100,000 shares were to be put in the treasury and the others were to be disposed of. George Forester was elected president; W. Williams secretary, and with the addition of Major Armstrong, the original syndicate members were elected the nine directors.

In 1891 the company shipped several tons of ore from **Le Roi** by mule-back, river boat, then railway to Butte, Mont., where returns in gold and copper were said to be the magnificent sum of $86.40 a ton. Due to transportation and treatment problems, regular shipments were not made for three more years. Durant, Harris, and Topping had established a camp at **Le Roi** by 1891 and development soon revealed ore that ran five to 20 percent copper, three to ten ounces silver, and from $48 to $470 a ton in gold. The secret could no longer be kept, and in contrast to earlier lonely arrival; as Moris and Bourgeois were leaving the area they were met by large groups of men hurrying into this new district.

Ross Thompson had come to the camp by way of Seattle, Butte, Great Falls, Bonners Ferry, and Nelson, and built his first cabin at Rossland in 1892. Pre-empting 160 acres for a townsite, Rossland would be incorporated in 1897 and named for that far-thinking gentleman, Ross Thompson. Topping followed Thompson's lead and pre-empted a townsite at Trail Creek landing where he shared a cabin with Frank Hanna.

As hard times hit and many mines closed in the Coeur d'Alene's in 1892, literally hundreds of men left there to prospect the Slocans. They ventured into other areas of British Columbia and as word got out about the Red Mountain discoveries, many headed in that direction. Some stayed and some even made large fortunes.

When **Le Roi** development continued to be slow, Joe Moris had sold his **Centre Star** claim to Durant. Part of this delay was due to many of the men wanting to prospect instead of mine for wages. The **War Eagle** was bonded to Durant and Tarbet in 1892 but was soon given up. In 1893 **War Eagle** was bonded by Pyritic Smelting Co. of San Fran-

Around 1893 – Sourdough Alley of Rossland, B.C. – pictured are a few of the bakeries that used sourdough, (reason for the name). Courtesy Rossland Museum.

condemned the whole camp as well."

In 1892 a wagon road was completed to Northport where a ferry connected with the Spokane Falls and Northern Railroad (later to become the Great Northern). In 1893 a good strike was made at **Le Roi** and a wagon trail was built to the Trail Creek Landing. Suddenly the camp was filled with all sorts of humanity. At the end of that year there were 99 claims staked and by 1895 the number had increase to 1997.

"Boomers" came from everywhere; tents were pitched and whipsawed lumber shacks spread up the steep hillside along a muddy trail soon labeled "Sourdough Alley." Dancing girls and Madames, shyster lawyers, stock manipulators, and claim jumpers swarmed the town that never went to sleep. Miners came – then families, and began to settle and build this "Golden City" to a population said to be 7,000 in 1897.

About this time a shrewd young man, F. Agustus Heinze, took notice of this boom in British Columbia. Heinze had been employed by the Boston Montana consolidated Copper and Silver Mining Co., large property holders around Butte, when he was ousted for company reasons. By finding all the legal loopholes in mining laws, he is said to have engaged in lengthy litigation often coming out the victor and making huge sums of money. Among those he had clashed horns with were the great Marcus Daly, and W.A. Clark, of Montana mining fame. "Fritz" Heinze made his move into the Rossland arena by first sending scouts James Breen, J.D. Farrell, and a mining man A.E. Humphrey to investigate the prospects of a smelter for Rossland mines.

With the cunning of an expert, Heinze had by 1894 acquired a third of Toppings townsite of Trail for a smelter site; had a proposed tramway site surveyed and had contracted with the **Le Roi** people to treat 75,000 tons of ore. Early the next year it was voiced around that he had obtained a charter for a railroad to the coast. This was a thundering blow to D.C. Corbin and his Spokane Falls and Northern; to James J. Hill and the Great Northern, and to Shaugnessy of Canadian Pacific. By 1896 Corbin and Heinze had settled their differences and railway rights and roads were being pushed from Rossland to the river.

In the meantime ground was broken Sept. 13 1895 for the $150,000 Trail Creek Smelter. On Aug. 10, 1897 the first gold brick was poured at his plant which at that time covered 10 acres and employed 150. With legalities cleared over the Trail

Trail smelter after it became property of the Canadian Pacific Railroad (S-R-1899).

Creek tramway, rolling stock was brought from Utah. A special car was said built for use by Brigham Young the Mormon leader. Old timers today say they remember old Jim Chase bringing in the first "shay." (Chase later moved to Colville where he, his wife, and daughter, Mrs. Clara McPhail lived until their deaths there).

Limited history on Moris' activities does credit him with building one of the earliest cabins in what would become Rossland (around 1893). Following a disappointing experience with Pyritic Smelting Co., at the **War Eagle**, Mr. Moris and Mr. Bourgeois are said to have bonded that mine for a sum of $17,500 to Patrick Francis Clark (Patsy Clark), in 1894. (Clark was former foreman for Marcus Daly at Butte). John A. Finch, Austin Corbin II, E.J. Roberts, W.J.C. Wakefield, and others were in on the deal. (Some of these men had earlier been interested in the Salmon River, Idaho mining district, and Clark, Finch, and A.B. Campbell were operators of the **Enterprise**, **Standard**, and **Gem** mines in Idaho).

Clark became president of **War Eagle**, and Finch was vice-president. Early in 1895 the **War Eagle** paid a dividend of $27,000, the first one reported from a Rossland mine. That same year the Centre Star Mining and Smelting Co. was organized by P.A. Largey as president; G. Larvell, vice-president, W.G. Benham, secretary, and T.M. Hodgens as treasurer – and Oliver Durant saw great hopes for his **Centre Star** mine.

Tiring of his lonely existence Joe Moris returned to Spokane to renew acquaintance with Rebecca Woodruff Trego, who he always good naturedly joshed about their first meeting by the "potbellied" stove in the Desert Hotel Lobby. Miss Trego was from Kansas City and had come to Spokane to teach at the Lincoln school. Moris had to go to California and Miss Trego joined him in Oakland where they were married Jan. 18, 1894. Different homes have been mentioned including the Auditorium building and the Felix Hotel. They did go to Rossland for awhile where Moris worked at the **Le Roi**. Col. Peyton had taken quite a liking to the little prospector whose only education had been through a squaw man at Colville; and assistance from his teacher bride. Col. Peyton hired a tutor for Moris, who rapidly learned all the social graces his good fortune was to lead him to enjoy in the coming years.

The couple returned to Spokane to farm on Pleasant Prairie until 1898 when Joe could not resist the urge to join a part of the Alaska gold rush. While there he mined for wages and prospected on the side. He made some locations in the Great Bear Lake region and as late as 1930 was being piloted back and forth to his claims by Nick Mamer, of Spokane. Mrs. Moris often went with her husband but they usually returned to Spokane for the winters. Transportation problems never permitted development of the properties he located. In 1901 he staked a gold mine near Nelson, on Hall Creek

A reprint of Rossland mines – taken 1895. The trail near center leads into the great Le Roi Mine, and on up the hill. (Courtesy Cominco Magazine Oct. 1962)

but this was reported as not turning out to be a rich find. His ventures led him to Montana, Idaho, Utah, and Nevada, often in advisory capacity but Moris made his last prospect trip in 1938.

Mr. Moris owned the Delaware apartments in Spokane and had built several homes in that city when Mrs. Moris died Dec. 18, 1940. For the next 16 years Moris made his homebase on North Nevada and in 1956 he moved to the Parson's Hotel where he lived until one month before his death in Spokane, Feb. 7, 1964 – missing his 100th birthday by such a few months.

In 1960 this stout hearted, but far from pretentious little prospector was honored by the Cominco people of Trail for his great discovery in the Rossland camp. A calendar device with **Le Roi** ore samples was presented him on the occasion of his 95th birthday, by R.B. Shelledy, now of Spokane, who had gone to the Rossland mines in 1909.

When Joe Moris died it was revealed that he had been the "Mr. X" to the Volunteers of America in Spokane, to whom since 1947 had secretly given five $20 bills each and every Thanksgiving and Christmas. When finally unveiled just before his death his only comment was that he could recall "missing a meal or two himself." As a prospector he had suffered many hardships but as his means increased he was forever generous and thoughtful of others. He was survived in Spokane by the Albert Beaudry family, and by a sister in Hillyard, Anna Beaudry. Mrs. B.B. Lyon of Spokane said he is also survived by 2 nephews and 2 nieces.

Among his possessions was a message he had left to those who might hesitate long enough to read. Imperfect spelling could not detract from the depth of meaning; **"My religion is as follows. First be true to yourself an to true to everyone else. And traet every body as you would like to be treated yourself. And if you can help cripple childern or anyone else do so. And put your Fate and love in the Lord Gease Christ. And ther is a Heaven your soul will be there but you have to die to find out what is behind the curtin no one has ever returned to tell us."** It was signed JOE.

Stiles

We have left the Rossland camp in the midst of it's growing pains to go on a few years and bid adieu to the great little prospector who made that possible. For the brief few moments we should pick up the trail of our other hearty prospector-miner, Al Stiles, who had headed for the mines of Philipsburg, Mont. in 1889.

Stiles found his first job for the **Granite Mountain** mine. "I was only 20 years old and didn't much

The Cominco people honored Joe Moris on his 95th birthday for his great discovery in the Rossland Camp. A calendar device with Le Roi gold samples was presented him by R.B. Shelleday who had been at the Rossland mines from 1909 to 1925. (courtesy the Beaudry's)

When Joe Moris died in 1964, his mining gear was found in clean, show condition and was photographed at the request of his nephew, Albert Beaudry of Spokane. It shows his ax and leather strappings for his donkey to the left; his knapsack to the right, his gold pan, pick, compass, and an ice bag which he used for ore samples. (Courtesy Beaudry's)

care about going into those deep mines so I got out mine stulls (peeled logs) for the company". He goes on to tell while he was there, of seeing stacks of silver bullion (bars) "as high as that heater stove," and valued from $800 to $1000 apiece. He says **Granite Mountain** paid out $55 million in dividends and supported Philipsburg; and recalls they used the old quick silver method now replaced by the flotation plants.

When Stiles was there, Philipsburg had become quite a civilized mining town. Another historian telling much of the earlier activity had this to say, "Nearly everybody drank and getting drunk was a venal transgression. The members of the Philipsburg Pioneer Association – composed of those who have assisted in opening up for settlement and civilization, California, Idaho, and Montana – in their resolutions – reserve the right to get decently drunk."

Prospect fever hit Stiles by 1892 and he moved to Henderson Gulch where he found some gold in the placer fields. Taking a job at a big tie mill he met two Butte miners who had given $5000 for a claim 15 miles below Butte on Flint Creek. They needed a new cabin chinked so Stiles stayed to help "mud it up for winter." He and Jack Walsh stripped rich lead and silver from the surface of the claim, hit a vein and sunk a shaft 75 feet but the mine never made good. He said checks received were signed by W.A. Clark (Senator Clark) who made a fortune in copper mines.

Stiles went on to the placer fields of Willow Creek, and in Miner's gulch was able to pan enough gold over a tub, that along with killing a deer a week for food he was able to keep "hide and soul" together. While there a Mr. Nordeen came out one Sunday and the two went prospecting up Princeton Gulch. They found copper showing, and silver that later reported out at 729 ounces to the ton. Talk around the fire-side led to his first personal loss through claim jumping. That night his "so-called friends" papered the claim and he said, eventually took out over $40,000.

Stiles brief brush with mining had "sort of" tarnished his dream and he returned to Spokane, then went to Davenport where he dug wells and later harvested on the Charles Buck farm. It was there he met the daughter Ruth who would become his bride ten years later. Stiles tells of taking up a section of land around Grand Coulee, plowing out

20 acres, and he said he had a pen of horses. Wheat dropped to 15 cents a bushel and in disgust he said "I just walked off the place."

Word of Joe Moris's find, and new strikes on Red Mountain in Canada was spreading. "Why not try it?" says Stiles, so taking a span of horses and his equipment, and along with J.W. Ponnell they headed up through Colville to the customs at Northport. Arriving on a Saturday night he said the customs man suggested they stay over until Monday and he'd see what could be done to save the 20 percent duty. On Monday, according to Stiles, he was told he could take out a declaration of settlers affects which would cost 50 cents but he must stay in Canada a year, which he agreed to do.

Stiles first job was at the **Le Roi** mine where he drove four head of horses on an ore wagon to the Trail Creek landing. He says at that time there were only two business houses at the landing; Steel's grocery, and Col. E.S. Topping's beer hall and lodging house. He said the ore was dumped on planks, then loaded on steamboats to be sent to the smelter. His remark "some of the richest I ever saw, they took $80-$85 million in gold and silver ore from those mines."

Stiles remembers seeing Joe Moris on the streets of Rossland but unfortunately the two never met. Among the many things he recalls about the Rossland camp was seeing the fellows standing around guffawing about that man Heinze building a smelter up on the side-hill; and where was Heinze going to get that 75,000 ton of ore he had contracted for? A couple of years later some of these same men probably stood in awe of the narrow gauge tramway that zig-zagged it's way upward for 11 miles to reach the smelter only seven miles away by wagon road. Stiles also recalled when Col. William H. Ridpath came to make the payroll. Giving the wrong Peone the wrong check, Ridpath reached into his own pocket to make up the difference. And Stiles remembered later when Judge Carey from Colville visited in the capacity of smelter attorney.

Stiles said he worked at the blacksmith shop when gold ore at the **Le Roi** was running it's highest. Whether it's a tongue in cheek story or not, he also tells when the 350 foot level was reached, the best ore on the dump ran as high as $1280 to the ton and he said he could still see Col. Ridpath, Judge Turner, Oliver Durant, and others walk around and around the pile and watch it grow.

When Roy Clark found that Stiles knew his "rocks" he put Stiles to work sorting ore along with three others. He said they hand picked 200 ton a day and received three dollars and fifty cents for a nine to ten hour day. He recalled that shares sold at the **Le Roi** in 1896 for one dollar and fifty cents and later went to $50.

Judge George Turner was already a famous man in the area. When "law and order" was in it's infancy in the vicinity of Colville, President Grant appointed Turner the eighth territorial judge. He was born in Missouri and had come to Spokane in 1884. As judge, he held court in Colville from November 1885 until June 1888, and was later elected Senator. Sometime later he represented the government at the Hague concerning the Alaskan Boundary dispute.

Stiles knew Judge Turner as a very humane individual. One day while the Judge was around the mine a little white dog walked in front of a Charles Smith who lifted it lightly with his foot but was just enough to accidently roll the dog down a 200 foot shaft. One of the miners brought it up and when Turner saw the dog limp he made a pallet behind the boiler and brought beefsteak to the little mongrel each day. (Interesting memories of a 96-year-old who recalled events he knew some 70 years before.)

In a camp of such great wealth there was bound to be men of various degrees of greatness – Olaus Jeldness was one of these. Born in Norway, he was onetime superintendent of a silver mine in Colorado, and came to Rossland in 1896. Jeldness is credited with opening the **Velvet** mine, but he is acclaimed for introducing skiing to Red Mountain which has continued to this day to be one of the most popular ski hills in Canada. Jeldness' prowess as a ski jumper was known far and wide but his ability as a generous host reveals one story that is too good to pass by.

The legend goes that he invited 25 guests to a feast at the top of Red Mountain to celebrate a $75,000 mining deal. Guests invited were to arrange for their own arrival but were well fortified for the trip up by pre-arranged rest stops. One guest said every time he fell down he found a bottle under his hand. The "thoughtful" Jeldness provided skis for the down-hill trip. Strapping these on after a "very large" evening the guests started down, and to where Jeldness had thoughtfully arranged that they be met by Dr. Bowes' ambulance. It has been reported Dr. Bowes had much business that night and that some carried scars of this notorious party for life.

Religious life was not overlooked in Rossland, many denominations were rapidly becoming active. One of the best remembered of the clergy would be Father Pat. If the situation called for fisticuffs he was right handy; if it called for compassion he was the first to give up his bed, the clothes off his back, and his strength down to the last breath of life in 1902. He is said to have helped celebrate the men's good fortune, to console those who had bad luck – and on occasion to hold service in the saloons rather than ask his congregation to come to him.

It was during the summer of 1895 that Rev. D.D. Birks saw the first section of the Methodist Church built and where the first school was held with the Rev. Birks as teacher. Since all mankind was interested in this earth-given wealth, it was he who

1. The famous Le Roi Mine and dumps of second-grade ore. 2. A group of miners at the Le Roi. 3. Assay office, boarding house and steam hoist at Le Roi (S-R-1899)

staked Al Stiles and W.C. Gerbeth to a prospect trip into the Colville reservation when it opened up Feb. 21, 1896. (Birks may be known in the Metaline Falls district where he was a teacher in 1928.) We will return to Stiles, and the reservation district after we learn of the disposal of the **Le Roi.**

As new strikes were made and the Rossland camp grew in importance the mines became grouped under holding and operating companies such as that formed by **War Eagle, Ironmask, Poorman,** and **Virginia; Centre Star,** and **Idaho; Jose, Monte Cristo, St. Elmo,** and **Mayflower.** 27,085 tons of ore was reported smelted by July 1896 from **Le Roi, War Eagle, Ironmask, Poorman, Josie, Cliff** and **Evening Star.** This produced 45,237 ounces gold, 67,793 ounces silver, and 265,362 pounds of copper valued at $1,007,007 – and average of $37.18 a ton.

A Chicago company offered to buy **Le Roi** for $500,000. A disagreement followed and company President George Forster resigned. He is reported to have sold 52,000 of his shares to the company for one dollar a share. Col. W.W. Turner was named president and John Maynahan became superintendent. Deeper development brought $225,000 returns to shareholders between October 1896 and October 1897, and new machinery was added. George Turner (by then a Senator) was named general manager and Col. Ridpath was company representative to Trail Smelter. Col. I.N. Peyton, who had resigned as board director in 1894, returned to the company. By 1897 the future of **Le Roi** looked bright and dividends of $25,000 a month were consistent.

Sometime during this period another legend has been told, this time about W.J. Harris (Billy Harris). It goes that he and a partner ran a bar in Spokane. As luck lagged with the **Le Roi,** stockholders traded shares for drinks and soon the wall was well papered. The two partners are said to have had an old gray racehorse. At the time they dissolved partnership they took their choice, Billy taking the so-called worthless shares, and his partner the race-horse. Billy is said to have realized $400,000 from these shares and from which he is said to have built the Victoria Hotel in Spokane.

Differences of opinion between the Peytons and the Turners, said over shipments to the Trail Smelter, has been given the reason for building the smelter at Northport. Herman Bellinger, manager of the Heinze smelter, and Heinze's man Breen, went with the **Le Roi** group to the Northport location in operation by October, 1897. At that time sale talk of the **Le Roi** was widely voiced.

In the meantime, Lt. Governor of the Northwest Territories, Honorable Charles MacIntosh had met in London with an old schoolmate, Whittaker Wright. They formed the British America corporation. MacIntosh returned to Rossland and is said to have immediately purchased the **Josie, Great Western, Columbia and Kootenay, Poorman, Nickel Plate,** and **West Le Roi.** Peyton and Turner were also in London and Peyton had located the Whittaker Wright interests and arranged a tentative sale of the **Le Roi** for three million dollars. Turner's offering price had been five million dollars, the difference was bound to lead to "bad blood."

Peyton, as representative, is said to have sold 284,000 shares of the company to the British American Corp., at six dollars a share. The minority group; the Turners, Major Armstrong, Bill Harris, and Frank Graves refused to sell at that price. The litigation to follow centered around the differences in British and American law. Washington's law stated no alien could hold property within the state and gave strength to the minority group's hold over the Northport smelter, and over the company offices in Spokane.

Almost fantastic events followed to include; Counsel for the majority group L.F. Williams was reported to have gathered documents etc., and fled to Rossland only to find that Billy Harris had outsmarted his move and removed the company seal. The Turner group obtained an injunction to stop the majority group leaving the country and hired deputies to enforce it by stopping all trains to Canada from Spokane. McIntosh, and some of the shareholders then were said to have boarded a special train which was told to stop at nothing – and didn't until Deputy Sheriff Bunce held a gun on the train crew. He had been refused entry to MacIntosh's car by reason of "trespassing in an Englishman's home and castle." Austin Corbin, president of the line, stepped in and gave orders for the train to proceed. Deputy Bunce persisted by hanging on the side from Spokane but had to give in at Northport when threatened with arrest if he carried a deadly weapon into Canada.

Other developments around the middle of 1897, sharing the limelight with the **Le Roi** squabble, included the erection of two 1,184 horse power hydro-electric units on the Kootenay River, with prospects of cheap power at the Rossland mines. Lorne A. Campbell became manager. This was to replace an earlier power company built by Spokane capital. The company then had been composed of W.S. Norman (who built the first dam in Spokane for the Washington Power and Light Co.), and who had interested Patsy Clark, J.A. Finch, R.K. Neil, and Frank, and A.M. Campbell in erecting Rossland's first electric plant.

The Canadian Pacific Railway was doing much business in the Slocan but their only rail connection with Trail and Rossland was over the Columbia and Western lines built by Heinze. They just never did reach a common ground for agreement between the great railway company and the smelter operator. By this date Heinze was faced with greater legal problems in Montana where he still "did battle" with Anaconda, and the Standard Oil Co., and he is said to have finally given up the battle in Trail and Rossland. His acquisitions over a brief four year period – the smelter, railroads and railroad charters, and mining interests were offered in a lump to Canadian Pacific Railway for two million dollars, of which he is said to have received in total $800,000.

A former classmate of Heinze at the School of Mine at Columbia University, Walter Hull Aldridge, undertook to settle with Heinze for the railway company. Aldridge had just turned down an offer from the Guggenheimers, of New York, who were in their initial stages of the huge mining empire they were building. Aldridge chose the west instead and was to become the first manager in March, 1898, of the Canadian Smelting Works, the home of today's famous Trail Smelter. Heinze is said to have returned to Montana where he amassed something like $50,000,000 which was said lost in one of the greatest debacles Wall Street has ever witnessed.

The **War Eagle** had been acquired by Toronto interests with W.G. Gooderham, president, T.G. Blackstock, vice-president; along with G.A. Cox, W.H. Beaty, and W.E. Gooderham. Stiles recalled this price as being $750,000. The mine was merged with **Centre Star** and at that time production was down. Old timers will recall the **War Eagle** steel hoist as one of the largest frames of its type in the world; and the similar hoist erected at the **Centre Star,** just to the east. **War Eagle, Poorman, Iron Mask,** and **Virginia** sold for $850,000.

Completion of the Crowsnest Pass road was imminent. Screened coal and coke would soon be available to the Trail smelter as fuel. By 1898 there were 30 beehive coke ovens in Fernie alone.

Le Roi troubles flared again in July 1898. Col. Peyton resigned, and by request of MacIntosh, Judge Spinks placed the **Le Roi** in receivership and W.A. Carlyle replaced Billy Harris as manager. The Turners (minority group) had the action reversed and Harris was reinstated. Large shipments of ore was being sent to Northport to the smelter.

Suddenly the whole affair was settled in Spokane on Nov. 22, 1898. The minority group is reported to have accepted seven dollars and forty cents a share from the British America Corp., and to receive payment for ore enroute, and matte at the Northport smelter. The figure has been reported in so many different ways. MacIntosh handed over one check of $1,042,054 as part payment for the **Le Roi** Mine reportedly sold for at least $5 million. This check was dated 27th August, 1898. (Historians have reported the total price being from $3 1/2 million on up.) The sale did represent one of the largest deals of that day and time and the Canadi-

Trail was still a pretty "rugged" town until some years later - Photo Sept. 30, 1929, (S-R).

ans were relieved to have gotten those from the United States out of the Kootenays. Spokane's economy benefitted greatly with the proceeds being spent on beautiful homes and large buildings. (Due to many irregularities in the disposition, Whitaker-Wright was reported taking poison at his trial in 1904 when he was convicted of fraud.)

Peak production at the Rossland mines was in 1902. After paying $775,000 in dividends, and more than $30 million in it's lifetime, **Le Roi** was liquidated in 1910.

In 1905 the **War Eagle** and **Centre Star** amalgamated to form the Consolidated Mining and Smelting Co., of Canada Limited; and in 1910 the **Le Roi** was acquired by Cominco. The Cominco name became legal in the 1960's. By 1929 production at these great mines was down and Cominco terminated operation. Rossland has become a comfortable residential city from which many commute to Trail to work at the smelter. As the great gaping holes stand on the side of Red Mountain as a warning that it's "no place to take a Sunday hike," new encouragement was being offered. What has been the outcome of a report of molybdenum on Red Mountain in 1964?

Advanced as a relief program during the depression years, Cominco entered a leasing operation from 1934 to 1942. Gold at $35 an ounce encouraged the working of many old mines to include: **Midnight, Golden Butterfly, Gold Drop, Snowdrop, Cliff, Evening Star, Georgia, Gold Drip, Hattie, Jumbo, Mighty Midas, Nest Egg, O.K., Silverine, Blue Bird, Lily May, Mayflower, Ural,and IXL,** plus all the former top producers we've just covered historically. Total output of several million dollars was the result.

Damage and Settlement Caused by Trail Smelter Fumes

Due to growth of metallurgical operations at Trail Smelter the district has been said once synonymous with the worst interpretation of "mining towns." An increased quantity of sulphur bearing smoke was the result of treatment of lead and zinc concentrates. The fumes or smoke had its effect on vegetation; on animal life (especially cats and dogs, according to the old timers who frequent the park benches now), and on the men who worked at the plants. An international tribunal was established to study the findings and settle claims, some on the United States side. The late Judge W. Lon Johnson, of Colville, was named American consul for Consolidated Mining and Smelting to settle such claims which occurred and were closed between 1927 and '37. With good management, and the elimination of smoke, Trail is today a model munity. The economy of the area has greatly been added to by the Cominco who employed something like 7,000 people in late 1960's to early 1970's.

Much Later, With Stiles at Northport Smelter

Mr. Stiles told one little story of the IXL when in operation in earlier days. A carload of ore worth $30,000 had been shipped to the Northport smelter from **IXL Mine** when Stiles worked there. A strike was in progress and the ore was dumped on the ground. This same strike had caused considerable hardship among the men. One Barney Rogers (whose wife was Anne Fields) gathered a milk pan of the ore and brought it to Stiles to melt out for what could be obtained. First they got a "collar button size", then a bar worth $500. "It was some of the best gold ore I've ever seen and that milk pan full sure helped that family through a hungry period."

It would be impossible to recall the names of all old-time miners now living in USA or who moved towards Spokane, that once mined at the Rossland camp, but one was "Bill" Cook, well-known prospector in the Northport mining district. After mining at **Le Roi** he located five claims called **"New England** in the Northport vicinity, and which Knob Hill Mining Co., had under lease in the early 1940's. At his death his wish was carried out to be returned to Rossland for burial. Frank Paparich, a Northport miner also worked the Rossland camp.

A Mass Return to the Coeur d'Alene's

The British Columbia mining history is akin to the history on the south side of the line and that relationship makes for interesting retelling. Because there has been so much association between those mines, and miners, and the ones on this side of the line, we have overstayed our time but have barely touched the surface of those great mining districts. We must now return to the Coeur d'Alene's and the rapid development there.

By 1891, there were 40 developed properties of which 26 were proven producers. 13 concentrators were said to be in operation. The cost of mining a ton of ore varied from two dollars and 50 cents to four dollars and 50 cents by 1892. Mine workers were only making three dollars and 50 cents for a ten hour day – a serious factor in conditions leading to labor war, even though the district was producing between ten and 15 million dollars annually.

Idaho became a state on July 3, 1890. Her last Territorial Governor, George L. Shoup, was chosen the first governor of this 43rd state to be included in the union. Idaho backed her request for statehood with a show of industrial responsibility – her mines were humming; agriculture and lumber industries were prospering and her future looked bright.

The first train had arrived in Wallace in the fall of 1887. Since then, freight rates, often added to by use of one of the steamships, was threatening mine owners' profit. Trains were met by such steamers as The Sherman; The Coeur d'Alene, The Kootenai, The Queen of the Lake, and the best remembered one, The Georgie Oaks. (The Kootenai was reported to have sunk with it's load of concentrates late in 1888 and was still being searched for in mid-1900's.)

As the towns developed, Burke aimed to be larger than Wallace. Strangely, Burke got its railroad in late 1887 – well before it had a decent wagon road. F.R. Culbertson, superintendent at the **Tiger** mine at Burke, made the first shipment from that district. Keeping in the race for town growth, Wallace was first with a public bathhouse in late 1888; and was requesting an equally needed facility – a jail. As Wallace continued to progress and eventually became the larger of the two cities, S.S. Glidden saw the need and started the building of his Tiger Hotel in Burke. To have 35 sleeping rooms and being built right down in the canyon, it was necessary to channel Canyon Creek under the hotel, and build the railroad and highway right through the hotel. This famous 1888 structure became one of the better known buildings in the entire district.

Ripley's Believe It or Not described Burke's Main Street as being so narrow the merchants had to pull in their awnings to let the trains go by. All that remains today of the Tiger Hotel is a concrete wall against the hillside.

By late 1890, the miners had formed four labor unions with some 1500 members representing different parts of the districts. These were based on unions in the Butte, Mont. mining districts and from where some participants in Coeur d'Alene labor wars would come. One aggravation to the miners was the need of better medical and hospital care; without what they felt was the "heavy hand" of the Mine Owner's Association, the other organized group in the district.

The Holland Memorial Hospital was built in Wallace in the summer of 1890. Reports indicate that a patient received all the medical care needed for a one dollar membership. That price was raised later. By summer of 1891 the union members got behind what became the Providence Hospital to be built in Wallace. This is sometimes referred to as the Sister's Hospital (for Sister Joseph who came from the Sacred Heart Hospital in Spokane to supervise), and sometimes the Union Hospital. (The one dollar membership cost was said effective there at the beginning.).

On the surface times appeared good, but there was an uneasy feeling in the district as serious problems arose between mine owners and union men. The miners claimed the owners refused safe-

Still said to be the world's largest smelter, this photo of Cominco was made August 26, 1963. (S-R)

guards to their health; and inadequate medical care, all for the purpose of making more profit for the mine owners.

Labor struggles are never one-sided and it depended which way the writer leaned as to his version of the telling. Miners were becoming irate over having to buy from and having to live in quarters owned by the mine owners. All of this provided fuel for the local news; and the subject of wages, along with union versus non-union provided headlines capable of starting plenty of trouble in the district.

The Mine Owner's Association was in the process of trying to obtain lowered freight rates. Said due to the high costs of operation of their mines, those mine owner members were also trying to lower wages and other costs. All of the many problems were causing great concern with the miners and was to become a major issue between those of the Miner's Unions and members of the Mine Owner's Association. Most of the owners of major mines were members.

The Mine Owner's Association membership was more or less secret but apparently included the following: from **Bunker Hill and Sullivan Mine,** Clement acted officially with Hammond, Bradley, and Jenkins, all Company men. Charles Sweeney spoke for **Last Chance;** George B. McAuley for **Sierra Nevada, Stemwinder, Inez** and **Granite.** Van B. DeLashmutt was also one of their men. Stephen S. Glidden officiated for **Tiger;** along with Frank R. Culbertson. Alfred M. Esler spoke for **Helena and Frisco;** and for **Badger and Black Bear** mines, and **Frisco Mill.** Patrick "Patsy" Clark was head spokesman for **Poorman;** Amasa B. Campbell and John A. Finch for **Gem, Standard** and **Union;** and C.E. Porter for **Custer Mine.**

That list covered the mines and mine owners that we will be concerned about. Our next concern will be an update of the specific mines involved in the 1892 labor war – to learn a bit more about their development and possible future.

The **San Francisco** mine near Gem (usually referred to as the **Frisco**), was owned by A.M. Esler. When war finally came, that mine was to be the first to suffer severe physical attack; reportedly because Esler refused to pay shovelers and car men three dollars and 50 cents a day. The Mine Owner's Association was holding out for three dollars a day.

Said to have the largest mill in the district, water for the mill was taken from Canyon Creek and run up to, and through a flume to the mill. Production

Said to be the only hotel in the world to have railroad tracks and creek running under it, the Tiger Hotel at Burke, Ida., built in 1888 over Canyon Creek had five trains daily rolling through it. The hotel was torn down by 1954 (Courtesy Wallace Museum, and S-R). (One release says the hotel was destroyed by fire in 1923 and never rebuilt, S-R, Sept. 11, 1983--Keith Goodman)

was so good that by April 1891 the company had paid out it's 12th dividend – a total of over $120,000. Principal investors were Helena, Mont. people John Murphy, A.M. Holter, and S.T. Hauser.

The combine of the **Helena-Frisco** mines were opted for 60 days, to be bought by an English outfit at a price of $1,500,000. Labor trouble news put a stop to that sale. Esler had wanted to ease away from the mine but when trouble did occur he fired Martin Kennedy and Billy Hutchinson and went back in full command.

The **Tiger** mine near Burke had been worked by around 30 men, with a few armed guards in evidence. The owners had battled water rights problems and in the summer of 1887 had accused the **Poorman** people of blowing up their dam. By December of '87, they had 100 men on the payroll and had a concentrator reported to be putting out 25 tons a day. By March 1888, lead prices were said the best in ten years – being quoted at five dollars and 30 cents per hundred.

The **Tiger** had installed a new drum hoist. Frank Culbertson, the superintendent, had noted that tension was bad all winter, but Glidden looked to a better future for his mine. Besides Glidden's new hotel, Burke was said to have large stores, other hotels and saloons and 60 men were supposed to be on a payroll.

The **Poorman** had been worked by Simon Healey in the summer of 1887 and by that fall, Patrick Clark (of Montana mining), had become manager. It was then known as the Coeur d'Alene Silver Lead Mining Company. Ore was being shipped to Omaha and some 70 men were employed. Stock went from 27 cents to one dollar a share by March, 1889, and some 30 more men found employment.

The **Poorman** had paid out $55,000 in profits to shareholders that summer and another dividend was paid that winter. The mine was mostly owned by Butte investors by 1890 and there was talk of a sale. By 1891 more dividends had been paid and Patrick "Pat" Clark was modernizing the mine by going electric. By the end of the summer the mine employed over 200 and by early winter is said to have paid out $310,000 in dividends.

(We note that by the end of 1889, many of the mines in this district were paying dividends.)

By January, 1890, the Coeur d'Alene Indian Reservation had opened. Hundreds – some say thousands of men rushed to stake claims and to obtain land. This provided new areas for discovery, but for the time, we will return to learn a bit more about the mining district that would one day become the greatest in the United States.

Gem Of The Mountains, or **Gem** Mines was also an 1884 discovery and was under supervision of A.B. Campbell, one of the principal owners. Campbell was also a member of the Mine Owner's Association. By mid-1889, over 50 men were working the **Gem** claims. The town of **Gem** was becoming noticed – with a postoffice, general store, saloons, and a request for a depot.

The **Union** mine on the extension of the **Tiger**

was building a 200 ton concentrator. The mine at Burke, then owned by Coeur d'Alene Mining and Concentrating Company was half owned by Finch and Campbell. Those two were also part owners of **Gem, Galena,** and **Banner Mines.**

The **Last Chance** headed by Charles Sweeney, was another "member mine". In the spring of 1888 that mine and the **Emma** had been sold for a reported $75,000 to A.M. Esler. That time he represented a Helena syndicate with Sweeney, and Frank Moore of Spokane.

The tramway of the **Last Chance** was damaged so badly by snow slides in the winter of 1890 that it had to be shut down along with a number of other mines in the Milo Gulch area. By early summer things were looking very good with an electric plant completed for power drilling. (The practice of hand drilling at all of the mines had caused silicosis death from rock dust, for hundreds of miners.)

Production at **The Last Chance** was at it's best with lead ore running 80 to 85 percent, in mid-summer 1891. By spring and early summer many of the mines would be on strike, pending labor problems between the Mine Owner's, and the union miners.

C.E. Porter's **Custer** mine was to feel the problems. The Gem and Burke unions had given Porter, and McAuley notice of intention to strike if wages were not raised at both **Custer,** and **Granite.** As early as winter 1889, these mines, along with others, had been paying dividends. It was on July 21, 1891 that over 200 men, mostly members of the Miner's Union, went on strike at the **Custer,** and **Granite** mine. They demanded that carmen and shovelers were to receive three dollars and 50 cents per day, a wage that was paid by some mines.

Apparently **Custer** handled the unions in a satisfactory manner because after settlement, and during the "War", the mine was little affected. Only four days after the vicious July 11, 1892, attack at the **Frisco Mill,** the **Custer** announced the biggest strike of rich ore so far discovered. They were shipping ore again by the end of July.

Granite made the earliest shipment of ore from the Wardner area in the winter of 1887. Owned by George B. McAulay, the mine was managed by Van DeLashmutt to being one of the first corporations to pay dividends. By mid-summer 1889, they employed over 50 men. When the mine was on a one week strike in July, 1891, the union won and both the **Granite,** and the **Custer** agreed to pay underground workers three dollars and 50 cents a day.

By April, lead had dropped to four dollars and 32 cents a hundred, and silver to 98 cents, creating costly decisions for the Mine Owner's. With so many of the mines on strike, or having serious labor problems, the **Granite, Gem,** and **Frisco** still tried to operate in early July, 1892, with non-union labor. The **Granite** was said to have 55 "scabs" and 20 guards, thereby creating more fuel for the bombing that was to occur. (Non-union men were often referred to as "scabs" in several area newspapers.)

McAulay also spoke for the **Stemwinder,** a Wardner mine. The mine had put the first aerial tramway in the district into operation in early 1888. By late 1889, the **Stemwinder,** and another mine of this conglomerate – the **Sierra Nevada,** had mills operating. During the extreme cold of the winter of 1890, the **Stemwinder's** tramway was damaged due to snow slides. The **Sierra Nevada** was not able to ship ore also due to the cold, a problem facing several of the mines in the district. As a result these problems added to severe hard times being faced by miners, and owners. Weather improving, by early summer these mines showed good profits.

By the end of 1891, **Stemwinder** and **Sierra Nevada,** were included with the larger shippers and there were close to a dozen other smaller shippers in the district. The Mine Owner's were still fighting for freight reductions and in January, 1892 had agreed among themselves to close down their mines for a few months with the exception of development work. Whether the **Sierra Nevada** was a member of the Association at that time seems uncertain because it was among the five producing mines still working.

As more demands were made by the Mine Owner's Association that miners were to work for as little as three dollars and 50 cents, and some said three dollars for a ten hour day, tension with the union men became worse. Some of the mines were using non-union, or "scab" men and importing more to break the union demands. Millions of dollars in wages were said lost with the mines shutdown. The railroads lost as much but were not willing to lower rates to meet the Mine Owner's request – and the Mine Owner's were losing another equal amount.

Times were getting worse in the district – the price of lead and silver was falling, and Idaho's population dropped. Mining ghost towns gave mute evidence of what was happening. Fires caused tremendous loss in Wardner in 1890, and again in the spring of 1893. Wallace's fire in 1890 almost stripped that "wooden" town but it was quickly rebuilt to something more substantial. The second fire in Wardner took an awful toll. Over 80 buildings were destroyed. Henry Day Sr. had become a prominent citizen by then and was appointed to dispense welfare of a sort to families who needed it badly. Even the unions from other areas, particularly Montana, were shipping in food and other needs of the families who were too long out of work.

The **Bunker Hill and Sullivan** had earlier been a "hold-out" from the Owners' Association because it objected to the new (Providence) hospital being built in Wallace. The Company had hauled it's first ore to the depot at Wardner in October 1887 and by

August of 1888 announced their first profit of $400,000 for the year. By early fall of 1890 they had 275 men working and it was estimated they would have over 400 very shortly. By spring of 1891 when their mill was ready to run, they also had a tramway over 8,907 feet long completed and ready to carry the ore to that mill. Robert Cheyne, from Helena, was mill superintendent and was said to know more about this business than anyone in the entire district.

The largest mine of the district, **Bunker Hill and Sullivan** was facing labor problems but did not yet see fit to join the Association. It was the spring of 1891 and it's new concentrator - said to be the largest in the world, made it's first shipment of seven carloads - over 140 tons. Silver was 97 cents and lead was four dollars and 22 cents per hundred so freight rates and labor problems were affecting this mine as it was the others.

Victor M. Clement was manager for the mine; and recent large investors included D.O. Mills, Silas McCormick, and the Crockers. Different versions of the Mine Owner's problems included objection to paying one dollar a month medical benefits for some 300 mine workers; and of course the ever present wage dispute. A strike did occur, closing **Bunker Hill** for part of the summer. In September that company faced serious litigation with owners of the **Emma and Last Chance** mines.

The entire district was reporting too many deaths and killings by the end of 1891 - mostly labor arguments involved. Strangely, an economic report stated the district had many new mills; and that nine million dollars in ore and concentrates were shipped from the Coeur d'Alene district in 1891 - plus $250,000 in gold bullion.

By December 1891 most of the mines were said shut down; some due to the weather, others to labor problems but throwing over 3,000 miners out of work, and some 500 common laborers. By January 1892 the Mine Owners were shutting down remaining mines due to the freight rates and by then, **Bunker Hill** was ready to join the "shutdowns", and the Owner's Association.

Nearly all the men at Burke mines were laid off along with those at the **Frisco**, and the **Gem.** Times were bad! Mullan was not so badly affected but many miners left for Butte, and others for the Slocan in British Columbia.

There was great bitterness in the district. Strikes were called at the **Frisco, Black Bear, Granite,** and the **Custer.** By the middle of 1891 mines of the district, excepting those at Wardner, had been forced to agree to the union wage scale of three dollars and 50 cents a day for all underground workers. (Shovelers and carmen were only being paid three dollars a day prior to strike settlement).

At the **Bunker Hill and Sullivan,** Simeon Reed is said to have disposed of his interest in the mine to California and eastern capitalists. Victor Clement continued at the mine as manager. F.W. Bradley was assistant manager, and Frank Jenkins was superintendent of the mine. High point of interest at the mine at that time was the tramway that carried ore from the mine to the concentrator on the hill and over a ridge at Milo. This was then better known as Wardner Junction, and now Kellogg. (When Kellogg tunnel was completed in 1902, the tramway was abandoned.).

A flare-up between Clement and the union men over the hospital fee, over which hospital was to be used and how services were to be supported, brought on a strike at the mine. The men added their demands for the same wages paid at other mines. The company met the demands of the men and for a brief time went back to work.

Association member mines closed by Jan. 16, 1892, said due to increased freight rates. Those rates are said to be six dollars a ton on second class concentrates to Helena, and $16 on first class concentrates sent to Omaha. Rates were lowered some and the mines advertised to reopen April 1, 1892 - but the wage bug-a-boo reared it's head in advance of that date. Demands on both sides were causing strong feelings and the date was passed by. The press, and the community began taking sides.

By early summer, non-union miners were imported into the district under armed guard. With these so-called "scab" miners, several of the Association mines re-opened. A relief fund was set up for locked out union members but only a few remained to do development work. By some loophole in regulations pertaining to shaft mining, Glidden and Patsy Clark found reasons for reopening the **Tiger** and **Poorman** Mines to union men. Clark is credited with the remark at that time, "We have not a friend in the camp."

Norman B. Willey had by then become the new governor of Idaho, and Shoup was elected senator. A complaint filed at Tacoma, by G.W. Dickenson, superintendent of the Northern Pacific Railway, is said to have propelled Governor Willey into an inspection of the problems in the mining district. Apparently Governor Willey termed the whole thing a wage dispute that would soon be settled. Suddenly deluged with mail from both sides, he went into action. Quietly, but with determination, he mustered state troops; and then he went so far as to request federal troops, but President Harrison refused him those at the time.

Captain Thomas Linn, of the 2nd Regiment of the National Guard said the unit was called to advertise Wallace. The presence of the Guard only tended to stir up more dissention. Wage scales dropped as mine owners let union men go and hired non-union from Michigan, California, or wherever the word reached about jobs available in the Coeur d'Alenes. Cases of guns and ammunition arrived, said for union men. Mines that worked were under armed guard and some killing was going on.

Lead was down to four dollars and 25 cents per hundred and silver 86 cents an ounce. Fewer mines were working and there were less jobs for the men, when in June the Governor came to Wallace to look the situation over. The Mine Owners had hired lawyers; and also a Pinkerton man named Charles A. Siringo, alias Leon Allison. In general, the area had the makings for an intrigue filled soap opera excepting this was about real live people with families to support.

War!! Martial Law!

Labor violence finally erupted at **Gem** Mine where on July 10, union men, armed with rifles gathered. Esler sent for troops but by early morning July 11, 1892, armed men were all over the hills. The tramway tracks were blown up at **Frisco** Mill; and dynamite was sent down the penstock and blew up the mill! Some $20,000 in damage was suffered; A.T. McDonald was killed and a number of men hurt. George Pettibone was accused of the dynamiting and was said captured later in July. (A new mill was built in September.) The governor declared Martial Law on July 13 and Shoshone County remained under such law for four months! - Until Nov. 11, 1892.

The Pinkerton man Siringo had posed as a non-union miner. Peter Breen, another trouble maker was also involved. He had forced A.L. Gross to surrender the **Gem** mine as union men and troops shot up the town and a saloon said to be a union hang-out. Guns and 2,000 rounds of ammunition were placed on a hand car and started down toward the **Granite Mill.** These were eventually taken over by union men. By the night of the 11th, squads of non-union men were seen leaving Wallace towards Wardner.

The union men took over the **Bunker Hill** Concentrator. Giant powder was placed under the mill and Clement given his choice - send non-union men away, or see his mill blown up. He gave in and 130 some non-union men were driven out of the district heading for Cataldo, and the Georgie Oaks steamship. There was panic and terror as these men were fired upon by union members. How many were injured, and how many were killed was never actually known. James Monaghan, foreman of the **Gem** Mine was believed killed but later showed up in Spokane.

In such a state of rebellion, Infantrymen from Vancouver Barracks, - 194 enlisted men and 11 officers, were said stationed in Wallace, 500 some troops in all in Wardner, and 250 at Mullan. Some 600 non-union men were said chased from the district.

Under martial law with by then some 1500 military present, several hundred union members and their leaders were arrested. Peter Breen, also arrested, was one of those from the Montana Union membership. Prisoners were eventually turned over to Federal control, the number diminishing to around 135. Twenty-five were taken to Boise, sentences meted out there by Judge Beatty.

Kellogg Tunnel built and used by Bunker Hill, from 1893 until 1902. (S-R)

The first of what would be many years of labor troubles saw non-union men returning to the district. Several of the mines reopened. The union had won one point - the companies could no longer tell the miners where to trade; and they also got better living quarters. By winter, many of the union men had been re-hired.

Gem, Granite, Custer, and **Sierra Nevada** mine's were shipping ore by the end of July with all except **Custer** hiring non-union men. **Bunker Hill** listed 365 men on the payroll the last of August and was hiring more. Their payroll was said $43,000. **Sierra Nevada** had 60 men working the end of August, and **Poorman** was touting their new manager Ben E. Thayer. He had come at a good time - going into 1893, **Poorman** was said to have hit the finest ore and had 110 men working until a December close-down due to price.

Esler was retiring from the **Frisco** and was working towards buying the **Argentine Mine.** Pat Clark had given the area the "heave-ho" and had headed for the Slocan. His **Poorman** was said to produce 700 tons of concentrates monthly by December, with 160 working. **Last Chance** was said to have 75 men working; **Stemwinder** with 100, and **Black Bear** with 70. But by February 1893 - **Bunker Hill** was shutting down - too high freight rates, and the price of lead and zinc down. With operating costs too high there was also the rumor of a sale of the mine. Lead was down to three dollars and 50 cents per hundred, and silver was 81 1/4 cents per ounce.

The military was in the district from July 13, to Nov. 19. To settle the many charges, a grand jury

Before – and After – July 11, 1892, armed men sent dynamite down the penstock and blew up the Frisco Mill. One man killed, several injured and $20,000 damage – and "War was on!" (Al Nugent).

was called at Murray the fall of 1892. Four major indictments were found against the Miner's Union leaders, members and sympathizers. District court followed through on the indictments after a change of venue to Kootenai County, those sessions held at Rathdrum, Ida. By March, 1893 the members of the Miner's Union charged with crimes were at liberty; basically because these charges could not be proven beyond a doubt. By July, legal sessions came to a close; 13 had been convicted of contempt and served out their time, as did four who were sent to Federal prison in Detroit.

Times were getting worse. By that spring, 1893, the bank in Wallace had closed, and the **Last Chance** mine was closed down. Wallace's saloons stayed operational, but there were said to be nine in the area closed rather than buy licenses. Of the 90 some saloons said licensed in the whole district by February 1891, one wonders how many business failures there actually were at that time. One comment was that "with jobs scarce," and the terrible fire in Wardner that had taken 80 buildings, there was need of relief "of one sort or the other." There was money and other help being sent into the district to care for the destitute and jobless.

The move towards unionizing resulted in a constitution presented by the Western Federation of Miners, and approved by the local unions in May, 1893. With the release of all the prisoners, union activity was renewed. Lead and silver prices were still falling and more of the mines were closing. Non-union men were leaving the district and union men took over. Strike demands were what closed the **Frisco** and the **Gem.** With the market in complete collapse, silver down to 62 cents, and lead still falling that summer, more of the mines were closed. In April of that year when Wardner had the fire, times were so hard there seemed little reason for rebuilding that town.

There were many mines in this district; some became great mines and others became parts of active mines – their tunnels, or shafts joined and being mined by one mutual corporation. Some of the mines not studied in depth include the **California.** It had ten men on the payroll by 1889 and was ready to ship in 1890. In the same area was **Black Cloud, Monarch, Panhandle** and **Yankee Girl.** Stewart Fuller was reported to have sold his ownership of **Monarch** for $30,000. **Yankee Girl** produced for many years for the Blake brothers who shipped a carload every two months averaging $75 to $400 to the ton. That claim became part of the **Sunshine Mine.**

Gray Eagle located in late 1890 above Burke, by Fred H. Harper, became part of the **Day Mines.** The **Mammoth Mine** near Burke shipped one carload in early winter 1890 expecting to make $2000. By July 1891 they had 25 men working, shipping 2,000 sacks of ore regularly and in August made a $10,000 ore shipment. The mine paid $5,000 in dividends.

Black Bear was discovered in 1885. John Bartlett and William Haskins were credited with that find and the claim was owned by O.B. Hardy in November of 1890. This mine was said to have the sixth concentrator built in the Canyon Creek area. The mine was the scene of tragedy when a gas blast blocked the mouth of the tunnel. It took five hours to dig it out and four men were found to have been killed.

Black Bear miners went on strike May 12, 1892, because all the employees were required to eat at the company boarding house – one of the factors that led to the 1892 labor war.

Killbuck claim was located in 1885 and years later became part of the great **Galena Mine.** It was leased to Mr. Cox in the spring of 1888 and was the scene of a new mineral strike in 1891. W.P. George and Lee George were said owners with H. Herrington working the claim. **Killbuck** was said located by Lee George when he sat on an uprooted tree to rest while prospecting. Kicking away, and picking up pieces of rock, he found shiny galena after breaking one piece apart.

Wardner and Blossom took a lease on the property and did a considerable amount of work. Luck was not with them. Later E.O. Cox leased the property and worked it with several men who shipped some good ore but could not make it pay. There was ore on every level but irregular. Assessment work continued until the George brothers went back to work where they found the underbrush so thick it was hard to find the ledge. 100 feet south of the original they found the right place and produced good ore.

The **Page Mine** was said operated from 1926 to 1969, producing 14.6 million ounces of silver; 270,000 tons of lead, and 27,000 tons of zinc. **Evolution Mine** operated from 1908 to 1948, produced 9,000 tons silver, 66 tons lead, and 71 tons zinc.

The **Silver Dollar Mine** owned by the Sunshine people was said to have been the last mine in this area to use mules underground. The **Tamarack** production was said to be 8.7 million ounces of silver, 169,154 tons of lead and 72,208 tons of zinc. The **Frisco,** where "all hell" broke loose in the labor war of 1892, had a reported production from 1897 to 1967 of six million ounces of silver, 106,000 tons of lead, and 87,483 tons of zinc.

The **Ore Or No Go** was located by Col. W.R. Wallace, on Canyon Creek. After his claim was jumped in January 1888, he reclaimed it in March of that year and went on to making it produce. That claim, along with **Lucky Friday** and the **Hecla** claim were combined to make the main portion of the Hecla Mining Co. The **Lucky Friday** was located by B.P. Potts who sold a quarter interest of the mine in the fall of 1887 for $750. J.F. Ingals was said hard at work there by June, 1890 along with two men. Bob Horn who also sold out too quickly at the **Golden Chest** was associated with his partner Alf Brile in this claim. He sold too soon to know the great wealth. The **Hecla Claim** was owned by Simon Healey, Patrick Clark and George Hardesty in April 1891. They hit a promising strike showing $20 silver and small amounts of lead. Ten men worked in 1892.

Other mines of interest were the **Standard** and **Banner Mines.** In August, 1891 Campbell and Finch had paid their last $23,000 on the purchase which included the **Sullivan Fraction** in Canyon Creek. They had 16 men working the **Standard** by April 1893. Another mine, the **Polaris** near Osborn, had some of the richest ore taken out of the entire district by December 1891. It also had the largest mining electrical plant in the world.

The **Gem** was said sold by Finch and Campbell to an English syndicate for $920,000 in June 1892. J.J. Monaghan was foreman. This mine, along with many others, was on strike a good deal into late fall of 1893. The "War" erupted at the **Gem** and Monaghan was listed as dead until seen in Spokane; later to be in Marcus and at the Cleveland Mine.

Custer, Granite, and **Sierra Nevada** were all shipping ore by the end of July, 1892. Some of these mines hired non-union labor. **Tiger, Frisco, Poorman,** and **Granite** were all on strike, or closed for repairs by fall of 1893. They had some hopes of getting back to work by winter. Times were very hard. Over 2,000 prospectors and miners had left the district and were scouring the hills in Slocan by June, 1893. Silver was down to 77 cents – then down to 69 cents. Many mines had closed.

The **Gold Hunter,** or better known as the **Hunter Mine** and **Mill** was also of the 1884-85 discovery period – the mill being built in 1889. Not listed publicly as a member of the Mine Owner's Association, the Hunter did lay off 60 men in February 1892, presumably to show support for the other "member" mines. By mid-April, some gossip connected the **Hunter** with withdrawing from the Association because it was a "wet" mine and would willingly pay the three dollars and 50 cents to underground men.

Times were so difficult it was hard to tell who was on the side of labor — and who was not. By June, 1892 a report stated that neither the **Morning,** nor the **Hunter** belonged to the Owner's Association. At least the **Hunter** mine and mill were working in late July, 1892, with some 100 men. With prices so low and costs getting worse, Martin Curran the superintendent of the **Hunter** had to close the mine down in September with little hope then of it's opening again. (It did again briefly in 1894.)

The **Morning** was little more than a prospect in 1887 when it was purchased by Charles L. Dahler, of Helena, and Charles Hussey. They were offered $70,000 for the mine within a year. Dahler accepted the offer but Hussey was against it so paid for the other share and also bought the **Evening** after it had been abandoned by a Helena company. He built the mill in 1890. When the bank failed it put Hussey's property into receivership.

The mine shut down, and opened again a couple times in 1893. In 1905 the **Morning** was reported purchased by John Rockefeller for three point two million dollars in gold coins. It produced 15 million tons of lead-zinc-silver ore. By 1949 it was a leading CDA Mine.

HERCULES and FIREFLY

The **Hercules** and **Firefly** were located August 24, 1889 by Harry L. Day and Fred H. Harper. It was near Burke and it would take Harry, Eugene, and Jerome Day more than a decade to put **Hercules** on a profit basis. Assisted by August Paulsen, Levi W. Hutton, C.H. Reeves, Frank Rothrock, Dan Cardoner and H.F. Samuels, all became wealthy with his share of the mine. Father Henry Loren Day came from Maine in 1854 by way of California, and Nevada. He had searched for gold and in hearing about **Bunker Hill** and others, he came north to Wardner where he started a general store and dairy business. Harry's two sisters, Eleanor and Blanche, shared in the profits from **Hercules.** The mine dump was up Gorge Gulch. The mine produced 30 million ounces of silver, 38,367 tons of lead, and 4,808 tons of zinc before it was idled in 1965.

From 1884 through 1985, four point seven billion dollars worth of metals have been mined from this region. The district has produced over one billion ounces of silver "enough to make silver dollars that would girdle the earth at the equator," says one publication. Silver production is believed greater than in any other district in the world. More than eight and a half million tons of zinc, one half million ounces of gold, and nearly 200,000 tons of copper have also been taken from these mines.

The same publication states production of over $120 million per year from at least five producing mines at that period. Future production of silver for the region is at 14 to 19 million ounces of known

The main winze at the MORNING MINE, Jan. 28, 1949 went to the 3850-foot level, with an extended shaft down to 4850 level, the vein opened to 6400 feet. There was 40 miles of underground development then, with 300 men working. (S-R).

The town of Wardner around 1907 – Bunker Hill upper level working is on the upper right; Last Chance is on lower left. (Courtesy S-R).

reserves.

Facts for much of the 1892 labor war years has been gained from a cross section of information from the Idaho Mining Association; from mining company publications, and some corroboration from a great book, **"Coeur d'Alene Diary"** written by Richard G. Magnuson, of Wallace. (The book is recommended reading.) Another source for much area history is from the **"Historic Wallace, Idaho"** publication picked up at their great museum. Information for that is from Norm Radford, and Pat and Sherrill Grounds, editors; with many good contributors; and from many other sources of history.

As we proceed to the labor war of 1899; the growth and decline of at least a part of the Coeur d'Alene mines; relate the history of some of the **greats** of that area, we eventually leave it for other mining districts that we will read about – but will find vastly interesting. But we have a few years to go including much activity before we leave the Coeur d'Alenes.

Still a hangover from the labor war of 1892 was John Kneebone, a strike-breaker and leading witness for the State in the previous trials. He had returned to the **Gem Mine** to work when trouble flared in July 1894. A band of masked men are said to have appeared at the mine and in his attempt to escape, Kneebone was shot down. A demonstration followed and a few weeks later a grand jury was called at Murray to investigate his death, but little came from it.

Heading into Warfare 1899

In the winter of 1897, uneasiness again prevailed in the whole district. Frederick D. Whitney, foreman for the **Helena-Frisco** concentrator was yanked out of bed, taken to the street and shot. He died as a result of the gunshot wound. A $15,000 reward was offered by the Mine Owner's Association for the arrest and conviction of the guilty man, and Governor Frank Stuenenberg reportedly offered $1,000 for each conviction. It did not help; fear hung heavy over the region and the labor situation was such that it made it impossible for anyone to speak up.

John Hays Hammond, president of Bunker Hill, left for South African gold and diamond mines in the summer of 1893, and reportedly took Victor Clement with him. N.H. Harris had been named president. Frederick Worthen Bradley became general manager, and went on to become president about 1897. (Bradley's administration ended in 1933 at his death.) It became Bradley's responsibility to ride out the economic storms, and turmoil, that continued to prevail. Frederick Burbridge, of Wardner, was named resident manager. The mine employed some 300 to 400 men in 1897, and built a large concentrating mill that year. Burbridge had been assayer for the mine in 1893. He stayed with the Company until 1901.

Even with labor troubles hanging over the area, Idaho mines were reported prosperous in 1898. $12,400,000 was said taken from Shoshone County mines. A second, and more conservative figure gave a total of 112,500 tons taken from the mines averaging $65 a ton, for a aggregate of ores and concentrates shipped bringing $7,312,500.

War – 1899

April 29, 1899 open warfare broke out again, with the strongest sentiment directed against the **Bunker Hill.** Historians tell of the wild capture of a train which was loaded with some 1,200 union men. Picking up a ton and a half of dynamite enroute, the union forces descended upon the Bunker Hill concentrator. The $200,000 mill was blasted to bits, the office, boardinghouse, and bunkhouse were said burned, and as the smoke cleared away – two men lay dead. Governor Frank Stuenenberg immediately went into action.

Shoshone County was put under martial law for the second time in seven years. Federal troops arrived and by May 7, 1899 some 518 prisoners were reported rounded up from the surrounding hills. The leaders were taken to Boise, tried and convicted, and sentenced to a Federal Penitentiary at Detroit. As the mines of the district re-opened, special permits for jobs had to be obtained from a representative of the State. These stated the miner had never been a member of the union, or that he had left it and would not join again.

When the prisoners were held prior to court, there were some in boxcars, and over 300 in a barn. These prisoners were transferred to a barbed-wire stockade, or "bull pen" due to the huge number. This treatment became the subject of a congressional investigation and more problems in the area.

Due to the 1892, and the 1899 handling of the union men, the martial law being invoked each time, the "bull-pen" treatment, and all the wage and personal problems imagined or real, threats against Governor Steunenberg were made – many made long after he had left office. On the evening of Dec. 30, 1905, as he opened the gate leading to the back door of his house, an explosion so mangled his body that he died as a result.

A professional dynamiter, Albert E. Horsley, alias Harry Orchard was charged and eventually confessed to setting the bomb. (Orchard was sentenced to life imprisonment and was still alive in 1962.) Clarence Darrow, a most famous attorney defended Orchard; also George A. Pettibone (of the 1892 labor war), William D. Haywood, and Charles H. Moyer. The three were officers of the Western Federation of Miners, (some from the union coming from Montana). Pettibone and Haywood were acquitted and charges were dropped against Moyer.

Labor troubles would flare again but until the end of World War I (1918), some harmony between

On April 29, 1899 BUNKER HILL'S $200,000 concentrator was blasted to bits. Two men were dead and there was said to be 1,530 miners out of work when Martial Law became effective. (S-R).

the Mine Owners and the union had been reached.

Colonel Wallace had fought many battles regarding his claim to his townsite. Though he had founded Wallace, there always were some legalities to disprove his claim. He finally moved out of the district and settled in Preston, Ariz. Later years he moved to Whittier, Calif., where he died in November, 1901.

Another area bit of interest – and sometimes contention, was the dredge operated by the Coeur d'Alene mine owners from 1930 until 1960 at Cataldo Flat. A dredge fund was set up to appease people who were concerned about the mine wastes getting into the lake. The "houseboat" machinery operated six months out of the year, 24 hours around, at a deeper and wider point of the river near the Cataldo Mission.

Because silver and lead were down so low in price, there were those who were still interested in gold mining. The **Nellie Wood** and **Alma** claims on Elk Creek was still working with 19 some men, increasing to 25 and 30 by early 1893, their stamp mill running full blast day and night. The claims were being operated by the Pandora Mining Company with Clarence Cunningham as superintendent.

These old claims were once owned by Jim Wardner (of Bunker Hill fame). The solely gold claim was averaging 12 ounces to the ton. In July there were said 18 men working on a cooperative plan. A reported discovery of another four foot vein of free milling ore assured their ten stamps of steady work for awhile.

Another gold property of interest then, and today, is the **Golden Chest.** Like several other gold properties in the area, this mine was paying dividends by late fall 1889. Said once owned by Bob Horn, a native of Main, Horn and his partner Alf Brile of Spokane, sold the **Golden Chest** claim in the Murray district early in this game. Horn was said to have moved out of the district by 1885.

Located near the **Mother Lode** that is said to have turned out a bar of gold every seven days with value of $1800 (we do not know for what length of time), the **Golden Chest** employed 22 men. By early spring 1892, the **Golden Chest** mine was said to match the **Mother Lode** output, with John Coumerilh the manager. The **Mother Lode** did clear $10,000 in May, 1892.

We do know the **Golden Chest** was producing

War II regulations, the mine was closed down at that time. Gibbs later came to Colville to mine the **Old Dominion,** and **Bonanza.** In August of 1987 it is reported that exploration drilling on a 3600 acre site is being conducted by Newmont Mining Corporation, at **Golden Chest.**

Shareholders and officers of the **Golden Chest** property include Minnesota Vikings tight end John Beasley, some of his teammates, and Brenda Kalatzes, a former Miss Utah who owns the **Vanadium King Uranium Mine** near Price, Utah. Beasley is president of **Tap** Resources Ltd., and Kalatzes is **Tap's** Corporate Secretary. Beasley played with the Vikings in Super Bowl IV in 1970 and retired from football in 1975. He holds a degree in natural resources. This is a joint venture into several old tunnels to include the **Idaho Vein,** the **Katie-Dora,** the **Klondike, Vik,** and **Dolly.** The mine's owner is composer Johnny Green, of Hollywood. He leases the mine to Golden Chest, Inc., and **Tap** Resources has a management contract, according to August 1987, and January 1988 news releases. Green died in May, 1989. In August, 1989, 200,000 ounces of gold reserve was reported discovered.

Some Towns, People and Points of Interest

Some mine owners tried to bring in Chinese, but in the early 1890's they were warned to stay away as in "Not Wanted in The Coeur d'Alene's." Strangely, up on the north-fork of the Clearwater at Pierce, the Chinese miners settled in droves. At first they were excluded by law, but with only 150 or so whites left in the district, a vote opened that area to the Chinese. By 1870 there were said to be 800 Chinese miners who did well on their findings, because they could survive on as little as five dollars a day.

Ignored at first, then hated, Sherry Devlin, staff writer for the Spokesman-Review states that Pierce owes it's existence today to the Chinese. She is quoting Darby Stapp, director of a University of Idaho "dig" that is going on in the area and will continue until 1992. One statement is that $400,000 in gold was taken by hundreds of Chinese where white miners were said to have taken one million dollars, then another $800,000. The Chinese continued to mine until a judge ruled their claims were illegal and in 1890 State court decisions barred the Chinese from mine ownership. More than half were gone within a few weeks and she states the last Chinese in Pierce was murdered in 1930. One mining report credited the Pierce area with $70 million dollars in gold having been taken from 1860 to 1870. Records were probably never kept from Chinese production, according to the Spokesman-Review story.

This report said 700 miners were held for four months in the "bull pen." The treatment became the subject of congressional investigaton and at the time three county commissioners and a sheriff were removed from office. (S-R).

Molly-B'Damm

The Coeur d'Alenes may not have been credited with Chinese residents, but by the 1880's, every town was known for it's saloons, and for it's red-light district excepting one (at the specific time of this writing). From Wallace Free Press came the best description of the house of ill repute. Stating that "Mullan was the one town in the Coeur d'Alenes which is not yet afflicted with those individuals of the feminine gender who build their houses on the way to hell, those beings who have been forsaken by the purity of the angels and who flaunt their festering identity in the face of public decency–"

One of the great historians of the Inland Empire, Rowland Bond in his tribute to early outstanding women gave cause for other opinions. He tells of Molly Berdan from Murray. Better known as "Molly-B'Damm," she died in 1888 at age 34. "On the slope above Murry is the grave of a woman whose life and profession coursed in such a manner that her death was mourned by every man, woman, and child in the Coeur d'Alenes."

She was a friend of Phil O'Rourke, Con Sullivan, and Dutch Jake Goetz, finders of the **Bunker Hill and Sullivan** . . ., and many others. Men sought favors of this blue-eyed Celtic lass who had been born in Dublin. She had been married to a man in New York who turned Molly into a thing for profit after his own money ran out.

Born Margaret Hall, Molly Berdan occupied cabin No. 1 on Gold Street in Murray's red-light district. Although her occupation was clear, she spent most of her time ministering to the sick and helping the needy. She was especially good with children. She was so well liked that when she contracted tuberculosis in 1887, the women of the town provided round-the-clock nursing service in cabin No. 1. She was refused absolution but three Protestant ministers and more than 1,000 friends gathered for her funeral January 18, 1888.

The courthouse and other outstanding places of business were closed in her honor and the Murray newspaper was black bordered. Her warm-hearted generosity was extolled and "she has drawn more public attention than any other woman in this part of the country for her many generous deeds towards others. We will never forget . . . Molly Berdan."

Another reminder in Murray, of days gone past is huge mounds of rock piled up by the gold dredge operated by the Yukon Gold Mining Co., which chewed down 30 feet to bedrock for some seven and a half miles each way of the town. Wendell Brainard tells that this dredging went on over a period of eight years. It was said that gold worth over $20 million (at modern day prices) was recovered. He adds that visitors are welcome to search for a gold nugget. (Courtesy "Historic Wallace, Idaho".)

Kingston and Osborn

Kingston with it's seven saloons, three bakeries, four restaurants, two livery stables, three barber shops, three general stores, plus hotel and rooming house seemed quite a village at that time. Osborn at the junction of Murray and Mullan road had a seven tent hospital by April 1891. More-often called Georgetown, The **Polaris**, and **Yankee** claims were nearby.

Delta, Mission, and Skiing

Other towns of the time include: Delta, said to have 3,000 population earlier and by 1889 was down to 300; and Mission, another "bedroom" town to several hundred working the mines. There was the town located and called by one - Jackass, later called Milo, then Kellogg. Wardner had lawyers and a dentist by July, 1887. The bad part was that travel as far as Mullan was under water part of the time. As **Bunker Hill** families came, coasting and the use of bobsleds started at the mine. Sleds ran almost to Wardner Junction, a mile away. By 1980 the hill became one of the better known ski-hills – "Silver Mountain;" and a committee was asking federal funding for a gondola to link Kellogg with the ski area in the late 1980's. (1991) operating.

Gem; Burke

There was Gem; earlier called Davenport, the town near the **Gold Hunter, Morning, and Evening, Black Bear,** the **Gem** and **The Frisco Mines.** With so much activity there, it does not surprise that an old newspaper picked up in Wallace relates: "Mayor Jackson ran a first class saloon in Gem and all trains stopped for 15 minutes on the run between Wallace and Burke." (One stop in a seven mile run wasn't too long to wait?)

Burke, up Canyon Creek was Andrew Prichard country. Col. W.R. Wallace was one of the Prichard party locating the **Evolution.** Prichard, Wallace, Theodore Davis, a Mr. Gilbert and a Kirby formed the Evolution Mining District. Each staked a claim. Wallace located **Ore Or No Go** May 10, 1884. That would become part of the Hecla that we know today.

The **Tiger-Poorman** mine; the **Frisco,** and eventually the **Star** would be the main-stay of Burke. All would remain open until the **Star,** being the last, closed June 17, 1982. Most of the residents of the town left long before that . . . with Keith Goodman stating in an article Sept. 11, 1983, there were 20 people left in Burke.

John Kennaugh, of Wallace, was born in 1903 at Burke. He remembers when that canyon was filled with people. Nearly 2,000 people lived in the narrow canyon through the 1920's. Goodman tells of a fire in 1923 that destroyed much of the town and killed 72 people (we have no verification of this). He said while the fire raged most of the miners worked at the **Hercules Mine** three miles up the canyon. That mine, along with the **Tiger** and **Poorman** were closed with only the **Star** working until 1982.

One loss in the fire of 1923 would be the famous Tiger Hotel in Burke, built with tracks running thru and over Canyon Creek. It has never been rebuilt.

Wallace

Wallace is one of the most historical cities in this "silver valley." It has approximately 1,800 people living in this beautiful setting formed by the junction of four major canyons, three leading to important mining districts.

Colonel Wallace built his cabin here in 1884. Six years of grubbing brush and cedar trees, and pioneers started building wooden frame buildings. By the time of statehood, the town had passed the log cabin stage and had some three story log buildings. It seemed unfair but Wallace had been beaten in his townsite case coming about with what was said the use of illegal tender in payments; claims that he did not receive notice of some ruling Jan. 24, 1887 – and the end result that his holdings in the town were not valid. John Mullan (of the Mullan Road) had served as one of Wallace's attorneys in this case.

The town was started as Placer Center. Mrs. Wallace came June 22, 1885 and soon Lucy Wallace became the town post-mistress. It was she who convinced the postal department the town name should be returned to honor her husband Colonel Wallace, on her appointment in August, 1886.

New businesses were coming about and John Cameron had the first saloon in the winter of '86. The town was incorporated May 2, 1888. A jail was badly needed so Wallace donated a lot for a city jail, and another lot for a school. On July 27, 1890 flames from a defective flue caught buildings afire and soon the entire business district was gone. Of great concern seemed to be the loss of 20 "dives" listing financial loss of some eight thousand dollars. The bank was burned but quickly set up in a tent next to the ruins so that the vault that remained could be used.

Food, lumber and help was quickly sent from Mullan and other neighboring towns. The mines were doing well so money was available for replacement of buildings and stock. It was said that a lot of 25 feet sold for as high as $2,200 cash.

Some newspapers had written of the debauchery that ensued from the terror of such a loss. The town was declared in a horrible state of drunkenness. Because there were families and people building towards a good life in a townsite they loved, we would come closer to believing this story: eight men and one woman rolled a barrel of whiskey up the valley and made a night of it. Some 50 men were said drunk in the vicinity of the Northern Pacific depot. And come morning they all went back to rebuilding the town of Wallace.

Before recent closures, the town was the center of where 40 percent of our silver is mined in the United States. It is true that it's local newspaper masthead gives the price of silver and gold alongside it's name **"The Wallace Miner."** The town itself built along Highway 90 brings in much tourist trade, partly because of it's buildings, a very large percentage of which are now on the National Register of Historic Places. Structures are beautiful, many dating from 1890. Yes, there are still said to be ten bars in a four-square-block area, with one restored to the gracious era of 1880.

It was the famous landmark the Old Northern Pacific Railroad depot that drew the movie film crew to make "Heaven's Gate," in Wallace. The film

Murray, Idaho in it's hey-day around 1890. (Al Nugent).

The Murray-Wallace Stage with driver said to be Robert Sage. The line also hauled freight in the 1890 period. (Al Nugent).

company was also drawn by the buildings, the history, and the people who are very proud of what they have developed as the hub of one of the richest mining areas in the world. A visit to their museum, and a tour of the Sierra Silver Mine will convince that all that mining heyday will return one day and the "Silver Valley' will once again be one of the most productive in the world. The depot was relocated in 1986 to make for Interstate 90. It is a marvelous railroad museum now.

Early Formation of HECLA MINES

Made up from some of the mines we have researched is the great **Hecla Mining Company.** As early as 1891 the **Hecla** claims was mined through shafts in Burke, and said to be the origin of this great company. The lead-silver mine was just south of the **Tiger Poorman.** The company was incorporated in Idaho by Patrick Clark, Finch, Campbell, George Hardesty, and Simon Healey in 1891 with capital of $50,000.

The same group also owned the **Katie May** claim which was worked intermittently to 1897, mostly by lessers. $14,000 worth of ore was reported extracted. The owners located ground for a prospective townsite and railway terminal at Burke. Showing poor prospects, the railway right-of-way was sold for money to do further development. Shares of one dollar each dropped to a cent or two by 1897.

The property was leased that year by Joseph Dolan, Dan Cardoner, and John Clark, but no improvement was done. In 1897 it became a Washington corporation, with capital of $250,000. Purchasing the Van Dorn property for a reported $27,500 helped push shares up to 25 cents par . . . First milling was done at the **Standard** mill, at Wallace, then later at the **Gem** mill which the company had acquired by then.

Finch and Campbell controlled the property until 1903, with E.H. Moffitt as manager. A young man, James F. McCarthy, who had earlier spent three years around the mines of Rossland, B.C., came to Wallace in 1899. He worked first at the **Mammoth** until it was taken over by the Federal Company, and soon found his talents in demand by the Hecla Company, which he served for the next 37 years.

First dividends were paid in 1900, a total of $100,000. **Mammoth** was said producing better than $89 silver and 55 percent lead. Lew (leu) H. Hanley had also worked for **Mammoth** in 1899. J.R. Smith, of Chicago, and his associates, had gained control. Former **Bunker Hill** man R.K. Neill was an early day superintendent, and McCarthy was manager.

By January 1906, Hecla company was said employing 125 men underground, and that $550,000 had been paid out in dividends. McCarthy saw great growth for the company as it shared a percentage of the **Star** zinc property with Bunker Hill. Then he directed the development of the **Polaris** silver property to the benefit of his company. He died in 1940 and Hanley became president of Hecla Mining Co.

The original mine was said worked out about 1940. This "Holding and Mining Company" is said by then to have certain ownership, and/or part control over **Ore Or No Go** (W.R. Wallace's claim at Placer Center); and the **Polaris,** both of the 1884 discovery period. They shared percentage with Bunker Hill over **Star;** a percentage over **Silver Summit;** a merge with Lucky Friday, and stock interest in Bunker Hill. As of 1963 they operated four mills, the **Star, Polaris, and Lucky Friday,** with the fourth in Utah, the **Mayflower.**

Because of Hecla's huge success with **Lucky Friday,** the silver mine that vies with **Sunshine** in a production race, we will relate a bit of that history prior to it's December, 1958 agreement with Hecla; and major stockholders of the mine.

LUCKY FRIDAY

Lucky Friday, and **Good Friday** claims were located in 1899. Nothing more than minor exploration was carried on until 1906. With very little return shown, the mine was sold at a sheriff's sale in 1912 to pay labor costs. By 1926 the Lucky Friday Mining Company obtained patents on the four original claims. Following mostly an idle period it became the property of Shoshone County in 1936 due to delinquent taxes. Still idle until 1938 one publication stated the mine was redeemed for unbelievable sum of $120.

One of Wallace, Idaho's buildings on the National Register of Historic Places. (S-R).

John Sekulic, of Mullan, is said to have obtained a lease and option to buy, at $15,000. Along with Charles Horning, they formed the Lucky Friday Silver-Lead Mines Company on March 30, 1939, that company coming about by lease and option. Some development was financed through shares but the Company was faced with discouragement again. A shaft was started in 1939 which led to ore being found by 1941.

Judge A.H. Featherstone, along with Sekulic and Horning entered into contracts with Golconda Lead Mines, and the first ore shipment of 478 tons was made in January 1942, to the **Golconda** custom mill. In 1946 working agreements were made with Hunter Creek Mining Company. By 1951, 80,000 tons of ore had been mined and profits permitted the first dividend.

The Hecla Mining Company took over management of the silver-lead mine in 1958; by 1960 had concluded a one million dollar plant expansion, and by 1961 had installed a $500,000 hoist. Development (and dividends) have been reported since. 1962 production was 700 tons per day; smelter returns to 1963 were about $30 million, and dividends about seven and a half million dollars.

Silver Summit

Historically the **Silver Summit** showed little cause for interest prior to 1930 when Harry Pear-

Wallace, Idaho railroad museum. Bricks imported from China in 1890 were used in part of the building meant then to be a hotel in Tacoma, Wash. Never completed as a hotel, it was built to serve as depot since 1902. Relocated to make way for Highway 90 in 1986, it is an outstanding museum, (S-R, and Tim Dunahee).

HECLA MINE with Burke in the distance (S-R).

Taken in Hecla Tunnel - shows a loaded ore train on a sidetrack waiting to be dumped on an ore skip and taken to the surface. An ore skip holds one car load. (Taken April 29, 1943. (S-R).

son and associates combined several claims, along with new ones they had located. The vein was said reached in 1931. In 1945, Polaris entered into a contract with Silver Summit Mining Company, and in the fall of 1952, **Silver Summit** was said merged into Polaris Mining Company, becoming a Polaris operation. From 1953 to 1960, Polaris, and the Defense Mineral Exploration Administration entered agreements for various explorations of the Silver Belt.

Silver Summit and Polaris into Hecla

On Oct. 31, 1958, Polaris Mining Company was merged into Hecla, after which Hecla owned all the interest held by Polaris including **Silver Summit Mine,** and the **Polaris** mill. (This mill was said by Lou Grant, to have been built in one of the coldest winters he remembers, in 1936. Since 1948, more than 16 million ounces of silver, and some copper, gold, lead and zinc, have been produced. The end has been predicted often but invariably new ore is found. Production was predicted for some time ahead in 1963.)

By way of this consolidation, Hecla combined the old **Silver Summit** and **The Polaris,** into properties they are presently exploring with Coeur d'Alene Mines. The two mines had produced until 1963, and 1969 respectively. A news release described the area as a ride; a mile into the mountain on a mine rail – the next leg of the journey is about a mile straight down the mine's shaft. The **Silver Summit** takes off at one level, the **Sunshine** at another. A man-way up the shaft is a series of ladders that could be climbed in case the hoists couldn't operate — the rungs are a foot apart. That makes 5,400 steps and he notes the 102 story Empire State building is only 1,250 feet high. Steve Bofenkamp tells of working in this mine in 1963 when it was called the **Polaris.**

Hecla's Mining operation near Burke, June 1, 1935 (S-R).

LUCKY FRIDAY shaft is 6,250 ft. below the surface, the deep mine is owned by Hecla. Taking 20 some years to develop, the mine has produced 4,700,000 ounces of silver; 2,400 ounces of gold, 35,000 tons of lead, and 3,900 tons of zinc up to 1985.

The purchases, leases, partial ownership, and mine operation of another mine becomes such a twist of details that it is impossible to follow all transactions. The fact that mines are being operated through other mine shafts; exploration is being done by one company for another, is all part of the mining game. As long as such operations continue, and there is production – and prices are right, this area will continue to be one of the world's best mining districts.

ASARCO – Early Organization

Returning briefly to Florence, Ida., we find a second boom there in the spring of 1896. By the fall of 1898 prospectors from around the west were gathering there. Among discoveries was gold quartz at **Big Buffalo,** and **Merrimac,** by Bert Rigley and Charles Robbins. The **Oro Fino** came into being, and Buffalo Hump became more than just a name. Clark and Sweeney, representing eastern capital, were said to have offered $500,000 for the three claims. The **Comstock** was located near Dixie, Ida., by Thompson, in 1898.

In early 1900, Buffalo Hump Mining Company bought **Tiger-Poorman** at Burke, and in 1901 these were said absorbed into the Empire State-Idaho Company. That company was said organized in 1898 to control the **Last Chance Mine** at Wardner, and to acquire additional territory west of Milo Gulch. (The **Last Chance Mine,** operated by Charles Sweeney, and F.R. Moore, was in reality the nucleus of the Federal Mining and Smelting Company.) It was formed September 1903 and took over Empire State holdings.

The **Morning Mine** was said sold to Federal Mining and Smelting in 1905. (The original **Noonday** vein of this old 1884 mine was said mined out in 1954 and the mine closed. A 1963 report stated Hecla was mining it through the **Star Mine.)** Federal was said to also have taken over Empire State holdings of **Standard** and **Mammoth Mines.** Lengthy litigation followed the transaction concerning the application of mining laws, between Last Chance Mining Company, and it's successors, and Bunker Hill and Sullivan Company. The U.S. circuit Court of Appeals are said to have rendered in favor of **Bunker Hill** in 1904.

The Federal Company was authorized capital for $30 million in 1906. One report said the final consolidation between Empire State and Federal Mining and Smelting was in 1909. (**Henry L. Day** son of **Harry L. Day**, president of the Day Mines, Inc., was general manager of the Federal Company

A generation later and we found ladies at the LUCKY FRIDAY – Marge White, Alice Wolfinger, Diane Maines emphasize the THINK SAFETY FIRST sign. Taken April 12, 1983. (S-R)

in 1913). Fred Burbidge was reported to be general manager from 1916-1930. **Morning Mine** was their heaviest producer. In 1953 a merger brought about the American Smelting and Refining Company – better known as Asarco. (A note of interest – Harry Day was elected president of the reorganized Idaho Mining Association in 1913 – first organized in 1903.)

The old **Standard-Mammoth,** once a well-known producing property, closed about 1925. Finch and Campbell were responsible for early development along also with the **Gem.** That same year the **Page Mine** was re-opened by Burbridge. It was reported the Federal Company chose to reactivate the **Corrigan** shaft on the **Curlew** vein of the **Page** property in 1924. All, or portions of these mines came under this operation. The Page property four miles west of Kellogg was still in operation in 1963 by the Northwestern Mining Department of Asarco.

Other Idaho operations of Asarco in the 1963 period included the **Galena,** one of the oldest mines in the district, and located one mile west of Wallace. (The property at the time was owned by Callahan Mining Corporation, and under long time lease to Asarco and Day Mines Inc.)

Asarco has continued to mine **Galena,** and **Coeur,** through failing times and on into 1989.

GOLCONDA MILL where the first shipment of LUCKY FRIDAY ore was sent January 1942. (S-R).

Tacoma Smelter

Their holding from 1911, the Ruston Smelter at Tacoma, Wash., was finally closed down June, 1985. Asarco had planned to close before that time but the plant was one of the few in the world equipped to handle copper bearing ores that contained substantial amounts of impurities such as arsenic. The last load of copper ore was processed through the blast furnace at the **Asarco** smelter March 26, 1985 closing a 74 year chapter of local history. 500 persons were on the staff and a clean-up crew of 150 were working for another six to 10 weeks.

It has faced the impossible to meet with environmental requirements. Also, the low-arsenic concentrates used in making copper came from the Philippines and were no longer available. The dispensed material was used in agricultural pesticides and in metallurgy. Because arsenic apparently wafted over the community the smelter received national attention in 1983 from the federal EPA.

The plant sits on 95 acres of some of the most desirable property in Pierce County, Wash. It is not clear what will become of it. Tom White, plant superintendent, says the company is unsure what can be accommodated on the site. **Asarco** plans to demolish some of the old buildings and the 485 foot smokestack, but problems arise there in that demolition might stir up some of the arsenic settled on the plant buildings.

Hercules

One of the great ore discoveries of the early period, the **Hercules** on Tiger Peak, just north of Burke was located by Harry L. Day and partners in 1889. The mine began to ship ore in 1901. Ray J. Horsman was chief engineer for a period. (Mr. Horsman retired in 1964 as project engineer for **Bunker Hill.** He had also been associated with **Homestake Mine** in South Dakota and with Phelps Dodge in Arizona. He was also a great advisor and helped in acquiring information for this book. This writer, and the readers owe him a lot.)

Gross production for the first three years from **Hercules** exceeded two million dollars in lead and silver. Crude ore was hauled by wagon, or sled to the railway at Burke,and monthly shipments were said from 1,000 to 1,200 tons by 1904.

New LUCKY FRIDAY mill Feb. 22, 1960 where more than 550 tons of ore a day was processed that week. The mill cost $400,000. DAY MINES-GOLD HUNTER property is in far background. (S-R).

Assisting in the efforts to "bring in" the **Hercules** were men from Spokane, including August Paulsen and Levi W. Hutton. It has been said many times that for a $500 investment, the dairy employee Paulsen became a very rich man. He did return much of that fortune to his home city by building an 11-story office building, followed by a 19-story medical and dental building. Mr. Hutton, a railway engineer, also brought his wealth home and built an office building and a group of houses for homeless children. His Hutton Settlement is a memorial to having been an orphan himself.

The **Hercules** was said closed down in 1915 (other reports say 1925). It was reopened about 1948 and in 1953 consolidated under Day Mines ownership. Their holdings include **Hercules, Tamarack, Dayrock, Sherman,** and **Monitor.** A 1947-1958 report shows 3,494,295 tons of ore produced, 30 million ounces of silver and named the **Hercules** as seventh in line of silver producer, and fourth in lead production in the district.

Of the two "silver wars" going on in the Coeur d'Alenes, in 1981, one was the hostile takeover of Day Mines when it was merged into Hecla Mining Company on the conclusion of business October 21, 1981. This included **Hercules.** The total figure accepted by Day Mines, and said agreeable according to Elmer Bierly was $105.8 million for Day stock. William A. Griffith, president of Hecla declared that a great day.

Caladay Mine

Callahan Mining Corporation reported a contract was let to sink the shaft and do all the drift work in another mine to remember – the **Caladay Mine.** Announced Dec. 28, 1980, Callahan estimated the total cost of the exploration effort to be at $26.6 million. The shaft is to be 5,100 feet.

The **Caladay** is located on the eastern boundary of the **Galena.** Wallace Diamond Drilling won the contract. Owner LoVon Fausett lists nine crews through the western states ready to drill as needed.

Canyon Silver Mine

Robert Dunsmore tells about the **Canyon Silver Mine,** formerly the **Formosa,** another mine of the early years. He says "when ore was struck in the mine all the employees and their families went to Hawaii at Company expense." Production was said from 1931-1974 to be 101,000 ounces of silver, 1,537 tons of lead, and 520 tons of zinc.

A report Oct. 16, 1983 written by Doug Clark, Spokesman-Review, tells of 11 men toiling in the mine trying to bring it back to life. Don Webster, Rich Hayman, and Ben Walters claim they haven't made a dime and that they are all broke. Two of the men are 24 years old and Walters is 53 with 27 years experience with **Hecla,** and five at **Canyon.**

The mine, said located in 1926 was called the **Formosa,** and a shaft of 100 feet was put down.

The mine was reported closed 1940; reopened and renamed in 1968 by William Morrow and his brother Roland. When bought, the site was to be a wrecking yard. An old time miner named Ralph Fritz talked the brothers into working the mine again, claiming there was still silver.

Using their own money they hired a crew; sunk a shaft to 800 feet and put it on the stock market where it was once listed at four dollars and 30 cents per share, then down to 35 cents. William Morrow was said electrocuted while fixing a pump at the 800 foot level. After his death, the mine was leased to a Seattle corporation. They abandoned it and the **Canyon** was said closed for another eight years. It was then that Ben Walters took a lease from Morrow. All of his help is "out of work miners" doing the hardest of work mucking the portal out by hand. They hoped the 600 foot level would pay enough to go on to the 800 foot level. The most serious problem they face is getting hurt. As these men are building experience that would take them onto a job anywhere, they are facing the same as so many mines of the Coeur d'Alene's – slow progress but which one day may make another top producer in that mining district.

Top producers of silver in 1982 as reported October 16, 1983 are;

1. **Asarco's Troy** in Montana – 4.2 million ounces
2. **Hecla's Lucky Friday** – 3.84 million ounces
3. **Asarco's Galena** – 3.8 million ounces
4. **Asarco's Coeur** – 2.5 million ounces
5. **Sunshine**
6. Four Mines in (Arizona, Utah, & Colorado) –
7. **Mapco's Delamar** (in southern Idaho) – 1.6 million ounces

Sunshine Fire

The great **Sunshine Mine** suffered one of the worst disasters in the Coeur d'Alene district when fire May 2, 1972 claimed the lives of 91 of their miners. 174 miners were on the job. Supposition was that the fire was possibly due to spontaneous combustion in a portion of the mine that was no longer in use. Flames, smoke and poisonous carbon monoxide swept through the mine. 81 miners reached safety. Rescue operations were started for 93 men still underground. As wives and families of the missing miners stood a grim vigil near the mine, each day brought discovery of new bodies and reduced hopes that men would be found.

Air was pumped into the tunnel and rescuers attempted all fronts. Six days after the fire and explosion, a new blast of heat and smoke turned the mine into an inferno. New hope was felt on the

LUCKY FRIDAY shaft. 1962 production was 700 tons a day with smelter returns of $30 million and dividends of about $7 1/2 million (S-R).

Miners change shift at HECLA'S STAR MINE. Notice sign "The best safety device is under your hard hat." Also notice ladies being part of the work force. Photo Dec. 1979 (S-R).

LUCKY FRIDAY miners on a shift change just days before the LUCKY FRIDAY closed because of low silver prices, April 1, 1986 (courtesy S-R).

Terrible hardships existed as prospectors stampeded into the Buffalo Hump country. By 1900-01 some large mines had been established. (S-R-1899).

7th day, May 9, when Tom Wilkerson, 29, and Ron Flory, 28, were found alive and in good health. They had survived by out-running the smoke and heat to a clear area, and by eating out of the lunch boxes of those less fortunate. All further hopes vanished when 40 bodies were found together – a total dead of 91. The investigation revealed the fire may have burned for months in the unused portion of the mine. The tragedy was that 173 children were left fatherless. **Sunshine Mine** is said to be the deepest below sea level mine.

Silver Summit

In December 1979, Alden Hull, president of Consolidated Silver Corporation, which now owns the **Silver Summit** shaft, announced that four mining companies – **Sunshine** and **Silver Dollar**, **Hecla** and **Coeur d'Alene** were forming a joint venture to lease the company's property and re-open the mine. It is Hecla Mining Company that will actually operate the mine and Dave Wolfe, Hecla's public information officer explained that the re-opening of the mine was based on silver price. A reserve of 67,000 tons of ore was said in the mine, with 20 ounces of silver to a ton.

The purpose of the project is to explore for more ore at deeper levels. The mine shaft is to be sunk an additional 1,000 feet from it's current depth of 4,590 to making it 5,524 feet deep. It was guessed at the time that five years would be needed and cost would be shared as to each company's percentage of ownership.

These massive joint explorations often bring about legal problems that have to be faced such as Silver Syndicate versus Sunshine. The case often referred to as **The David and Goliath** relationship between Sunshine and smaller companies was in court on Oct. 7, 1979. It involved the "Apex Law"

The happy face of Floyd Longley who kept track of miners' lamps at Galena Silver Mine through the crash of 1981 that did not stop this mine. (S-R).

which states that whoever owns the highest portion of an ore body has the right to pursue that vein wherever it goes, even if the vein goes under somebody else's property, as reported by Barton Preecs, Spokesman-Review writer.

Finding that "Apex" is not always so easy, so the two companies negotiated a contract covering the **Chester Vein** agreeing to share the ore on a 50-50 basis. The contract was not clearly written and when three more veins were discovered in the same area within 200 feet on either side, trouble arose. District Court Judge James G. Towles ruled in favor of Syndicate in 1976. The Supreme court affirmed the decision in October, 1979.

A dispute between Sunshine and Metropolitan Mines follows similar lines with some differences. Sunshine filed suit in March to force Metro to allow Sunshine to pursue development of an ore body called the **Copper Vein,** at the 5,000 foot level into Metro property. Then Metro filed a countersuit denying Sunshine the right to mine the ore in question and filed other charges. Metro has no shaft and has Sunshine mining it's ore and sharing the profit. Metro was asking $50 million in damages as a result of the action it claims Sunshine has against Metro. Claims were not settled at date of

GALENA MINE, July 22, 1933 - still in operation by Asarco into 1988 and beyond. (S-R)

COEUR d'ALENE mines corporation at Wallace, Ida. Jan. 30, 1945 and still operating in 1988. (Courtesy S-R).

91 miners lost their lives in the SUNSHINE MINE fire May 2, 1972. 173 children lost their fathers. (Courtesy S-R).

the publication in 1979, but is an example of mining at less cost to come about the very best production and more profit, but once in awhile becomes a legal snarl.

Before we leave the Coeur d'Alene district there are a few more developments that are pertinent. We must again drop back in time to catch up on portions of our history otherwise lost.

The scars of two labor wars were healing and a mill at **Bunker Hill** was under the direction of a colorful gentleman from Italy, Gelasio Caetani. He remained as mill superintendent from 1905 to 1910, during which he conducted various experiments concerning milling, and laid the ground work for the west mill which commenced operation on Nov. 9, 1909. This romantic young 1901 graduate of the University of Rome was a civil engineer who turned to mining and metallurgy. Following a year's study in Liege, Belgium, he graduated in 1903 from Columbia University as an engineer of mines. With letters of introduction to Stanly A. Easton, then general manager of the Bunker Hill and Sullivan organization, Prince Caetani headed for that far western mine. (Easton became vice president in 1925, and president in 1933.)

At the end of Prince Caetani's highly successful service with Bunker Hill and following World War I, he was appointed Royal Ambassador from Italy to the United States. During the three years in that office he was accorded many honors from the mining association in the United States. Royal S. Handy succeeded Caetani as **Bunker Hill** superintendent in 1910, remaining until 1942.

The stories Robert L. Anderson could tell! A 47-year veteran of Sunshine Mining Co., May 11, 1989 says "I get no salary, and they get no work. It can't be beat." A mining engineer, he is an unofficial consultant for the Company. His grandfather G. Scott Anderson came to Wallace in 1886 and was manager for several mining companies, including YANKEE GIRL. (Courtesy S-R).

Erection of the **Bunker Hill** lead smelter was important to that company's expansion, and was operational by July 5, 1917. The Sullivan Mining Company was formed in association with Hecla Mining Company that same year. Along with purchase of the **Star** mine, a zinc processing way had to be found and Wallace G. Woof was hired to do that. As a result of his knowledge, the three million dollar **Sullivan** Electrolytic Zinc plant went into operation Nov. 6, 1928. Woof was reported to have retired as general manager of Bunker Hill in the 1960's.

World War I had little impact upon these mines and the Korean War imposed no federal controls – as a result, the Coeur d'Alene district prospered. World War II did cause the closure of gold mines in the area.

Numerous expansions were reported by the Bunker Hill management and in 1966 Hecla's interest in the Sullivan Mining Company was secured for a reported 275,000 shares of the Bunker Hill stock. This brought the zinc plant under direction of Bunker Hill, along with the **Star Mine.** Hecla had operated it under contract. (A recent change to Hecla made them 30 percent owner of the mine, in lieu of operating contract.)

1955 negotiations with Hecla brought Bunker Hill 495,000 shares of the Pend Oreille Mines and Metals Company stock. (We will return to Washington to learn about that mining district, at a later date.) That transaction increased Bunker Hill ownership to 36 percent of the stock in the zinc-lead mining firm in Pend Oreille County.

J.B. Haffner retired as president of Bunker Hill in 1955 and Charles E. Schwab became president. In 1956 the Sullivan Mining and Concentrating Company name was dropped; the company officially became Bunker Hill.

Don "Whitey" Leonard and son, Phil at work in the SUNSHINE. Part of Whitey's legacy to his son will be to quit at retirement – "before something happens." (Dec. 27, 1981 S-R).

In 1960 a 130-ton per day Phosphoric Acid Plant was constructed between the smelter and the zinc plant. That same year a few slow-downs occurred due to a strike which lasted eight months. At the Lead Smelter and Refining works it was said about 500 men were needed to keep that busy unit operating 24 hours a day, seven days a week. Some 2,200 tons of ore was said processed at the concentrator daily, producing 214 tons of lead and 85 tons of zinc concentrate. The zinc plant was said to have an annual capacity of 73,000 tons of 99.998 per-

Very early photo of BUNKER HILL and SULLIVAN (S-R).

Stacking one ton refined lead pigs at the BUNKER HILL lead smelter at Kellogg, Idaho (S-R).

cent pure zinc. Considered the finest in the world, the storage room sometimes held one million dollars value. Two new products had come into production in 1959.

According to the 1960 report it is of interest to note the type of use of Bunker Hill products – such as; the automotive industry was one of the largest users of the lead and zinc. In 1961 the average car had about 87 pounds of zinc in die casting, in and on it. Some 65 percent of all the Special High Grade produced by Bunker Hill goes into the nations' autos.

In that report, sales amounted to $50 million, and 2,600 people were employed. Plants were located by then at Kellogg, Seattle, Portland, San Francisco, Oakland and Los Angeles. The Company at that time reportedly supplied about 16 percent of primary lead refined in the United States, and 20 percent of the high grade zinc. Production was from the **Star Mine** at Burke; **Crescent Mine,** at Kellogg; **Pend Oreille Mines and Metals,** at Metaline Falls, and the **Reeves MacDonald Company,** in Canada – all of which Bunker Hill had an interest in besides owning the largest single lead mine on the basis of production, in the United States.

The **Bunker Hill** smelter produced from 1917 to 1981, 20 percent of the zinc used. The mine itself produced from 1928 to 1981 when records were kept – three million tons of lead, 160 million ounces silver, and 12 million tons of zinc. Their greenhouse 3,000 feet underground produced a quarter million tree seedlings which were planted in the district from 1977 to 1981.

Since work began in 1886, 25 million tons of ore had been extracted from **Bunker Hill's** famous lead-zinc mine. Value of the metal refined would be over three quarter billion dollars. (To recall – the Coeur d'Alene district is one of only eight districts in the world where more than one billion dollars worth of ore has been removed from the earth – silver, copper, and small quantities of gold.) The **Bunker Hill Mine** used the largest underground hoist in the United States, capable of lifting one and one quarter mile. Eight million board feet of timber went into the mine every year, never to come out again – so **Bunker Hill** maintained it's own sawmill.

In the 1960's Gulf Resources and Chemical Corporation became **Bunker Hill's** parent company. Prices were falling for their products; costs were going up and EPA was imposing millions of dollars in requirements to correct emission standards. It had to happen! – **Bunker Hill** was closed down! – in the fall of 1981. That closure was followed in the next few months by **Lucky Friday** and by **Sunshine.** One by one the mines of Coeur d'Alene were down – except **Galena** and **Coeur.** Harry Magnuson said that unemployment could be as high as 6,050 jobs lost with a payroll around $109.2 million in Shoshone and Kootenai counties. The ripple effect was felt by Spokane, and the Inland Empire.

What will develop from those terrible days forward will have to be another chapter but Harry Magnuson again put it well; "Hope springs eternal from the breast of a hard-rock miner."

Montana

Because our prospectors have gone from placer to quartz, from mining camp to mining camp – and because we have seen some of those camps develop into vast districts, it has been difficult to leave an area to stay in chronological order. We have gone far ahead of the Montana development and must return there.

We may have almost overlooked one danger these wandering prospectors faced. Through the entire history of mining, the prospector and miner has been exposed to the worst element of humanity. Highway robbers, horse thieves, gamblers, desperadoes, and ex-convicts drifted along with those "seekers of fortune." Some even traveled with the law abiding families coming west to join their men. Gangs of outlaws, often organized groups of thieves and robbers, watched for gold shipments. They held up travelers and robbed ordinary folk and often this was done in the name of the vigilante – special groups formed to provide what law and order existed then.

Diesel-powered front-end loader picks up waste rock in driving tunnel deep inside the BUNKER HILL MINE (S-R).

Henry Plummer was said to be one of the shrewdest and least merciful of the cut-throats. He was said to have operated around Elk City, Lewiston, and Pierce, in Idaho. He had made a name for himself, being accused of killing a man – and running off with someone else's wife before he arrived in Montana. Then he became sheriff of the gold camp area at Bannack. Vigilantes were generally the last word of law but one questions how brave these so-called law men who reportedly shot 100 bullets into one accused of breaking "their" law.

Whether Plummer had a good effect on his surroundings — or was even to be considered a "good" man, or a bad one, is being researched by two fine local Spokane historians. One historian believes Plummer's gang killed around 125 people before he was lynched Jan. 10, 1864. Good, or bad – Plummer's name comes up in Montana law history, often considered a hero. This type of criminal element – or worse, and this type of law, or lack of it, was part of the prospector's problem which he faced in our entire area of history now being told. Plummer's type was the rule, not the exception.

Butte- Anaconda

We learned earlier that Butte, Mont., came about in 1864-65 as miners drifted in from the failing camps at Virginia City. The **Summit Valley Mining District** was formed in 1864, one of the greatest mining districts the world had seen. It turned out in 1864 to be an insignificant gold camp and as gold played out, Butte almost became a ghost town. A few hardy prospectors did stay on and in the early 1870's large deposits of silver were located. Butte boomed again!!

As the miners followed the deepening silver veins they discovered the immense underground bodies of copper ore. Butte emerged as the most fabulous copper district in America. The "Richest Hill on Earth" became a fact.

With more and more new ore findings, many immigrants came to "make their fortune" – including the Irish miners from Nevada; Cousin Jacks or Cornishmen from Michigan, and from Cornwall. Marcus Daly was one of those immigrants from Ireland. Differing historians gave Daly's arrival in Montana country as of 1878, and 1881. Born in 1841, Daly was only 15 years old when he arrived in America. He went to California first, where he evidently learned something about mining, before going on into Montana some 22 years later. As he continued to learn about the mining around Butte he is credited with proving that silver could be extracted from below the water table. One of the results was the finding deposits of great copper wealth soon to become known as the **Anaconda Mine.**

Daly bought the **Anaconda Silver Mine** from a Civil War Veteran (said in 1880). Michael H. Hickey was credited with finding what would become the great Anaconda, and which Marcus Daly was said to have bought for $30,000. With 55 percent deposit of copper, that metal became the richest in the mine and thus a Montana copper industry was born. Over the next 100 years, 18 billion pounds of copper was taken from the hill. Copper wire for electrical use expanded the market. Butte was booming – little ethnic villages (neighborhoods) were springing up including Corktown, Finntown, Dublin Gulch, and miners and prospectors poured in.

Backed by California financiers, the Anaconda

Shift off at BUNKER HILL Feb. 28, 1961, (S-R).

Mining Company bought other claims and built a smelter town. It had added coal mines and timber holdings. In the meantime William A. Clark became interested, and F. Augustus Heinze had come to town.

Heinze was born in Brooklyn in 1869. He had earned a degree in mining in Columbia University, then studied in Germany. He arrived in Butte in 1891 where he is said to have owned a copper smelter. We do associate him with the Boston Montana Consolidated Copper and Silver Mining Company, large property holders around Butte. Anaconda Copper Mining Company gained control of the Boston Montana Silver and Copper Corporation's smelter-refinery at Great Falls, in 1910. It was said built by the Heinze group in 1892. (We are told Augustus Heinze was ousted for company reasons. By finding all the legal loopholes in mining laws, he is said to have engaged in lengthy litigation often coming out the victor and making huge sums of money.) He clashed with Clark and Daly and left Butte by 1894 – going to Trail, B.C. There he built up a fortune which we read about in conjunction with the **Le Roi Mine** and **Cominco**, and from where he eventually

BUNKER HILL greenhouse 3,000 feet underground. (Photo April 6, 1981, S-R).

Photo Jan. 28, 1976 shows tree seedlings planted in the BUNKER HILL underground greenhouse used to reclaim environment above the mine. (S-R).

Miners on the hoist at BUNKER HILL Feb. 13, 1957 (S-R).

leaves, and returns to Montana.

In his early years in Montana, one historian wrote that Heinze employed as many as 32 lawyers and had as many as 40 court actions pending at one time. At first Heinze allied himself with Clark, furthering Clark's political ambitions, while Clark aided Heinze in his battles with Anaconda and Daly. Some of those battles were with Standard Oil after Daly's death and that company took over Anaconda. Heinze was accused of bribing legislators and corrupting judges with large fees. His miners were said encouraged to fight underground with dynamite, with other chemicals, and with water hoses.

Heinze returned to Montana and in 1906 he was said faced with millions of dollars in lawsuits. By some "hook or crook," he sold what he could in Butte for $10.5 million and went east where he is said to have died broke and in disgrace in 1914.

The three "Copper Kings" Daly, Clark and Heinze became vicious competitors purchasing huge holdings such as timber, ranches, railroads, farmland, hotels, commercial buildings, and newspapers. Until 1959, Anaconda Copper Company owned five of the state's six daily papers including The Montana Standard, in Butte.

Patrick "Patsy" Clark whom we have met in the Coeur d'Alene district as owner of **Poorman** and other mines, had earlier been a mine foreman for Daly in Butte and Anaconda for seven years. He was also a native of Ireland. (In later years he left Coeur d'Alene and headed for the Slocan – and eventually became involved with the mines of Rossland and Trail, B.C., and with Republic, Wash.

Following the "war of the copper kings" it was Daly's **Anaconda** that finally bested the others. As the area developed there were as many as 20,000 workers in the 150 mines that honeycombed the hill beneath Butte.

Transportation of his ore from the mines at Butte, to the smelter at Anaconda brought about his own railroad. The Butte, Anaconda and Pacific Railroad was the brainchild of Marcus Daly, who wanted to avoid other railroad's high shipping rates. His miners built his railroad in 1892 and the 25 mile run became one of the busiest short lines in the nation. In 1912 the line was electrified and 28 locomotives hauled up to 1,000 cars of ore a day to the smelter. The payroll for this enterprise was then to more than 500 employees. The line once boasted eight passenger trains a day making the trip to Butte in one hour. That last run was in 1955, according to information from the columns of Dorothy Rochon Powers.

Recently down to 21 full-time employees, Ray Lappin was the sole remaining engineer. His job in January 1985 was to make the once-a-week run to Butte and back, hauling only three cars and coming back with a caboose. With both mining and smelting shutting down there was nothing for the little railroad to do.

The railroad better known as **"BAP"** was operating at a loss of one million dollars a year. Owned by the Atlantic Richfield Corporation **(ARCO)**, in recent years it was offered for sale for two million dollars and was well worth $20 million. On April 11, 1985 it was reported the Railroad had been bought by management employees of the BA&P, with John Greene (former president); Dave Bisch, and William T. McCarthy the new officials. The sale also involved other arrangements with the state of Montana.

There is said to be 25 miles of mainline track and a total of 115 miles including the yard and siding. The line was used to carry scrap metal out of the area to meet other railroad lines, and hopefully would soon be contracted to haul 45 million tons of copper slag which was stored at Anaconda. Called the Rarus Railway Company, a name in memory of the old **Rarus Mine** at Butte, the new corporation hopes to see "BAP" used for tourism.

Probably one of the most determined defenders of "needs of the people" is Dorothy R. Powers, recent retiree as associate editor for the Spokesman-Review in Spokane, who grew up in Anaconda. One of the most beloved columnists in the Inland Empire, her articles relating to the "Little railroad that could" have helped encourage the "Little railroad that will!" and her support will help in developing tourism interest. Some of our information has come from her actual knowledge of the line.

Butte has been said to be one of the brawlingest mining camps in the west. Dennis E. Curran, writing an article for the Associated Press in June, 1986 told how at the turn of the century "Uptown

One troy ounce BUNKER HILL silver, 1981 (S-R).

Butte" was considered the most cosmopolitan in the northern Rockies. At that time Butte had some of the west's first skyscrapers, up to ten stories high. He also tells that the pollution from Butte smelters killed virtually every tree and blackened the skies at noon – but never stifled Butte's desire for a good time. There were over 500 bars, many never closed; along with dozens of gambling halls and brothels. The population peaked during World War I, with over 100,000 of which nearly 20,000 were underground miners. (One of the better histories of Butte is written by Neil J. Lynch, former Montana senator; a local historian.)

Daly was known to be a warm and generous man. Unlike the other two "Copper Kings", Daly made many improvements and left many things to the State of Montana. It was said that Clark and Heinze did not leave a thing for the betterment of the state where they made so much money.

Early in this Montana mining period Daly was reported to have allied with James Ben Ali Haggin and George Hearst, both mining men from San Francisco. Just before Daly died in 1900 he sold out his Anaconda holdings to Henry Rogers and William Rockefeller of Standard Oil. That included the mines, lands, timber, and mercantile establishments.

In the period of the "War of the Copper Kings", Marcus Daly had a mansion built in Hamilton. It was a three-story structure of Georgian revival design built similar to a summer home he had built in the 1880's. The mansion was 42 rooms, the largest home ever built in Montana. In 1909, nine years after Daly's death, his widow hired a Missoula architect to redesign and enlarge their Queen Anne Style home. The present mansion is the result. Daly had become one of the richest men in America and spent every summer at his estate from 1886 to 1900. It was boarded shut after his widow died in 1941.

The property went to Daly's granddaughter, Countess Margit Bessenyey and at her death in 1984 it went to her stepson, Francis Bessenyey, who was said to owe millions of dollars in State and Federal Income taxes as a result. The town tried to make a deal but the mansion's furnishings were sold at auction. Now a local group is trying to raise money to buy the mansion and to buy back the

BUNKER HILL zinc plant. This photo Dec. 6, 1956 well before 1981 closure. (Courtesy S-R).

Before the closure threats; SUNSHINE Mine's 1,100-ton mill and mine buildings cover the canyon floor east of Kellogg as early as Jan. 28, 1949. Famed Jewell shaft is at the center top. (S-R).

furnishings. Daly's rise to one of the most powerful men in Montana history was far from unblemished but the mark he left on the state is indisputable. Tom Brader, local bank officer and chairman of the preservation trust fund said, "Marcus Daly, for good or bad, was a part of Montana history. He did a tremendous amount of things that made Montana the state it is today. After seeing the 19th century artwork and treasures that filled the mansion – all hauled out on the front lawn and sold for auction, we realized what we are losing historically."

Brader is vice-president of the bank in Hamilton that Marcus Daly founded in 1895. He has seen the 50 acres of the once elegant gardens turned into horse pasture. The mansion hadn't seen a paintbrush in 50 years. To solve these problems over 400 Bitteroot Valley residents have volunteered labor to restore the mansion and the grounds. In the fall of 1987 scaffolding covered the front of the mansion as work progressed. The grounds and the estate were open for public tours by August, 1987 and 10,000 people paid one dollar to roam the ground since May of that year. 4,500 had paid nine dollars for a guided tour of the house. Others have attended many events in behalf of the mansion repair.

It was said that businessmen from Missoula and

as far away as Spokane have offered help and material. Of $200,000 for the preservation trust, about half has come from outside Montana. The mansion and 50 acres are now owned by the State of Montana. The local trust that operates the place came about following negotiations with the New York banker Francis Bessenyey who inherited the property from his step-mother.

Many great homes were built as a result of these huge mineral earnings. Senator Clark's mansion is built in Butte – a 32-room red brick mansion of Victorian style and elegance. An intimate billiard room, and the 62-foot long ballroom are just some of the features.

Clark was born in 1839 and began his working life as a store-keeper in Ophir City; then he became a wholesaler in Helena, Mont., and had more stores before going into banking. He was said to be detached, arrogant and uncompromising. A saying that followed him was that "not a dollar got away from him except to come back stuck to another."

Clark was accused of spending $431,000 to bribe legislators to vote for him for United States Senator. Daly challenged Clark constantly revealing much that was not honest in Clark's campaigning. Daly died before Clark finally made the office after four tries, but Clark was soon removed from office for illegal campaigning. He died in 1925.

A tour of Butte would include the World Museum of Mining. There are 35 acres and is the site of an inactive silver and zinc mine named **Orphan Girl.** There are thousands of articles on display and a 100-foot steel headframe is astraddle the three compartment 3,200 foot shaft. Wandering through Hell Roarin' Gulch one would see the general store, assay office, saloon, drug store and soda fountain of that heyday – then continue to walk further over cobble-stone streets to over 25 other businesses.

Much of the earliest mining in Montana was around the Libby district, Troy and Sylvanite – and in the big Yahk district; placer finds calling for most attention. Large hydraulic operations were in use and the **Snowshoe Mine** at Libby was the biggest of the area. An English company purchased it for $150,000, the final payment made in May 1899. Concentrates shipped ran 65 percent lead, 26 ounces silver, and eight dollars in gold.

The **Buzz Saw, Silver Cable,** and **Silver Crown** were being developed at the time. Quartz mining was taking over requiring large capital and showing slower mine development. At Sylvanite the **Keystone** and **Gold Flint Mines** were on the same hill and owned by Spokane Companies . . . the **Keystone** being the first quartz location in the camp. It was staked in 1895. A ten-stamp mill was on the mine and gold bricks were being brought into Spokane – the ore running around eight dollars per ton. The **Gold Flint** had a 20-stamp mill and was producing some 60 tons per day.

Anaconda Copper-Silver smelter from the air. Probably taken September, 1956. (S-R).

From the railroad museum at Anaconda, Montana. The photo taken June 6, 1985 is of engine 97 of the B A & P line. (S-R).

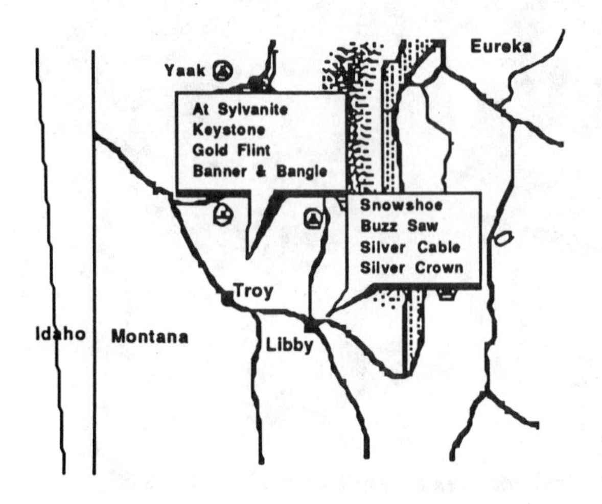

The **Banner & Bangle** was the main mine near Troy – a silver-lead mine showing values of 25 ounces of silver and 60 percent lead. These mines all came about prior to 1899 and now nearly 100 years later there are some of the State's best mines coming into operation in near locations, mostly silver; some gold – and even the start of a rare platinum mine.

Total dividends reported in Montana at some mines as of 1899 included: **Alice – $1,075,000; Anaconda Copper** – $9,750,000; **Colorado** – $1,945,000; **Montana Ore** – $1,120,000 and **Parrot** – $2,690,898 – just to name a few. Of course there were other districts reporting heavy dividend payments by that year to include the **DeLamar** in southern Idaho – $2,346,000, and the **Standard** in northern Idaho at $1,745,000.

In earlier Montana history the big stack that was "blown" was said to be in Anaconda – instead of correctly being in Great Falls. It did belong to the **Anaconda Company** long since taken over by Atlantic Richfield Corporation **(ARCO).** The stack was built in 1908 to disperse poisonous fumes and when the plant at Great Falls was closed in 1980 everything was being dismantled for salvage. Said at different times to be either 585 foot or 506 foot high, it was constructed of brick and mortar and weighed an estimated 36 million pounds. It was said the Washington Monument could fit inside. The demolition Sept. 10, 1982, took two tries when dynamite smashed the base and a great shower of dust exploded from seams between the bricks. It collapsed straight down rather than toppling as planned and as dust drifted away, portions of the giant stack stood out against the skyline. Five hours later this mighty beacon of industry lay level on the ground.

Mining has continued to be a leading industry in Montana for over 125 years. A directory list in November 1974 names 300 mining enterprises in gold, silver, copper, lead and zinc. It lists 25 gold placers, four tungsten mines, three iron mines and several other mines including many non-metallic operations. All of these have been developed and have produced ore.

Madison County listed 40 gold, silver, copper, lead and zinc mine operations. Jefferson County-36; Beaverhead County-33; Granite-34; Lewis and Clark-23 and other counties less. Names of many mines would fit right into "skin novels" of today but for a history read by families we will only mention some of the "unusual" Montana claims – "Scalded Cat," – "Dirty Muzzler," – "Spoofer," –"Green Lizzard," – "Scratch Awl," – "Hard Luck," –"Moose Trail," –"Sky Baby," and "Freezeout Pass." These were in the 1974 directory, some rather obvious, and all very mild in contrast to some names given as claims were staked in our area of interest in past years.

Earlier construction of smelters included one built by Parrot Silver and Copper Company around 1881. That new smelter at Butte was said to have produced the first blister copper in the district. Also at Butte, copper matte was produced from a blast furnace at the Bell Smelter. The Utah and Northern Railroad completed a narrow-gauge line from Ogden, Utah to Butte – all according to a county agent's report.

When Marcus Daly had his smelter built at Anaconda, W.A. Clark built his near the townsite of Meaderville, in the Butte district, these both in 1884. W.A. Clark also built the first smelter in the Coeur d'Alene region near Wardner, and is credited with the Butte Reduction Works on Silver Bow Creek, south of Butte, both were said in 1886. In 1888 the Helena and Livingston Smelter Company acquired the **Wickes Smelter** and constructed a lead smelter at East Helena, with a capacity of 12,000 tons of ore per month.

At Butte the **Berkeley Pit** opened in 1955 and was at one time the largest all truck mine in the

Entering Butte, Montana, "The Richest Hill On Earth". Photo taken Feb. 7, 1946. (S-R).

"Copper King" Marcus Daly in retirement at Hamilton, Mont. (S-R).

world. It was discontinued in 1982, the work moved to a smaller operation the Southeast **Berkeley.** As the first huge pit grew, underground miners were phased out and all of a sudden the pits were closed. The smelter at Anaconda and the refinery at Great Falls were closed down representing around 1,500 jobs. Despite many statements as to what might still operate, by 1983 all of Atlantic Richfield's Anaconda Copper Company was closed down. It was due, according to James L. Marvin, president for the company, to the fact they could not stand paying upward of $400 million cost to meet federal clean-air standards – and Montana's even more stringent rules. The EPA statements issued said the early closure had not been required, that the company could have operated under existing laws until January 1988 and had until then to comply with at least a portion of the requirements.

The union squabbles were said a big part of the closure. Townspeople stated the unions ripped off the companies and fought each other. Part of that problem may have come about when huge shovels and ore trucks were put to work at the **Berkeley Pit** instead of miners. Fewer were being hired and many jobs were no longer there. The tourist of today is not being overlooked – the viewing stand for the huge **Berkeley Pit** in Butte is open during daylight hours at no charge.

On June 30, 1983 Anaconda Company suspended mining – some 3,500 people lost their jobs – many lost their cars, their homes – and there were more people out of a job than with a job. The entire area is trying to come up with new jobs for the people. Mayor Don Peoples is one who is doing his best for the area. He still has to combat this attitude that "I'm a union man – we have the Local #1 here," – and those who say "I worked for $22 an hour in the mines – I won't work for less!" That does not hold true for everyone who has seen what loss mine operations can cause, but also see that Butte, Anaconda, and Great Falls have to go on. Taking advantage of every loan, grant, and assistance program in the book, Peoples sees growing high-tech industries, motels, restaurants and an ambitious plan to expand the Port of Butte into a trading center.

As Atlantic Richfield (or Anaconda) developed the pit that was more than a mile wide and a half-mile deep; that swallowed entire communities like Meaderville and 50 blocks of Uptown Butte, Anaconda phased out it's underground mines leading

Restoration of the Marcus Daly mansion at Hamilton, Mont. could cost $1.5 million as reported May 26, 1985. It is said to be 42 rooms and the largest home ever built in Montana. (S-R).

A view of the leveling of Butte, Mont. by open pit mining. Photo July 21, 1962, (S-R).

DeLAMAR Mine in Southern Idaho once belonged to a company from London. It paid well in dividends from it's gold and silver ore. Photo in the 1899 period. (S-R).

to many miners out of work. Then it closed the smelters in Anaconda and Great Falls and by 1982 the **Berkeley Pit** was closed. By the end of June everything was closed and Peoples, along with others realized the district could no longer survive on a one-industry economy.

Scars of mining days are everywhere including the abandoned **Berkeley Pit** filling with acid water, padlocked gates around mine headframes and the empty union halls – and if some of the people of the area have their way, those union halls will stay empty. With all the reminders – there is a bit of new interest in mining but on a different scale – and different wage scale. Montana Resources, Inc. a subsidiary of Washington Corps, a Missoula-based construction company, bought the mine in December 1985 and hoped to resume copper and molybdenum mining by summer 1986. Reverend Joe Warren said "there's a spirit in Butte that will always be there, whether that new mine comes in or not." That company expected to hire some 300. By July 1986 prices were falling and were down to 68 cents a pound from the 78 cents for copper, when Anaconda quit.

Washington Corp is better known for building roads but is one of the largest conglomerates in the northern Rockies with contracts totaling more than $100 million a year since 1982. It excavates everything from dams to gravel pits and operates one of the nation's largest open pit phosphate mines near Soda Springs, Idaho. Dennis Washington is head of the company.

In July, 1986 the new company planned to begin shipping copper and molybdenum concentrates to smelters in Japan, Korea, and Taiwan, which commenced in August. They hoped to reach full production of 40,000 tons of ore a day by August 11, and hoped to employ 315 by the end of summer – about half what Anaconda had when they suspended operation. Ore was coming from the **Continental Pit.** The reopening of the mines has generated controversy over Montana Resources' refusal to negotiate contracts with unions before resuming operations. Butte has been a hotbed of union activity since copper mining began and some union men have been critical of the new hiring practices. The new company refuses to follow Anaconda Minerals, who had agreements with 13 different unions. "Now we will let the workers decide," says Ray Tillman, human resources manager who adds "We don't vote on that, the employees vote on it. We are pro-people, let the people decide." (Remember it was union men from Butte that came into the 1892 labor wars at Coeur d'Alene – possibly from Local #1.)

By Oct. 19, 1986, three years after Anaconda Company unceremoniously shut the gates on operations – the "richest hill on earth" was back in the mining business. Some 300 were back drawing a paycheck – maybe 50 percent smaller; and only half the force had been re-hired. Non-union hands were producing the concentrate. Montana Resources Incorporated (MRI) proclaimed it's intentions to operate in a union-free environment in what was once America's strongest union town. Wages at five dollars to nine dollars and 40 cents an hour with tops at $10.40 in February 1987 look some different to old time union men who drew $13 to $15 an hour — and more. Joe Maynard, business agent for Butte Miners Union, said the MRI operation was union-busting and claims workers got sold down the river by labor and the mayor.

From the **Anaconda Mine's** start, over 20 billion pounds of copper was extracted from the underground mines – with over 20,000 men employed. The 13 separate guilds that represented nearly every worker in the mammoth operation came to friction with one or more of the unions causing strikes. In 1977, **Anaconda** was purchased by the Atlantic Richfield Corporation who very soon decided to close their Montana operations. That was complete by 1983. They threatened to tear down the ore concentrator so it could reduce the property tax bill, but Mayor Peoples formed a committee who saved that operation. Then plans proceeded to entice MRI to opening the pit and restore some jobs.

"The union, they cut their own throats," said the bartender at Helsinki, which still offers steam baths as well as beer, as it did to the Finnish miners who were once it's most loyal patrons. "They just ripped the company off like crazy, and all the unions did was fight with each other." "People think that the unions killed the golden goose years ago, that's the general perception," Mayor Peoples said. The feeling is strong among some that miners are too afraid to organize again, but that they will. MRI, however, seems determined to keep the unions out and sponsored a "Thank you" barbecue for it's employees and families. "It was not a hot dog and hamburger affair but baron of beef was served," Tillman commented.

It happened in CDA

With so many miners out of work in the Coeur d'Alenes and in Montana, it was good news when the **Lucky Friday** and the **Sunshine** reopened in 1987, both without union contracts. In Butte it is not strange to hear that many workers say they don't see the need for a union. "If they treat us fine, we don't need a middleman," said Ed Konda, a mechanic in the mine. His expression may apply to today's problems more than the old "hardline thinking." At least with the elaborate three-story Butte Miners Union hall still vacant in March of 1988 it indicated some changes in peoples' thinking about the union.

Butte's mayor, Don Peoples, was said by Oct. 7, 1986 to be doing his best for the area. He was taking advantage of loans, grants and other assistance to bring about new industries. (Courtesy S-R).

Changes unforeseen did occur in Butte, and Al Hooper says Butte has "never been worked out – never will be." Bert Caldwell of the Spokesman-Review tells us July 16, 1989 that New Butte Chairman Alan Richardson looks for extremely high potential for gold and silver production at their **Rainbow Claims.** Taking the name of the **Lexington,** the mine is just on the edge of the old **Kelley Mine,** at Walkerville, Mont. Richardson is a former Pegasus Gold Corp. vice-president and manager for Inco Ltd., said to be one of Canada's larger mining firms. Some claims were from Montana Mining via Dennis Washington and Washington Corp.

Jardine Gold Mine

There is much other activity in mining in Montana. An announcement April 20, 1986 is that the American Copper and Nickel Company, and the Homestake Mining Company have joined on a venture to reopen **Jardine Gold Mine.** This is two miles north of Yellowstone Park, and is an old mining town along Bear Creek. It is also a spawning water for cutthroat trout; a prime grizzly bear habitat, and adjacent to a major elk migration route. A prospector, Joe Brown, had first discov-

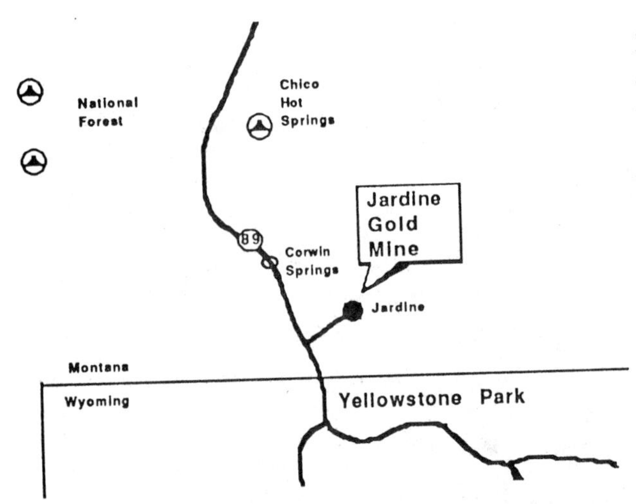

miles south of Jardine and Gardiner. From 1920 to 1948 an underground mine was operated in Jardine. It has been dormant since then with interest renewed in the 1970's.

The schist is three point three billion years old; making it some of the oldest rock on earth. One cubic yard of the schist weighs two point six tons and it will take two to 12 tons to produce a single ounce of gold. (It is the kind of crystalline rock that splits easily into layers.) Because the area is environmentally sensitive, every precaution known is being taken. The townspeople question there being a mine opened here but the companies hope to begin in 1988 with 100 employees.

"We may get a new mine every five to 10 years over the next 50 years," Buster LaMoure, regional director of mineral and geology in Montana, told Larry Wills of the Spokesman-Review in an interview in January, 1985. LaMoure looked for an explosion of mining activity along a rich vein of silver and copper ore between Troy and Noxon, in Western Montana. (As we will note – the principal mining in Montana has been on the western side and down toward the northwest side of Yellowstone Park, taking an east-west swath in that portion of the state.)

Asarco Troy Unit

At the time of this interview, Asarco Incorporated was proposing to mine it's claim on the southern edge of the wilderness area; and U.S. Borax was planning a mining operation next to the **Asarco** mine, both on northwest Montana's Cabinet Mountain Wilderness. Both mines were said to be about the size of the **Asarco** operation at **Troy** which employed then about 350 people and had a $25 million payroll.

Impact on wildlife is a major concern and in the case of these two potential mines, would affect (according to the article), – 20 grizzly bear in the 91,000 acre wilderness. (A documentary from Yellowstone Park states that each grizzly there ranges some 1,000 square miles for food.) These mines would be located in spring bear habitat and the Forest Service is studying ways to mitigate the harm to the bear. U.S. Borax has spent $50,000 on grizzly bear studies and **Asarco** has hired a biologist for it's project.

LaMoure said the only visible mining impact on the wilderness would be a ventilation shaft – with no other surface impact. The **Asarco Mine** would be 1,100 feet below the surface and the ore would be brought to the surface at a mill outside the wilderness area. The massive mine would use 100 ton trucks and other equipment in giant caverns below the surface.

Asarco Rock Creek

John Gatchell of the Montana Wilderness Association remarked in December, 1987 that "it's the steepest range between the Cascade Mountains and Glacier National Park." He worries about the grizzly population up there being on the verge of extinction. **Asarco's** new **Rock Creek** project is five miles northeast of Noxon on the edge of the wilderness. Terry Erskine, manager of the project, said the mine is expected to produce about five point three million ounces of silver a year. U.S. Borax is studying whether to put it's mining complex in the Rock Creek drainage about three miles from **Asarco's** – or over the mountain on the other side of the wilderness.

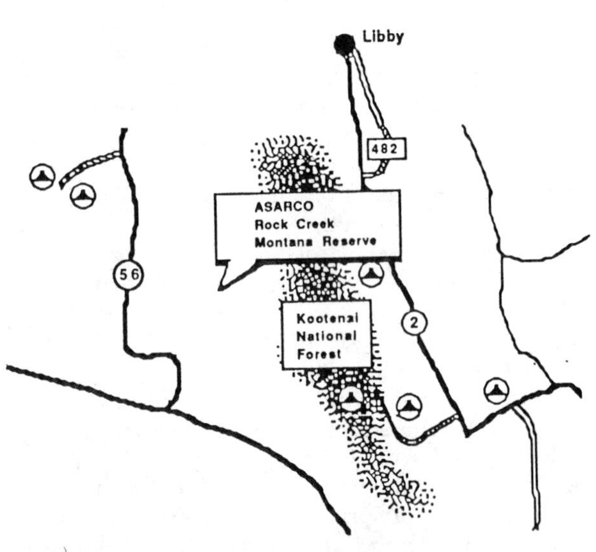

Montana Silver Venture

In March, 1988 a report by David Bond of the Spokesman-Review said that Harry Magnuson and Frank Duval had purchased the Borax property – the purchase of the **Montana Silver Venture** near Noxon, was for $94 million. The transaction for the undeveloped mine is between U.S. Borax and Chemical Corporation of Los Angeles; and the Montana Reserves Company. The property boasts reserves of 280 million to 420 million ounces of silver. It also contains significant reserves of copper ore. Magnuson said "it's going to be a major low-cost producer and the world's largest reserve of silver." He claims this to be even larger than

Asarco's Troy Unit mine . . . a five million ounces per year producer.(estimate 2.5 billion lbs. copper reserve).

Borax was said to have sold 55 percent interest in the Montana Silver Venture Reserve for $55 million in cash. The two men announced they had bought the other 45 percent interest from Jascan (or Jacan) Resources Inc. of Toronto, and Atlantic Goldfields Inc. of Halifax, N.S. According to the news, the two companies are to receive $25 million in cash and $15 million in production-income bonds for a total of $40 million. Borax said it has spent five point three million dollars exploring and developing the property. **Asarco's Rock Creek** development flanks the **Montana Reserve** property.

Industry insiders say the three mines – **the Troy**, the **Rock Creek** development, and the **Montana Reserve** property together could rival production from the Coeur d'Alene mining district with it's total of about 15 million ounces of silver per year.

Harry Magnuson is president of Clayton Silver Mines and many other mining and non-mining enterprises including a connection with the new Bunker Hill Company with high hopes of that mine opening again. Frank Duval was a founder of Pegasus Gold Company Incorporated in Spokane.

Proving that not all is "wealth and riches" in the mining game, Harry Magnuson was recently rejected on a request to lower property taxes on his **Clayton Silver Mines** in Central Idaho. The mine, some 60 miles northwest of Boise, has been closed for three years according to a July 14, 1989 news story. The mine shaft is said to be filled with water but the stock still trades on the Spokane Exchange. The assessment had been lowered from $691,454 and Magnuson was asking it be down to $160,000. He claims extraordinary financial hardship. The Company is said not having paid taxes for 1986, '87, and '88 of $27,291.

Montana's mining industry may never recapture the glory days when it was the state's primary industry, but it appears to be rebounding from its slump in the early 1980's, writes Dennis E. Curran of the Spokesman-Review, in December, 1986. With it's oil, coal, natural gas, and a diversity of minerals, the "Treasure State" will never be a one-industry state, according to Gary Langley, executive director of the Montana Mining Association. He anticipated at that writing that in coming months the mining industry would pump more than $400 million into Montana's economy to construct precious metal mines that eventually would employ 600 people.

In Jefferson County, Centennial Mining Company was just developing the **Montana Tunnels Mine**, an open pit operation that hoped to extract many millions worth of gold, silver and lead left when the area was honeycombed with mines a century ago. In the fall of 1984 the mine had a construction force of nearly 300 workers and was said would employ 200 miners. It expected to yield more than 100,000 ounces of gold a year, making it the sixth largest gold mine in the United States.

The Mine is located four miles west of Jefferson City in the old Corbin-Wickes mining district, a rich silver mining area in the 1880's and 1890's.

In September, 1985 Pegasus Gold Corporation, of Spokane, a subsidiary of Pegasus Gold Incorporated, of Vancouver, B.C., made a tender offer for the outstanding shares of Centennial which was developing the **Montana Tunnels Gold** project near Helena.

Still in the feasibility stage at the time, Pegasus said in 1984 that deposits at the site 20 miles southwest of Helena, holds 52.5 million tons of ore containing an average 0.033 troy ounces of gold per ton plus other precious metals. Hobart Teneff, Pegasus president said that when the acquisition of **Montana Tunnels** is complete and the project in full production, Pegasus expects to be among the leading gold producers in North America. Norman Thorpe of the Spokesman-Review also told that the company operated the **Zortman** and **Landusky Mines** in north-central Montana, through the Spokane subsidiary. During 1984 it produced about 70,000 ounces of gold and 152,000 ounces of silver at those mines.

In June, 1987 a report stated Pegasus Gold Incorporated, along with U.S. Minerals Exploration Company, of Denver, officially opened the **Montana Tunnels Mine.** It was said such a gala affair that some 500 guests attended the opening. This is the third mine Pegasus has placed into production in the last year. It developed the property, then bought it from U.S. Mineral Exploration Company, which maintains 50 percent interest in the mines profits after development costs are paid.

Going into production in March, it expected to reach full production in August. Designed to process an average 12,500 tons of ore a day, Pegasus projects it will produce 95,000 ounces of gold, 1,430,000 ounces of silver, 46 million pounds of zinc and 15 million pounds of lead in 1988. In July 1987 the company was offering three million shares of common stock in the United States and 900,000 shares in Europe, at $21.125 a share, said to repay outstanding debts and provide capital for acquiring gold mining companies and properties. (A news report in early August stated less production than anticipated at "The Tunnels.")

The three mines were producing well when in October, 1987, it became known through the Spokesman-Review that Pegasus Gold Incorporated accused several former officers and directors of fraud, racketeering and breach of fiduciary duty in a federal lawsuit that had been filed in February. The suit was immediately settled the next day after the complaint was received by the U.S. District Court in Spokane. Pegasus said the suit had been settled Feb. 11. The quickness of settlement and the fact it was kept secret from the public was to protect stockholders according to one official. Bill Sallquist, of the Spokesman-Review reported the case as; while being on the Pegasus payroll, the defendants failed to tell company directors about a promising mining property in Nevada, and instead acquired a stake in it "for their own benefit,"the suit claims. Then, without disclosing their personal interest, they sold part of that stake to Pegasus "at great profit to themselves," according to the complaint. An earlier 1981 interest in the **Florida Canyon Mine**, near Winnemucca, Nev., was involved.

Besides former Pegasus Chairman Hobart Teneff, and Milton Zink, other defendants in the civil suit were American trading and Investments Ltd., Translantic Investment Company, Carl Toporowski, Frank and Janice Duval and Zink's wife, Linda. The two corporations are foreign, based in the Cayman Islands. Teneff and the Duvals are Spokane residents, Frank Duval being a former Pegasus consultant. Toporowski and the Zinks are British Columbia residents, Milton Zink previously served as Pegasus senior vice-president and legal counsel. Teneff retired as Pegasus chairman and chief executive officer less than a week before the lawsuit was filed; and Zink was said to have resigned as a Pegasus director about the same time. Duval's consulting arrangement had been terminated in January (according to the report.) Teneff and Zink both denied any interest in American Trading and Investments and Zink was pursuing a claim against the company. After being kept under wraps for such a long time it is hard to guess whether either side would ever bring this case into court again.

James H. Foreman, recently retired Pegasus president and chief executive officer, said the settlement saved the company the expense, delay and uncertainty of litigation. (As mentioned earlier in this history, the building up of inventory and adding to a company's or individual's holdings often brings about lawsuits of one kind or another. It is called progress and growth. The old time prospector might even call it greed. His major worry in most cases was having his claim "jumped.")

Production from this company through 1988 includes from holdings with **Florida Canyon** and **Relief Canyon** mines in Nevada; it's development of **Montoro Gold Company** property it acquired in late 1984, and results from signing the merger with Gold Reserve Corporation. **Pegasus Mines** are principally of the open pit heap leach process to extract the minerals. In January, 1988 the company released figures setting a new record for gold production exceeding it's target of 1987 by 3,000 ounces. They produced 228,000 ounces of gold, 782,000 ounces of silver, and 14.4 million pounds of lead. Vice-president Alan H. Richardson said Pegasus goals for the year 1988 are 270,000 ounces of gold and one point five million ounces of silver.

Other Montana Mines

Still another gold and silver mine went into operation in the Tobacco Root Mountains, near Pony, Mont. The mine is operated by a West German firm, Denimil Resources Incorporated. It is using a chemical leach process to extract gold and silver amounting to around 30 tons of ore a day.

In being interviewed Gary Langley also did state that "Mining has somewhat of a sordid heritage in Montana, the companies that operated in the past were irresponsible with their care of the environment." He added, "We've got to live that down and I think we are."

The Associated Press tells us in February, 1988 that 300 people turned out at a hearing on a proposed gold mine near Lincoln, Mont. . . . most of them in support. Sunshine Mining Company wants to build the **Big Blackfoot** Gold Mine a few miles west of Lincoln, Mont., near the banks of the Blackfoot River. Lincoln is about 45 miles northwest of Helena. Mark Hartmann, Sunshine's project manager said the mine would bring about one million dollars in annual payroll to the community and pay wages of more than $10 an hour.

Stillwater Platinum

Possibly one of the most interesting projects was reported by David Bond in March, 1987 concerning the opening of a platinum mine. Back in 1930 chromium was discovered along the vein of the now named **Stillwater Mine.** Some chrome was mined under contract with the federal government during World War II and again in the early 1950's. It remains stockpiled in rock form at Columbus, Mont., beside the railroad track. Geologists suspected there was platinum at **Stillwater** as early as 1935 but it wasn't until 1964 when a Manville Corporation exploratory diamond drill bit strayed off it's intended course that the platinum reef was intercepted, Joe Dewey, project manager explained. Manville and Anaconda performed additional searches for platinum at opposite ends of the "reef" until 1979 when Manville assigned an operating interest to Chevron, who drilled on Manville's "Frog Pond" claim.

Platinum around $525 an ounce 1989, was at $1,040 an ounce in 1980. Canada is a minor source, some is imported from South Africa, and the major supplier is the Soviet Union. The Chevron Resources and it's partners have gambled they can profitably produce the rare metal from their mine 60 miles southwest of Billings in the Beartooth Mountains, just outside the town of Nye, Mont. If successful, the **Stillwater Mine** will become the Western Hemisphere's only primary platinum mine. Production is anticipated to be 25,000 to 35,000 ounces per year with room for expansion. That is only three and one half percent of the United States' consumption. It is used from electronics to automotive catalytic converters, for jewelry, for tools in surgery and dentistry, for powerful magnets, and is even being tested as a cure for cancer.

In addition to platinum, the mine will produce palladium, a less expensive but similarly useful metal. Chevron is the managing partner and the unrefined metals are to be shipped to a Belgium, or Norway smelter, where the concentrate is reduced to it's final product.

Bulldozers, tractors, and trucks are all around the mine, grading roads and putting the finishing touches on the tailing pond while hundreds of deer graze nearby apparently unconcerned about the intrusion. Bighorn sheep gaze down from the cliffs above and are often seen near the deer and by the workers. Along with the work done, the rubber-tired equipment is doing the underground work – electric ore trains will be installed as the mine extends deeper into the mountain.

Some 250 have gone to work for the **Stillwater Mine** – many from the **Bunker Hill, Crescent,** and **Sunshine Mines.** Project manager Dewey was being interviewed by David Bond and told him how rough it's been for these fellows with the Coeur d'Alene shutdowns. "A lot of these guys coming in here were flat broke." To help them, Stillwater guaranteed local bank loans on mobile homes, and developed a trailer park near the mine. Mine Superintendent Cherie M. Tilley said, "I don't think we are abusing the labor market." They require drug testing; they also offer safety measures and an open-door-policy. Top scale miners earn $13 an hour but have medical, dental, vacations and pension plans. The company is non-union – and working to stay that way.

Mine and mill production will be about 500 tons of ore per day, the mill running seven days a week. **Stillwater** has indentified 400,000 tons of proven and profitable ore reserves. Ore grade runs point eight ounces of combined platinum and palladium per ton of rock with traces of other minerals. The Stillwater Mining Company has obtained permits to operate for 29 years – and Dewey and Tilley hope that's just a fraction of the mine's life.

Bonner County, Idaho

Before we leave this area of our history we must join mining fields in Idaho with those of Montana by investigating claims of mineral wealth in Bonner County, Ida. It has been proven that thousands with mining fever came north from the Boise Basin, Ida., strikes around 1862 and headed for the new camps at Virginia City, Mont. Yes, some did go through Idaho by way of Salmon, and on into the Virginia City area but with a portion of the Rocky Mountains ahead, there were those who headed into the mountains by traveling north. To a prospector this offered more potential.

Hope for wealth ended in Virginia City around 1867 but it brought many hearty men through an area that for the past 100 years, and especially in more recent years, has shown some prospects. Many of those prospectors had come through the Coeur d'Alene district some 50 miles south but until 1878 that district had little to show for itself. They boarded steamers at Lakeview on the south end of Lake Pend Oreille (then called Ponderay) and went up the lake as far as Clark Fork or Hope. From there they boated the Clark Fork River and it's tributaries, or more often they just walked the long trek south east to Virginia City.

When nearing Sandpoint, Ida., the prospectors had located a number of claims, three districts being the better recognized, Talache, Lakeview, and around Hope. Beginning in 1917 more than two million ounces of silver was extracted from Talache Silver mines. Silver down to 32 cents in 1926; the crash, the market slump of the 30's and early 40's all made it discouraging and unprofitable to mine in Bonner County.

Iron Mask was discovered nearly 100 years ago by Robert Rennie and Josiah Williams, according to the geologist Donald D. Kotschevar, of Sandpoint. The men tunneled 300 feet through quartzite but with the hand drilling and hard work, they were forced to abandon the mine – just 30 feet short of the riches they tried to reach! That was reached more than a century later.

In 1984, Clare Nichols writing for the Spokesman-Review says the Iron Mask Mining Company is six months into the extensive exploration on Blacktail Mountain. With the lake behind it and facing the Schweitzer Basin ski area, **Iron Mask** director G.E. "Red" Sarff calls this the "most beautiful mining country in the world."

Company founder George Watt, formerly of Sagle; Harlan Churchill, and Robert Evans picked up where the old timers left off and began turning up silver bearing mineral. Before incorporating in 1957 the miners had sunk a winze, a steep shaft to the vein, and discovered some of the rich silver ore prospectors dream of.

Sarff recalls one piece of ore that was big enough and heavy enough that he couldn't turn it over – and it was good looking ore. When that one specimen was assayed it ran 1,100 ounces of silver per ton. A channel sample one foot wide was assayed at 141.4 ounces of silver per ton and Sarff said Watt was ready to buy them all Cadillacs when that ore discovery was made. **Iron Mask** must now determine whether the silver exists in large concentrations. Kotschevar said "We don't have a producing mine yet but we have good showings."

The mine had shown good profit in one shipment in 1979, to a smelter in Trail, B.C., which brought the company $12,500. The high-grade ore averaged 83 ounces of silver per ton – and there have been other shipments.

At the **Iron Mask** Robert Evans Jr. is mine foreman. Evans; miner Gene Driggs, and miner's helper Dave Watt have been spending long hours in No. 2 – drilling, mucking, and tramming out the rock. A rail has been built through the tunnel but the heavy ore car must be moved manually back and forth on the track. Evans says the character of the rock is showing good promise. The geologist said the tunnel is in the Revett, considered the best host rock and the same type as the Coeur d'Alene mines. (This type of host rock is prevalent in Bonner County.)

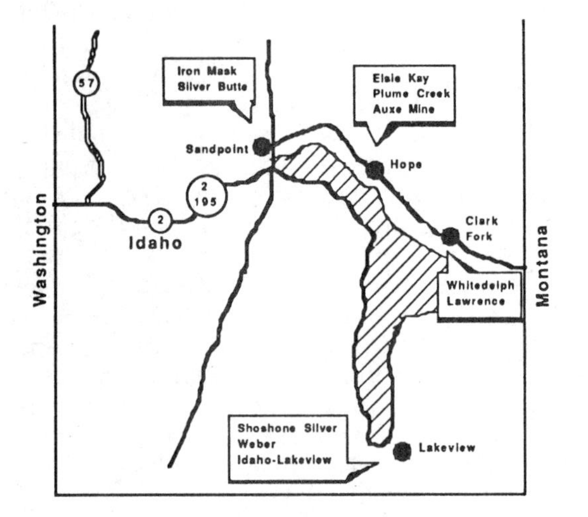

The **Iron Mask** raised $500,000 for it's exploration through stock offerings. A 700 foot hole was to be drilled at the site to study the rock. Assays recently had run from 34 ounces of silver to the ton to 146 ounces. "Faith seems what will make this mine a winner," Nichols wrote.

The neighbor mine **Silver Butte** is involved in exploratory work. Secretary-treasurer Harlow McConnaughey, of the Sandpoint area, said **Silver Butte** has discovered some good ore on a raise – an inclined shaft. Some of that ore assayed at 95.6 ounces of silver to the ton and point zero two ounces of gold.

The two mines from 1917 to 1926, during all the low prices for silver, were some of Idaho's leading silver producers grossing two million dollars in silver when it went from one dollar an ounce down.

In the Hope district Hope Silver and Lead Mining Company has been in business for years. Charlie "Gunny Sack" Johnson sat down for lunch one day and his hound "Big Nose" dug in a hole near him. The dog hit a silver and lead float that has kept **Elsie Kay** in production for a long time.

Also in the Hope district, Roy Anderson says he was operating a grader during the Plume Creek fire in 1967 when he had a time keeping his eyes off what was being revealed. As his rig's tracks cut into the bank where water leaked he saw shiney surfaces and guessed it was ore. He tried to keep others from seeing what he had uncovered because he wanted to be the only one who knew. He was sent out of the district before he could do anything about staking his claim and was greatly relieved when he returned and found no one else had seen what he did. Since then he has put over 14 years work up to 1980, and was planning a mill at his **Plume Creek Silver Mine** (Sometimes said Plumb Creek). He said that would employ eight to 10 men. He did say he had shipped $60 thousand to $80 thousand in mostly silver ore.

Since 1948 Pete Kiebert has worked the **Auxer Mine** in the Hope district. When he first started drilling ore, gold was selling at $70-$80 an ounce. In 1980, after 32 years of work, he hoped to take out four or five ounces of gold to the ton, paying something over $500 an ounce. His mine is some 25 miles up the back roads from Hope.

Another mine reported in 1980 was that of Irv Scheller of Post Falls. It is in the Lakeview area where the most activity occurs in Bonner County at the time. He was in his 7th year of production at his **Shoshone Silver** and employed 10-13 people. In 1979 he claims he grossed about $300,000 from ore produced at two mines. He has done surface exploration at the open pit **Weber Mine** and some underground work on **Idaho-Lakeview Mine.**

Between 1913 and 1943, two and a half million dollars worth of silver and lead was produced in the Clark Fork area. Compton White holds the then dormant **Whitedelph** and **Lawrence Mines** near Clark Fork and pointed out that although the price of metals has risen considerably, so has the cost of mining equipment, labor and energy. "Exploration and development is a fantastically expensive venture." **Whitedelph** was said discovered by Jack Pugh when a tree blew over after a strong wind and exposed the ore. There has been limited development.

While most of these reports were made around 1980 concerning the mining picture in Bonner County, silver had gone up to $19 an ounce and great interest was again shown, but White was trying to explain to Janet Jensen, a Spokesman-Review correspondent that "any profit only comes about after nearly $10 million is expended and six hard years of work is put into the claim before there is any production or profit."

A favorite remark of miners but said well by Compton White is that "Most prospectors think they're five feet from a million dollars, when they're really a million feet from five dollars." He added that "Out of 1,000 mineral occurrences only two or three become good."

An Idaho Bureau of Mines and Geology geologist said "There are literally thousands of recorded mineral claims in Bonner County and 99 percent of them are worthless."

"Someday the old mines will spring back to life – it's just a matter of economics," predicted Harlow McConnaughey of Sandpoint. "Someday Bonner County will be just as viable a mining area as Kellogg-Wallace," says Kermit Kiebert, of Hope. It may be true that Bonner County is as long on dreamers as it is on mineral claims, it is also true that there is more going on than just dreaming. It shows when more than six million dollars has been extracted from Bonner County hills.

One of the more interesting restorations in this area is the Jeannot Hotel, near Hope. It was built by a mining man and gambler Joe Jeannot in the 1890's and catered to the most refined traveler. Teddy Roosevelt, J.P. Morgan, Gary Cooper, and Bing Crosby were among the elite who stayed here before the hotel fell into total disrepair.

Guy and Helen Neyman saw the "For Sale" sign in 1971 on what then was one of the oldest buildings in the Panhandle. They bought the Hotel and have gone to great lengths, and spared no expense to rebuild it and add their collection of antiques to it's decor.

Restored to it's original style and class, it has 19 rooms upstairs and a dining room and lunch counter on the ground floor. One elegant piece is the rich mahogany and mirrored bar that was featured earlier. There is also a general store including old tools, teapots, china and notions.

The hotel was still minus the veranda and landscaping so the date of opening was unsure when Janet Jensen visited.

As she put it – Mrs. Neyman said they were approaching the light at the end of the tunnel – which may have a double meaning. After cleaning away dirt and mud in the second basement, Guy Neyman discovered a tunnel running down to where the train depot used to stand along the water's edge. The tunnel was used in the past to haul supplies from the depot to the hotel, but for the Chinese Coolies who worked in Jeannot's mines, it also supplied access to a meat cooler turned into an opium den. This area was one of the last traces of the big Chinese population.

Spokane

Just across the border from Bonner County to the southwest was Spokane. It was still far behind, but Spokane House had been located in 1810 by the Northwest Fur Company. Into 1865 Spokane Falls (as it was first known) was only a crossroads. One story told by an old timer; a sawmill and a discouraged homesteader was the sum population of Spokane Falls. He claimed the homesteader offered to sell J.T. Logsdon, his father, that homestead for a shotgun and ten dollars – which his father turned down. That land today would contain most of the downtown area of Spokane.

In 1878 James N. Glover secured title by government deed to the village after buying out the only other three people living there. Glover proceeded to populate his village with professional people, a lawyer, merchant, doctor and missionary, and to encourage business. Spokane County did come about in 1879 and from that time, Spokane City mushroomed until the tragic fire that nearly leveled the entire city on Sunday, Aug. 4, 1889.

John Welty, the first to make great discoveries in the Eureka, or Republic mining camp on Feb. 21, 1897 when he staked Blacktail. (Courtesy S-R).

Into Ferry County and Republic

To follow the prospector we have had to cover, and learn about this large portion of the northwest. At present we must catch up with those who have gone ahead into Ferry County, Wash. It would still be three years before Ferry became a county and four years before it's county seat at Republic would be official – but there was activity. Many had scoured the Slocan and had been involved in the Rossland-Trail area. Some had edged southward through Camp McKinney and other Canadian camps towards what was then Colville Indian Reservation on the Washington side. In 1870 President Grant had established that reservation to be 100 miles wide and 200 miles long. There was supposd to be 300 Indians from eight tribes located there. It included land in Chelan, Okanogan, Ferry and a part into Stevens County. Grant was not complimentary when he said, "The best use of this portion of the earth's surface is to cut it up into Indian Reservations."

There was so much objection to the area Grant had designated for reservation that the lines were re-drawn to include the Okanogan River on the west, the Columbia River on the south and east, and the Canadian border on the north. From 1872 until 1896 there was limited interference from the white man. There were some who found mineral locations, including "Tenas George".

Eureka Camp, 1897 (photo in Totem July/Aug. 1982).

North half of Reservation opened to mining

On Feb. 21, 1896, the north half of the reservation was thrown open for mineral exploration. The major access in Washington to that part of the Ferry County was by way of Colville, Marcus, then into Grand Forks, Canada, and south by way of Curlew into what would become Republic – first called Eureka mining district. Some of this journey was made by horse, some by wagon train or mail carriage and much on foot. The Sherman Pass across the mountains from Kettle Falls to Republic would not come about until 1955, but the majority of prospectors in 1896 were coming from Canadian mining camps.

There are several stories of the actual first discoveries but the one I use is from the family of John Welty, said to be the first white man to make a location and did so a few hours before the actual opening. John and his brother, George M. Welty had camped out on San Poil River most of the winter leading up to their findings, according to John's daughter, Lillian Welty Mathews. (She and her half brothers were residents of Colville up until the early 1960's). One history said John Welty spent that winter with a squatter O'Brien. By being "on the spot" John staked the **Black Tail**, and his brother George was credited with locating the **Quilp** and other claims three days later on Feb. 24, 1896.

The area was called "Cleopus" by the Indians, meaning high mountains. What followed in this great stampede, and the tremendous production, brought about the name Eureka which it carried until wanting a post office and finding there was another by that name. It was changed to Republic in honor of what was to become the best mine in this earlier history.

Philip Creasor and Thomas Ryan were said to have left Rossland where they had been grubstaked by L.H. Long, Charles P. Robbins, and James Clark, and headed for the Reservation country. The men were old-time miners, Creasor was from Ontario and had come west in 1885 and located in Colville. He had mined most of his life and at the time of meeting James Clark, he was prospecting Slocan. Ryan was said born in New York and had followed mining through almost every field of discovery. He had mined in Silver City and Coeur d'Alene, Idaho, among other places and was mining in British Columbia when he, Creasor and others got together. Robbins, one of their grubstakers was also an old timer at this game, having been born in Silver City, Ida., and having mined through much of the country.

From varying versions, one has to choose and the story that Ryan and Creasor had first headed for LaFleur Mountain where they had heard of great showings but found the locality already staked, sounded logical. Their first recorded locations were not until the last days of February, and the first week of March.

Disappointed, they had returned to the little town of Nelson, later named Danville, where they met Alan Blackburn, John and George Welty. Weltys had a wagon load of supplies and were heading back to do development work at **Black Tail.** Creasor and Ryan were going placering but because it was bitter cold, 20 below zero and lower, Weltys encouraged them to come back to camp with them. On the way they described the area and told Creasor and Ryan where there were some big quartz ledges.

From camp, Ryan and Creasor went up Granite Creek where they first located **Copper Belle;** then **Iron Mask, Lone Pine** and **Micawber,** on Feb. 29. On March 5, they located **Republic** and **Jim Blaine.** They were also credited with locating **Iron Clad, Surprise, Pearl,** and **San Poil.** Disregarding who made each discovery, it is worthy to note that among the locations made in the first two weeks of prospecting, most of those claims became leading mines of the district.

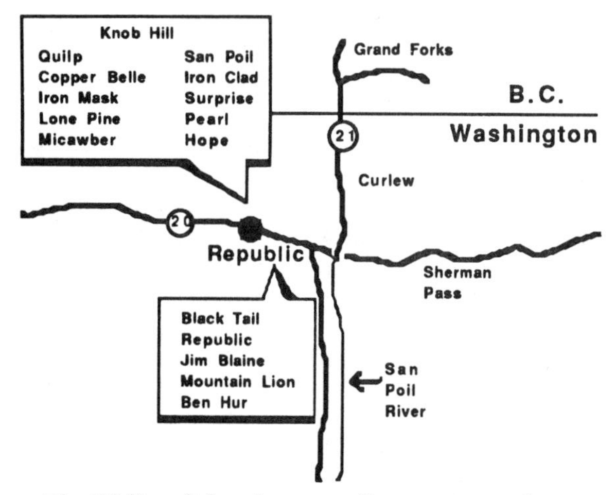

The Weltys did make more discoveries and more will be told of their claims. When mining began to take a real slow-down in the Republic area, John Welty returned to Stevens County where he will be remembered as fire warden around 1917. He died in 1930. No more has been learned of George's later life.

Creasor and Ryan had found a vein 15 feet wide that was said to assay from $72 to $300 mostly gold, in the **Iron Mask.** By early April it was down to a yield of only four dollars to the ton. In June the two men parted and Creasor was said back at the **Republic.** By October he was a bit discouraged there and sold one eighth interest of his shares in **Republic;** also in **Jim Blaine,** to Dennis Clark. Creasor retained a like interest for himself. In the following December, Dennis Clark came to the new camp to meet Thomas Ryan. Seeing so much production and development in this booming field, he was amazed. With the prospects of so much future, he and one of the grubstakers James Clark, who was a brother, rushed off to Rossland, (or Spokane) to interest their wealthier brother, Patrick "Patsy" Clark in this new field.

With plenty of money behind him from his association with Marcus Daly at Butte, Montana; his **Poorman** and other mines in the Coeur d'Alenes, and his **War Eagle** among other claims in British Columbia, Patsy Clark was in a position to spend huge sums of money in Republic. In March, 1897, probably by reasons of wealth, and as being the oldest brother, he bought the first of the 50,000 shares of treasury stock offered for sale for the **Republic Mine.** Arriving in Republic not long after, he was named president of the Republic Gold Mining and Milling Company. From that day dated the industrial growth of **Republic Mine,** and of Republic camp.

Republic Mine

By April 18, 1896 there were said only 64 men in the entire camp but 24 got together and formed the Eureka Creek Mining District. Activity was very heavy at both north Republic and south Republic. Most historians credit Tom Ryan with having staked the **Republic** and **Jim Blaine** while his partner was off to Nelson recording other of their locations. Ryan was also busy at his **Lone Pine** and reportedly struck a rich vein assaying $320 to the ton, mostly gold. By the fall of 1896 he had been taking money earned from the **Lone Pine** to continue development work on the **Republic.**

Patsy Clark bought Ryan's interest in **Republic** amounting to 200,000 shares at seventeen and a half cents a share, or $35,000. He bought Creasors 100,000 shares for 20 cents each, or $20,000. Ryan thought he should have had the better deal but he was thankful for his $5,000 house with "all the comforts of home." Creasor was credited with all the locations of their mines that would show a value of around $10 million. Ryan's discoveries would total well above the three and a half million dollars the **Republic** sold for. Both men benefitted greatly in their affiliation with the Clark brothers, Patrick, James, and Dennis, and both Creasor and Ryan became wealthy men.

On Oct. 31, 1898, Patsy Clark and a Major Kingsbury were said involved in the production at **Republic** of some high grade ore being sent to smelters. They reported a strike giving $300 to the ton, and had built a 50-ton mill. In a later period a 200-ton roasting and cyanide plant called the Red Mill was built. Up until 1899, 20,000 tons of ore

Eureka Camp, 1897 (photo in Totem July/Aug. 1982).

had been milled under unfavorable operating conditions, which did not improve.

The **Republic Mine** was located half a mile southwest of what became the railroad station. It had produced sufficient gold to pay development expenses by early 1899 and had paid $260,000 in dividends. There was estimated to be three million dollars in gold in sight when Canadian investors took over on the basis of $3,500,000 capitalization. Until January 1898 the mine was only surface work, but now a tunnel of 1,000 feet was being opened.

The new Republic Consolidated Gold Mining Company had seen production of $300,000, from 37,000 tons of ore – one sample running $180 in gold and five to six dollars in silver. By 1901 total dividends of $625,000 had been paid but there was to be a reversal of good times. Failure of the mill, and the closing down of the **Mountain Lion** mill caused serious problems. On May 13, 1901 the mine lodging house used by 60 employees, and the bunk house burned with a loss to the **Republic** Company of over $10,000 only covered by $4,200 insurance.

One report said the **Republic** operated into 1903 by shipping ore to one of the smelters. Most reports indicate that until the railroads came in 1902 and 1903, that the majority of the camp shut down including the **Republic.** The county was said to have taken over the **Republic** during this period due to unpaid taxes. Bad recovery of the metals with the mills in use was cause for the **Republic** to close down followed by the **Mountain Lion** but with available smelters at Greenwood and Grand Forks, there would seem to be a market for the low grade ores of the Eureka Gulch if they could be shipped. Any ore moved up to that time was by wagon train and it was not profitable to do so.

Sadly, by the time the railroads were completed by 1903, demand for such ores had dropped not only at the Canadian smelters, but also at Tacoma mill. That temporarily defeated the development of the Republic area.

Smelters available to Republic ore included the Cariboo at Camp McKinney, B.C., opened in 1896; the one at B.C. Copper Company in 1896, the Old Ironsides at Greenwood in 1897, and the Knob Hill one at Greenwood in 1898. The LeRoi mill at Rossland was also taking ore by 1890, along with the Hall Mines mill, but without railroads, it was too costly to transport the ore. Day's mill at Northport would soon be on the list.

Very little was done at the **Republic** until around 1909 when it was opened by New Republic Company. (Improved life was shown at the **Republic** – not to be confused with the famous **Knob Hill** of today.)

In the next two years the output was large ,with some high-grade ore. On one assay a five-day sample showed $445.80 to $685.17, with the lowest $303.40 in gold to the ton. Stock had been sold in May 1899 for $100,000. It ran from 10 cents a share up to four dollars. There was said to be a 24-foot vein of gold running $100 to the ton, reported in 1899 when the mine was paying $35,000 a month dividends. With the new com-

THE REPUBLIC MINE was so great in this early time that it paid $260,000 in dividends to 1899. There was estimated to be $3 million in gold in sight at the time. (Dept. Nat. Resources).

pany developing and working the mine, it was not unusual in 1909 and 1910 to hear the **Republic** mentioned as a five million dollar mine.

A 1909 geological report stated there were more than 100 men employed at the **Republic Mine** with a payroll upwards of seven thousand dollars per month. The mine was said a regular dividend payer and ore was averaging better than $70 to the ton. Many claims were made that under Clark management they had gutted the high-grade ore with an output set at $1,500,000. Under the new Canadian management, that much again was expected to be produced according to an engineer's report.

One quite reliable report to 1910 gave such impressive figures as **Republic** having produced $1,400,000, and **Quilp, Mountain Lion,** and **Lone Pine** having made up the remainder of two million dollars in all. M.H. Joseph gave his report of the first 10 years production of Republic camp at $3,051,000. He said that in 1910 there would be another $813,686 produced. He estimated that figure would go up to one million dollars produced in 1911.

In 1911 **Republic Mine** had control of Pearl Consolidated Group including **Pearl, Surprise,** and **Lone Pine,** all leading producers. Other important producers by 1913 were **San Poil,** where the only mill was said in operation; **Ben Hur, Knob Hill, Republic Mines Incorporated, Hope,** and **Quilp.**

Republic's best production was said in 1902 but with only half hearted work done at all the mines until 1909 there was little output reported. In 1937, **Republic** had 2,757 tons smelted and in 1946 there were some shipments made but by then gold production figures were seldom told. Day Mines acquired **Republic Mine** in 1951 along with many other claims.

By going so far ahead in explanation of **Republic,** the biggest mine in the area at that time, we have missed much of the early development of the town. By early summer of 1896 it became evident that the needs for supplies, for groceries, and other merchandise and services must be met. W.C. Otto had heard of the stampede to this new gold camp. In May of that year he abandoned the Almira market and came to Republic with full stock. He pitched a tent on Eureka Creek and presented his wares for sale. Phil Creasor secured a townsite and built a two-story frame hotel in July, 1897. Harry Kaufman had built a log structure earlier and John Stack opened a tent store July 22, 1897, followed by the building of his two-story store and office building in Oct. 30, 1897.

By March 1898 the town was large on new additions – everyone had their idea and many used claim-sites. James Clark, then president of The Blue Jacket Development Company laid out a a townsite March 22, 1898. It was combined with another May 21. The original one was said platted April 29, 1898 by Tuesday Development Company, Patrick Clark president. There were many more locations presented and argued before a decision was made. Within two months nearly 2,000 people came to Republic and put up canvas tents, or shacks; hasty architecture being the rule. By May 6, 1898 phone service had been connected and reports of new strikes were daily reported to the stock companies in Spokane, with stock quotations given in return. There was also a newspaper.

The south half of the reservation was thrown open for mineral research on June 30, 1898, bringing more needs for supplies the length of the county. The **Pioneer** reported on May 14, 1898, "Large quantities of whiskey, flour and other necessities arrived in Republic during the week. When the camp is older, some luxuries will be on hand. The wagon road for 85 miles was the site of freight teams coming and going from Marcus to Republic. Some merchandise was coming from Seattle on a steamboat that came up the Okanogan River and a number of four horse teams were seen waiting to load. L. Hallenbeck, steamboat man said there was 11 acres of freight at the landing and more arriving every day – he said 75 freight wagons could not transport all of it."

By June of 1898 there were schools, justice court rooms, a jail, and the first church – a Roman Catholic was built by December 1900. A Baptist Church followed in 1902. The post office from April 1 to June 30, of 1898 had 1003 registered letters sent from Republic – more than any place in the state, and by 1899 it was busier. The office did more business than ones in Los Angeles, Seattle and Portland.

Among the many men who had hurried to the Eureka district was our prospector-miner Alfred Stiles, who had been still mining around Rossland at the time. He and his friend C.W. Gerbeth were grub-staked by D.D. Birks. Stiles said they located 16 claims showing copper and gold in the vicinity of the U.S. Canadian line near Patterson. NO LUCK! From there they moved out of the district and on to Grouse Mountain, then to Squaw Creek above Northport, Wash. Here Stiles staked other claims but nothing of consequence ever came from any of these.

There were so many claims filed in Republic; to name some of the others in the vicinity include the **Ben Hur, El Caliph, Morning Glory 1898, Little Cove, Insurgent, Princess Maud, Tom Thumb, Admiral** and many others.

The **Quilp** located by George Welty was earlier called the **San Poil.** Up until 1906 the **Quilp** had produced 2400 tons, assaying zero point four ounces gold, and five ounces silver with a value of $263,000. Along about 1911-12 it became the property of Imperator Mining Company and total production to 1920 was $720,938. In 1937 it produced 22,402 tons, and in 1938 another 9,828 tons, and continued to produce into 1940. In 1951 it became the property of Day Mines, along with **Imperator, Surprise,** and others.

The **Imperator** was an important producer in

The beginning of Republic or Eureka camp. (S-R-1899). 1.–a swell barbershop–50¢ for a shave. 2.–small building on right was first bank. 3.–Spokesman-Review headquarters. 4.–Part of the town in 1899. 5.– Republic's Clark Ave., looking north.

Photo of BEN HUR MINE at Republic during the early days of that great mining camp. (Stevens County Historical Society).

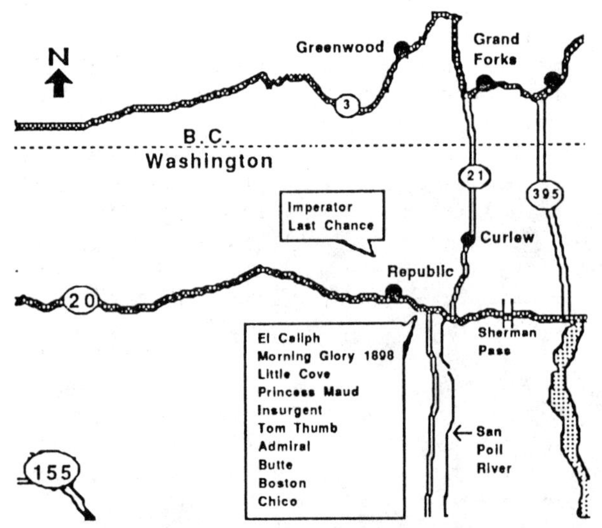

1899. The **San Poil** which became a part of this combine was thought to be the southward extension of **Ben Hur.** It had produced 200 tons prior to 1902 and in 1911 it reported zero point seven ounces gold and four ounces silver to the ton. Continuing to produce through 1921, and on to 1935 the highest test showed $400 to the ton in gold and $15-$17 in silver.

John Welty's **Black Tail Mine** was opened with a short tunnel and a main tunnel. Production was not reported heavy but some single showings were $200-$300 in gold to the ton. By 1911 it became the property of Hope Mining Company and was considered a part of the **Surprise** claim. This property also became one of Day Mine's.

Another Ryan and Creasor location included **Lone Pine** a producer right from 1896 and from which Ryan and Harry Kaufman took earnings enough to work the **Republic.** By Jan. 1, 1898 the men had a short tunnel and by 1899 were going into heavier production. **Insurgent** was a producer by 1911 and was believed to be an off-shoot of **Lone Pine.** Adjoining **Lone Pine** is **Pearl** another of these prospector's finds.

Slower to produce, **Pearl** was said to have yielded 2400 tons to the end of 1909. Charles P. Robbins reported values of 12,194 ounces of silver, and 2,301 ounces of gold. Up to 1912, several hundred thousand dollars worth of ore was shipped. It has been said owned by Republic Mines Corporation from 1910 to 1912, and by Republic Consolidated Mines from 1915 to 1922. Production reports of **Pearl** according to a later date were 8,924 tons with 30,256 ounces of silver and 5,436 ounces of gold with a yield of $124,000. It was reported to have produced $137,000 to 1910; over one million dollars from 1910 to 1923 and **Pearl** was still producing up to 1950, when it also became a Day property.

Mountain Lion and **Last Chance** were both said located by "Tenas George" Runnels in 1898. A 50 foot shaft and 100 feet of the big tunnel were finished by 1899 and was a good early day producer. $200,000 reported earnings were from smelter ore assayed 1904 to 1905 at an average five dollars for gold and two ounces of silver to the ton. In 1938 the claim came under **Knob Hill** ownership and was mined by 1941 along with their open pit operation. (Two "squaw men," Best, and ST. Peter also take credit for staking **Mountain Lion.)**

South Half Reservation Opened

Not all activity was in the north half of the Ferry County when the south half was thrown open for mineral exploration the summer of 1898. There was reported to be 46 miles of nothing but trail dust as more than 60 horses thundered south from Republic. Some 400 men were prospecting who were out of the Republic camp. There were also prospectors from many other areas and directions and one week later there were 5,000 claims located.

It was during this time that Tenas George was said to have stood guard with his Winchester in defense of his **Iconoclast** on the Toloman Mountain. He had found the claim earlier and believing this "the greatest mine I have ever staked," he had told his wife he was prepared to die and wanted to be buried at the claim if anyone intruded with serious intentions of taking the claim from him. There were but a few other locations of import and most prospectors and miners returned to the north half to work.

Back to the claims around Republic that became good mines would include **Ben Hur,** an 1898 find. Some ore was shipped then the mine was closed until 1909. Production to then was said $65,000. The mine continued to produce through 1911-12 with a considerable amount of ore running six dollars to $10 a ton. There have been a number of different owners and the mine was producing from 1923-34. Yield was gold at six dollars to $15 to the ton with high at $130 to the ton.

Tom Thumb, Morning Glory and **Butte** and **Boston** are a few more of the claims – mostly in 1899 finds. **Tom Thumb** was said in the advance stages of development and ready to produce early in it's life. By 1909 it had included six patented claims and was owed by Midget Gold Mining and Milling Company. The mine had been closed and was being unwatered in 1909 when it became a property of New Republic Mining Company. Some ore was shipped in 1909-10 with value said from $10-$15 gold to the ton. That assay later ran as high as $23 to the ton and at one time the gold and silver was figured at one gold to five silver.

Morning Glory was incorporated in Nov. 1898, and held in high esteem due to native gold at, or near the surface. In 1899 it was bought for $2,000 by ten Spokane men. That company received

TOM THUMB SHAFT #2 at Republic camp around 1898. (S-R-1899).

$35,654 for 55 tons of ore shipped to Granby smelter. They claimed ore on the dump was worth $25-$26 to the ton as an average. Rich ore was found and the mine price was set at $150,000. By August 15, 1903 there was a sense of permanency when the mine produced 500 tons a day and could have put out 1,000 tons a day.

It was rapidly being developed, then closed down. It reopened in 1911 with owner R.R. Pence and R.M. Skidmore, said from Spokane. It produced over $100,000 and carried values up to $400 gold and silver. It operated through 1936 and was still in operation in 1940. Both **Tom Thumb** and **Morning Glory** became Day properties.

Butte and **Boston** once owned by John McCann and others, was just east of **Princess Maud** and was filled with water in 1911. Assumed to be on **Princess Maud's** vein, it was owned by Blaine Republic Company in 1932. It assayed as high as $16 gold, and $40 silver.

Princess Maud – an 1899 find was being opened to 240 feet that year. By 1901 some ore was being shipped to average $25 to the ton. The mine was closed in 1906 and reopened in 1909, the later ore running from $12-$20, 60 percent silver. **Chico Mine** came about then and was probably an extension of **Princess Maud.** It produced through 1932 and was predominantly silver.

From Jan. 1, 1898 to June 1, 1899 there were said to be 40,000 feet, or eight miles of tunnels, shafts and drifts run in Republic Camp. Construction was beginning on railroads. The most traveled route from Colville, took in 80 miles via Marcus, Bossburg, Grand Forks, and on to the reservation – and still they came, and still more claims were filed!

Valley, Golden Valley, or **Lame Foot** – are probably three claims owned by Everett and I.G. Haugland since 1939. In 1941-43 these produced 1,994 tons, 5,800 tons, and 1,302 tons. Operating up until 1950 these are some of the "new-comers". They did produce prior to 1902 under Hillside Mines Incorporated. Haugland also owns **Ida May** and **Iron Monitor** claims.

El Caliph belongs to Haugland. Up to 1902 it was said to have produced $9,000 with an estimated earnings between $15,000 and $20,000 to the end of 1936. To the end of 1939 there were reports of an average $125 to the ton. C.M. Trevitt was connected with the company and Hillside Mines Incorporated became the predominate owners.

Trade Dollar is an early one but from 1909-11 it's total shipments were said to be 1,000 tons averaging $17 a ton with silver one to 12 ounces to the ton. Said to have produced $25,000 to the early 1930's, some picked samples showed $3,120.92 in gold.

There are so many other claims of note to include: **Bryan** and **Sewall; Lost Lode,** and **Iron Mask** – at least one was Ryan and Creasor's find. In 1941 **Iron Mask** was owned by B.B. Chapman, of Republic. **Macawber, Gold Ledge, South Penn,** and **California,** were also noteworthy claims.

The newcomer **California** was bonded in 1900 for $5,000 and was later given up. E.J. Delbridge purchased the claim for Apollo Gold Mining Company of Connecticut because it had much richer showing than other mines of the camp. The first ore shipped by rail went on the Kettle line and 105 tons went to Curlew by wagon, then by rail to Granby mill. The three carloads of that shipment ran samples of $10,000 to the ton. Over $100,000 worth of ore was said shipped by 1903. On August 15, 1903 after costs on 1,300 tons were figured, the shipment showed a profit of $54 to the ton, or $70,238.

Little Cove was south of **Knob Hill.** Prior to 1934 it produced $1,450 for two carloads with an average of $11-$12. It became a Day property. **Last Chance,** one of the better known Day properties produced three million dollars by the end of 1923. From 24,000 tons of ore it averaged about $13 to the ton showing 7.85 silver and one ounce of gold. It produced well to 1940.

With many more mining locations to be made in this fabulous gold district, we return to the matter of organization of Ferry County. It came into actual being on Feb. 16, 1899. Records were sent from Stevens County to the new county seat at Republic but unfortunately they were placed in a temporary courthouse which burned June 3, 1899. Many mining claim records that had been filed were lost in that fire but luckily the majority of records were still in the courthouse in Colville. A new courthouse was finished in Republic by January, 1900.

Colville's first white child, and one of it's most beloved citizens, Clara Hofstetter Shaver, worked in the Stevens County Auditor's office for a number of years. Broadminded, but totally a lady, she found transcribing mining records quite revolting at times. Her comment to this writer regarded the vulgarity of names given to some claims by men she termed "smart alecs." Her thought was "how awful" that these names had to continue on the records in use. She transcribed a good many of the records that were sent to Ferry County when it was separated from Stevens County.

In the second year after the first discoveries, a Miner's Union was organized by members of the Butte, Mont. union. Edward Boyce was said came to Republic Aug. 11, 1898 to form the new union. Minimum wage for miners was set at $3.50 an hour.

North 1/2 Opened for Settlement

Along with the opening of both halves of the county for mineral exploration, the north half was opened for agricultural settlement Oct. 10, 1900. In the guise of prospecting, many had settled into temporary quarters ready to spring onto their chosen location. Before, and after the opening, there was pitched tents, partially erected cabins; and after staking their land the rush to establish ownership required making improvements.

The new county had been named for Elisha P. Ferry, the first governor of Washington State 1889-1893. With the mining boom, and now the settlement for homes, Republic was suddenly the sixth largest city in the state. The county population for 1900 was reported to be 4,562.

There was said such a "sea of stakes" that first day of opening for land. Over 500 locations were made where there could only be 128 valid ones. Numerous stakes would be placed on the same land, and as many as half a dozen attached to the same tree designating the chosen homesite. It was said a real surprise that no gun play came into the decisions as to whose stake was legal. In time the stakes were limited down to the future owner and due to regulations calling for overhead protection by nightfall, shacks and other buildings sprang up in the next three hours.

One brave lady named Miss Elizabeth Beecroft told the Republic Pioneer (the second newspaper in Republic), three days after the suicide race for ownership, that she had also been involved. She had sent out a load of lumber to be left near her prospective ranch site. Shortly before noon on the 10th, the day of opening, she rode onto the land, posted a notice and made the mad race back to town to record her location. It was a hard horseback race; her trusty steed was white with foam after racing from outside the Curlew area, somewhere around 20 miles south to Republic. She had to reach Commissioner Stockers' office before several men who had frightened her by being on her property that morning. Possibly her fright helped but she had won her race – and her land, and will be ready with a house and developments to show the United States Land Office representative, at the proper time.

The old history books tell that the Indians were allotted 80 of the best acres. The newcomers were allowed 160 acres to a married man, and timber land to total 480 acres. This was to cost one dollar and 50 cents an acre to be paid for within five years.

Knob Hill and Others

W.H. "Billy" Kells is credited with staking the **Knob Hill** claim early in 1897, with others joining him in ownership. Billy was born in New York, had gotten into this mining game and had arrived in Republic by March, 1896. The Eureka Gold Mining Company came into existence in 1897 as a result of Billy's locating the **Knob Hill.** By 1903-04 reports indicate a shaft and two tunnels had been opened with average ore running $18 to the ton, one shoot running $30 to the ton in carload lots. (It is unfortunate that some of this company's files were lost in the fire in 1899 shortly after Ferry became a county, making it more difficult to give us exact details.)

In 1902, before mining activity south of Republic ceased, or at least slowed considerably, properties north of the town or in Eureka Gulch were opening in anticipation of the coming of railroads. It had been nearly impossible to realize a profit on ore shipped by horse drawn wagon to the railhead at Grand Forks, or to Marcus.

The Republic, Kettle River Railroad arrived in 1902. The company had done so much bragging that it was termed the "Hot Air Line," even to the delight of the residents of Republic ready to celebrate it's arrival April 12, 1902. It did NOT arrive on that date but one of the most rip-roaring celebrations was held including the golden spike put in place to represent the line's arrival several months later. This railroad went bankrupt by 1920.

The Spokane, British Columbia Railroad arrived shortly after, making it's way from Danville to Republic through prime grouse territory. This story has been told and re-told so many times, and there are many versions, but it was said the conductor sat in a chair in an open car and shot grouse as the train chugged along. When he hit one, the train backed up so he could retrieve it. (We have heard this story told about venison during fall hunting for deer. True, or not, it took care of a few miles of boredom.) (Yes, there was a game warden for Ferry County – George F. Baisley.)

Kells' need of capital, along with the market drop, has been given as his reason for selling his property to Jonathon Bourne, Jr., a former United States senator from Oregon, in 1908. Kells had been reported as having netted between eight thousand and $10,000 on ore shipments from a 40 foot shaft sunk on the outcrop. Payment to Kells was in stock in the Eureka District Gold Mining Company, with holdings listed as: **Knob Hill, Mud Lake, Mammoth, Little Cove, Gold Dollar,** and

Looking south at Eureka Gulch showing LONE PINE and SURPRISE ore bins. G.N. track on left and "Hot Air" track on right. (photo Republic).

Alpine.

One report said that Mrs. Jonathan Bourne was given half interest in the **Mountain Lion** as an act of gratitude by the original owner whom she had nursed back to health. Shares for that mine were reported selling June, 1899 on the basis of $2,250,000 for the property.

By July that year, the **San Poil** "patch of ground" was estimated worth $350,000 with stock jumping from five cents in 1898 to 70 cents a share in 1899. Stock was sold in the **Black Tail** in 1899 on the basis of $250,000 for the property, and on **Lone Pine** on the basis of $500,000. That figure represented a drop in value of $100,000 from the year before's figure.

There were "goings-on" in this mining camp that did not make for pleasure for those trying to make a fortune – or at least provide for themselves. Travel in, and around the area required the use of a good horse; and for family type people there needed to be a cow, or more. Gangs of horse and cattle thieves found "pay dirt" when they raided camps such as the one at Eureka and Republic. They would generally come in by night and drive off everything that could be moved without facing a gun themselves. Often they would hide this stock in a ravine outside the area and actually had the gall to return some of it to it's owner for a huge reward.

One notorious gang was said headed up by the vicious character Charles McDonald. Thought to be the head of a gang, that person was shot down by Deputy Sheriff Griswold. McDonald's partner, Frank Draper was taken into custody.

Weather caused some of the worst problems and in the winter of 1899, called the "rainy winter," mud made the roads impassable. As teamsters became more discouraged, and many afraid to take such chances, some refused to haul the freight. Marcus, Bossburg, Grand Forks, and Wilbur reported 100's of tons of freight piled up. Even light loads were said to take nine to 12 days to make the trip to Republic. Freight rates went up to four cents a pound, or $80 a ton tempting the teamsters but many still refused to haul. Failure of much needed equipment to arrive did slow mining developments, and prospecting considerably.

Up to the early part of 1900 there were 12,500 locations recorded. The type of "close to the surface" ore found in the area, brought about one of the greatest "booms" in history from 1896 to 1899. Surprisingly, by 1901 there were mills closed and too many claims and mines shut down. There was actually no sustained producer until after 1935.

Production reports, and population figures were given in many different amounts by differing historians. Even state reports varied. The legal population count 1900 was at 4,562 for Ferry County. That year when Republic was incorporated there was reported by news and by mouth to be 12,000 citizens in the midst of the gold rush. One state report in 1910 gave the early district production from 1903 to that date as five million dollars, with 83 percent gold. Another gave total production for the district at 2,500,000 tons, with gross value at $50 million. And so it went in boom time!

My favorite parable to this kind of reporting, each one having their own truths is; From the Statesman-Examiner at Colville for whom Saga Joe sez: "Washington with it's millions of employees is best summed up by Mark Twain's old story 'fifty thousand ants going down the Mississippi on a log and every blamed one of 'em thinks he's steerin'." From

The mountain showing above the present day KNOB HILL Mine, was where Billy Kell made his first location in 1897 and called it the Knob Hill claim. Nearing 100 years this claim is still the basis of the mine operated by Hecla today. (Courtesy J. Maurice Slagle.)

frantic activity often comes frantic reports of output and growth. By cross-checking we hope to come about the one that is the most accurate.

The 1909 report on other mines in the area were **Black Tail** assaying up to $100 a ton and **Surprise** assaying as high as $300 to the ton. This mine belonging to Seattle men was considered one of the best in the area. **Quilp** was well developed and had produced more tons of ore, and paid more dividends than any mine in camp except **Republic. Mountain Lion** was said to have shipped many tons of ore, some assaying up to $57 a ton. **Southern Republic,** (former **Princess Maud**), reported values up to $600. **Ben Hur** along with **North San Poil, San Poil,** and **Lone Pine,** was generally termed the **Republic** ledge. **Morning Glory, El Caliph,** and **California** had produced high-grade ore but at the time of this 1909 state report were only considered prospects.

Ferry County has the longest recorded production of gold in Washington State and has led all counties in production during 28 of 50 years. Chelan took the lead in 1938 as a result of production from the **Holden Mine** which lead it held until 1953 except 1947. Stevens County had produced the most gold from 1905 through 1908, and again in 1922. From 1900 to 1982 there has been approximately $90 million in gold, and $15 million in silver produced out of the Republic district.

A state geological report said there were daily trains to the camp by 1903, both lines having reached there by then, and that there were five smelters that could be reached. This report also included the Hall Mines Smelter at Nelson. (This is not the Nelson that was changed to Danville by 1902, the town just north of Republic.)

Knob Hill

The **Tom Thumb** was mentioned in the 1909 report. It also said **Knob Hill** was mined through tunnels on a limited basis. In 1910 the Knob Hill Mining Company was organized by J.L .Harper, Leslie Anderson, John W. Lloyd, Bob Mabry, and George Hurley. They purchased the **Knob Hill** and **Mud Lake** claims from the Eureka Gold Mining Company, for a reported $125,000. New tunnels were added and in 1915 the recent mining engineer, S.H. Richardson, took over management of the mine for the Knob Hill Company. He negotiated the purchase of the **Alpine** claim from Jonathan Bourne for $20,000. At that time, Bourne still owned the **Mammoth, Little Cove,** and **Gold Dollar** claims, all adjacent to the holdings of Knob Hill Company.

During the years 1916-1918 the **Knob Hill Mine** shipped a large tonnage of siliceous ore to the Trail smelter for wartime needs. Late in 1919 a district wide strike by miners virtually closed the entire Republic camp. Mining supplies were increasing in cost but gold remained at $20.67 an ounce. Two shifts still worked at **Knob Hill** in 1920 with operations closing there before the end of the year.

Production reports from the **Knob Hill** Aug. 1, 1910 to Jan. 1, 1912 included 7,142 tons with a gross value of $224,795. The ore yield later was reported as 10,850 ounces gold, 31,914 ounces of silver with net returns of $149,005. To the end of 1951 the mine was said to produce more than $10 million. Major operation until Days took over the mine was between 1937, and 1955.

In 1922 the Knob Hill Company sold it's holdings to the Balcalala Corporation of California. Development was renewed under Steve Holman as manager, and E.O. Blades, as superintendent. In that year, **Knob Hill** became known as the largest producer of gold in the state and continued production under the new management until 1928. Lack of demand for silica by smelters at Trail and Tacoma again brought about a shut down that October. Except for some leasing operations by Claude Trevitt and L.J. Plant, in 1931-1932, the property remained inactive.

Gold price went up to $35 an ounce in 1933 and most of the mines in Eureka Gulch reopened. Steve Holman put **Knob Hill** back into production for Balcalala and in 1934 the mine was once more one of the principal shippers of smelting ore.

Resulting from renewed interest in gold mining, an option to purchase Balcalala's holdings was acquired by Mountain Copper Company in June, 1935. R.A. Burgess was resident manager of this English concern which proceeded with extensive exploration, particularly in the **Mud Lake** claim. The option was relinquished in February, 1936. Through Holman's efforts, the property was again optioned. This time to a group headed by A.O. Stewart, of San Francisco, E.E. Check, and F.L. Metler took over exploration resulting in a corporation which purchased the property June, 1936 for a reported $75,000. The new owners, Knob Hill Mines, Incorporated added a new cyanide plant with construction beginning in July 1936, and which was still in full production in the middle 1960's. Sixty-five were employed at this 400 ton mill.

While foundations were being poured, a water line was laid to bring water from the San Poil River four miles distant; and a power line was constructed by the Republic Power Company to bring power from Tonasket, 40 miles away. It furnished power for the mill and provided a dependable source of domestic power for the town. Overburden was stripped from much of the **Mud Lake** claim and breaking of exposed ore commenced. First ore delivered to the new mill was on May 10, 1937.

Many changes of management and operation took place over the next years. A.R. Patterson became mill foreman and was eventually vice-president and general manager. In 1936, A.O. Stewart was president; Walter Lyman Brown, vice-president and general manager; Henry N. Kuechler, Jr., treasurer. In 1948 Patterson became general superintendent; J.E. Davis, mine superintendent, and Louis Lembeck, mill superintendent. L.R. Atwater, R.M. Grantham, R.E. Blain, and H.M. Miles were also on the staff of those from Republic.

A power line from Tonasket, 40 miles away furnished power for the KNOB HILL mill and the town. The mining boom town of 1897 shares the economy with logging and ranching today – and a new discovery of fossils drawing wide-spread interest. (Courtesy Byron Larson, Portland, and Statesman-Examiner).

There were other San Francisco officials, and titles kept improving as the mine grew in importance.

In 1942 most of the mining was done underground, and "only through the introduction of the contract system of wage payment was production kept up under the increasing shortage of labor caused by the demands of the war effort," according to Patterson at a later date. Since the start of World War II, **Knob Hill** had been the only consistent producer of gold and silver in the Republic camp. A 1955 report said the company owned 13 patented claims, **Knob Hill, Mud Lake, Alpine, Lone Pine, Rebate, Mountain Lion,** and others.

Day Brothers enter Eureka Camp

The Day brothers, Jerome and Harry first came into Eureka Gulch in 1916 when they purchased J.L. Harper's Republic Consolidated Mines Incorporated, with assets including the **Surprise, Lone Pine,** and **Pearl Mines.** George S. Bailey, of Republic served as Day's local representative, and in the capacity of ore buyer for the Northport Smelter, where the Day brothers had partial interest. The principal Day interest was in the Coeur d'Alenes where they had been prominent in development of the **Hercules, Tamarack,** and other mines. The Republic subsidiary the Aurum Mining Company was merged into Day Mines Incorporated in 1949 according to one report. Aurum was the acting company for most of Day's Republic mines up until then.

In 1931, Days purchased the **Little Cove, Mammoth, Baby Franction,** and **Gold Dollar** claims in Eureka Gulch. In 1932 Bailey was said to have added **Tom Thumb** to the Day account, and in 1934 the **San Poil** mine. In 1935 the Aurum Mining Company acquired **Trade Dollar** and **Ben Hur,** followed in 1940 by **Quilp.** The last was said a county tax sale due to the bankrupt Eureka Mining and Milling Company. Days bought **Last Chance,** formerly owned by Charles Robbins, and around this time they acquired the **Jim Blaine-Republic.** They were reported to have at least 92 patented claims.

Prior to World War II, the Aurum Company carried on development work at **Last Chance** and **Little Cove,** and were reported to have shipped substantial ore. Leasing operations at the **Ben Hur** and **Surprise** mines also accounted for some production. Such activities ceased for that company with the coming of the war.

Around 1944, Aurum Mining Company, and Knob Hill Mines, Incorporated entered into a leasing arrangement. Knob Hill leased a block of Aurum ground and mined from Knob Hill's workings paying royalty to Aurum from profits. This lease lasted two years but another lease was written in 1953 on the **Gold Dollar** by Knob Hill with what was then the Day Mines, Incorporated. Crediting respect to Knob Hill's management and ability to treat Republic ores, the lease was still effective in the mid 1960's. The two companies continued joint exploration, development, and operation,and together owned most of the properties of Eureka Gulch.

Around 1953, Knob Hill Mines, Incorporated operated theirs and the Day company's properties. To properly carry on the joint operation Knob Hill improved it's facilities through a major rehabilitation program. In 1955 a new steel headframe was erected to replace the original wood structure serving **Knob Hill's** main shaft. They installed a modern hoist and nine modern steel buildings were constructed. In 1959 the mill and nearby buildings were equipped with automatic sprinklers for fire fighting, and a complete system of fire hydrants were installed. New steel tanks for greater water storage were also built.

Open pit mining on the **Mud Lake** ore body, and the **Mountain Lion** which started in 1937, was discontinued in 1942 due to scarcity of parts and increased costs making low grade ores unprofitable. Underground mining was begun about 1940 and is still carried on.

The **Knob Hill Mine** is known as the biggest and most consistent producer of gold and silver in the history of the Republic camp. Several times the mine was thought worked out but in 1965 it celebrated it's 29th year in continuous operation since purchase in 1936. It's program of diamond drilling and long range development assured ore in sight for many years to come.

The Knob Hill-Day Mines combined operation was said the largest gold and silver producer in Washington and was reputed to be one of the three or four largest producers in the United States. A 1946 report stated $20 million in gold and $3 million in silver had been taken out of Ferry County by that year.

(We owe special thanks to J. Maurice Slagle of Republic, who provided much of this information on **Knob Hill** to this writer in 1964. It was verified and added to by A.R. Patterson, then vice-president and general manager of the Knob Hill Mines, Incorporated.

Boundary Mining Districts

The Eureka, or Republic mining district has so overshadowed the remainder of the county that it is easy to forget there are other Ferry County mining districts, and also important camps just to the north. Discovery, results came about when important locations were made in the Boundary Creek district as early as 1862. Until 1891 not a lot of attention was paid to this part of lower British Columbia, nor to what was then reservation on the south side of the border. It was following discoveries in the Rossland district in 1890 that prospectors spread over a wider country; many would prospect the reservation.

In 1890 and '91 the Deadwood and Phoenix district of British Columbia, became important as prospectors made new locations. Many had come via the Rossland camps, or by way of Marcus in Washington State. Some came by pack train following the Kettle River Valley to connect with the Dewdney Trail at Grand Forks, British Columbia.

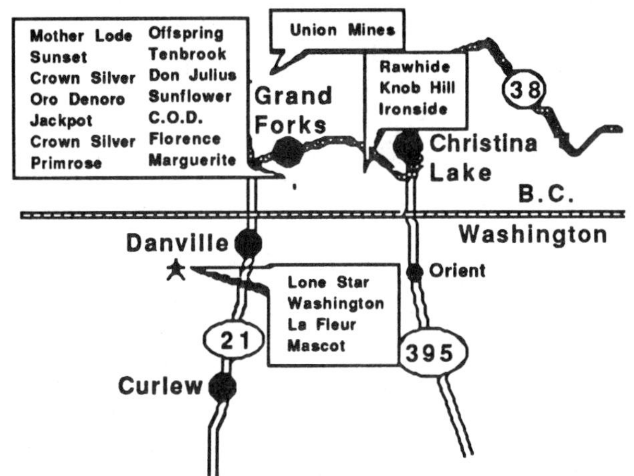

Branch trails were built by the prospectors and the **Mother Lode** was discovered in the Deadwood Camp on May 23; 1891 by William McCormick and Richard Thompson. The **Sunset** was located on June 2, 1891 by John East, and the **Crown Silver** the same day by William Ingram.

We are indebted to the Canadian Department of Mines, 1913 publication for giving us the details of those discoveries. Both the Deadwood and Phoenix districts were said to be of some disappointment due to low grade but there were large bodies of ore found.

The **Mother Lode** was bonded for $14,000 June, 1896 to Colonel John Weir of New York, who formed the Boundary Mines Company. In 1898 that became the property of British Columbia Copper Company. A smelter was built to the south of Anaconda (adjoining Greenwood), and a spur line was built from the Columbia and Western railway to the mine. The smelter furnace was blown Feb. 18, 1901 with a second one following in 1902 and a third sometime later. The company expanded and the plant became capable of treating 2,400 tons of ore per day.

Production of the mines at Deadwood up to the end of 1910 was said no less than 2,114,481 tons of ore, of which 2,014,481 tons were credited to **Mother Lode.** The recovered values amount to 37,648,281 pounds of copper; 93,924.2 ounces of gold, and 336,889.5 ounces of silver. Average assay was 1.1 to 1.3 percent copper, $1.10 to $1.20 gold and silver. We have no report as to value received if any, but garnet is found in most of the mines in this area and just to the south side of the border.

The **Sunset** and the **Crown Silver** mines were reported sold in 1897 and by 1909 came under the holding company New Dominion Copper Company. They also eventually came under the control of the British Columbia Copper Company.

In the mining game you will find many duplications in names of prospects. In this district there was a **Knob Hill** and **Ironsides.** We have no knowledge of association between that **Knob Hill** and the one in Eureka camp at Republic.

Mother Lode ore runs to copper-gold-silver. It was the chief mine of the British Copper Company who was also holding company for other claims including **Primrose, Offspring, Tenbrock, Don Julius,** and **Sunflower.** The company also controlled **Sunset,** the **C.O.D.,** and **Florence** claims.

After **Mother Lode** came under control of that company the **Sunset** was sold to W.L. Hogg of Montreal. Taken over by the Montreal and Boston Company, they also obtained the smelter of the Standard Pyritic Copper Company at Boundary Falls. With other changes of ownership, the mine came under the British Columbia Copper Company.

This company's principal mines in 1910 were the **Mother Lode** at Deadwood; the **Oro Denoro** at Summit; the **Jackpot** in Wellington Camp, and the Rawhide **at Phoenix Camp. They were also developing the Lone Star and The Napoleon** mines on the Washington side of the line. Their smelter was at Anaconda and later proved very beneficial to developments in the Eureka Camp.

The smelter at Anaconda treated 1,800 to 2,400 tons a day. The matte was shipped direct but in 1904, a Bessemerizing plant was installed to convert the matte to blister copper 99.3 percent pure and carrying the gold and silver values. Blister copper was shipped to New Jersey to complete refinement. Some 120 men worked at the smelter.

Mine production in the Boundary district in order of importance placed **Knob Hill-Ironside** mine at Phoenix Camp at the head with **Mother Lode** coming in second.

Development of the **Mother Lode** consisted of a series of quarries forming a large "glory Hole" (a form of mining to reach the highest most valued ore sometimes said to reach the "jewel box"). Horse haulage was used on the 60 and 200 foot levels and electric haulage on the 300 and 400 foot levels. The mine was lighted with electricity by 1910. About 235 men were employed.

The **Sunset** mine is southeast of the **Mother Lode.** It commenced shipping ore in 1901 but became idle in 1908. Something less that 100,000 tons was shipped in that interval.

Crown Silver mine is east of **Mother Lode.** The **Marguerite** mine owned by the Quebec Copper Company was farther east. Assays reported good returns in copper-gold-silver and considerable work had been done by 1910.

Danville Mines

Prospectors knew the reservation would be thrown open for mining and most had knowledge, and some had even ventured into what would be Ferry County to examine the ore potential. In the meantime they worked impatiently, north of the border. We have already read the "rush to riches" that occurred when the area was opened to mining.

Danville was still known as Nelson until 1901. Just outside of Danville, the British Company's **Lone Star** and **Washington** mine was reported active in 1913; also the **Napoleon** at Orient. Ore from the **Napoleon** was sent to the British Columbia Copper Company's smelter at Greenwood to supply the smelter's need for fluxing ore. To retrieve this ore economically an aerial tramway was installed with one terminal at the main level of the mine, another terminal at the railroad siding more than 1,000 feet lower and about a mile west of the camp. Garnet was found in irregular patches at the **Napoleon,** as it was in several other of these border mines. It was a yellowish-green, light brown, or yellow. Run of the mine ore at the **Napoleon** was said to contain 33 percent iron, 12 percent sulphur, 10 percent lime, 30 percent silica, 0.3 percent copper, a trace of silver and 0.5 to 0.1 ounce of gold to the ton.

The **Lone Star** and **Washington** claims, one half mile south of the border were said on the extension of the **La Fleur** and were under development of the Reservation Mining and Milling Company. Edward L. Ensel, of Seattle; J.N. Scott and W.W. Hawks of Everett made up this company in 1897. They did report 1,700 tons of production at that time. The company was said to hold the **Mascot** some mile and a half away which consisted of 20 prospects. The British Columbia Copper Company controlled the **Lone Star** and **Washington** by the 1909 report and leased to Reservation Mining.

Danville and Orient District

Toroda Camp some 15 miles northwest of Republic was intersected by some of the largest veins in the district. **Sally Ann** was a claim owned by a Minneapolis Company and assayed over $70 to the ton. By 1899 it was worked more than other claims.

Sheridan Camp was almost overlooked due to the rush on Eureka Camp. The **Zella M.** was said the best developed mine in 1899 with values running from $200 to $1,000 a ton.

The **La Fleur** on Goosmus Creek was one of the more interesting discoveries. An English concern, probably well aware of it's content, had been keeping an eye on the **La Fleur** for several years. So had certain prospectors, and so had native Indians. The beautiful purplish hue of "peacock copper", along with other values, had been one more encouragement for the early opening of the reservation to mining. Moments after the news came over the wire that President Cleveland had signed the necessary documents, two men on horseback

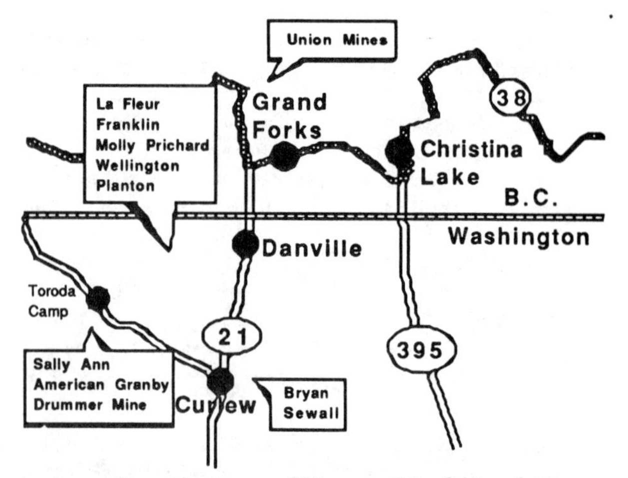

took out from the town of Marcus, Wash. In a bitter race to be the first to stake this claim they covered the long, snowy and very cold miles but arrived almost simultaneously.

Angry in defeat in this extremely cold February race, it was only by one of the men's cool head and hasty retreat that the snow covered purple vein was not stained red with blood. He backed off, withdrawing stakes already driven in place. Legal battles ensued and in 1897 the **La Fleur** (known as the **Butte**) along with the **Comstock**, was said held by Comstock Mining and Milling Company. By 1897 ore was averaging $75 a ton, 30 to 54 percent copper, the remainder in silver. By 1899 the company was reported under a British American Corporation, the same company that owned the **Le Roi** mine at Rossland. This mountain was the same that Philip Creasor and Thoms Ryan were heading for when turned back by other prospectors.

Irvin Ballew had the **Franklin** claims west of Danville and Millard F. Crounse had 480 acres of placer ground on Gold Creek, both residents of the area. Even the postmaster was drawn to this gamble and owned the **Molly Prichard** and **Wellington** Camp near Danville. John W. Siebert was offered $4,000 for half interest in his **Planton** mine.

Orient District

South of Danville and on the east side of the Kettle River, is the **Hercules** claim. Also in the Orient district is the **Faithful Surprise** which along with the **Lucille Dreyfus, Mineral Hill, Virginia,** and **Morning Star** made up a portion of the ten claims owned by Morning Star Mining Company in 1936. This group of claims paid $15,000 before 1910; $27,000 in 1917, about $15,000 in 1935 and produced another 790 tons between 1940-1943.

The **Surprise** prospect revealed coarse yellow gold, some high grade and averaged $17 a ton in the 1914 report. The **Laurier** and **Jenny** prospects were also in the Orient district, neither showing much value in 1910. Of those claims that did produce later, one has to realize that Orient district is 400 square miles with very little over one half of the claims in Ferry County. The majority of claims in that district will be included under Stevens County.

Curlew, Toroda

The town of Curlew was said founded by trappers who built cabins on the site in 1880. Many of those structures are said standing today. When the reservation was thrown open the first white man to establish a place of business was Guy S. Helphrey along with partners Henry Nelson, and Andrew Seymore. The Helphrey family business approached 100 years operation in 1988.

In 1903 Toroda Store was built as an extension of an earlier established business in Curlew some 30 miles distant. Mountain top location was selected to serve some 200 miners and hangers-on who clustered around the flourishing **Sheridan** Mine. The store was said serviced daily by freight wagons from Curlew and it was abandoned some five years later. The structure remained in 1965.

Within reach of the then booming town of Curlew was the **Bryan** and **Sewall.** In 1897 it was reported that F.C. Robertson, J.M. Hamilton, and R.F. Rogers were showing returns of $45 gold and seven dollars in silver at that claim. On the headwaters of Toroda Creek the Gold Dust Mining Company boasted two prospects on a ledge of free milling quartz assaying from $14 upwards to one as high as $2,673. Little is known about the **American Granby** said to contain some platinum, and less is known about the **Drummer Mine.** (One wonders about this sleepy little village and the effect that **Echo Bay's** fantastic mining developments of 1989 will have.)

Belcher Mountain Mines

The **Iron Mountain** was closer to the **Belcher Mine** and in that district. W.C. Zutt of Republic was said owner. From four prospects, ore had only been shipped from the **Blue Horse.**

In the early 1900's Belcher Mountain Mining Camp (described as on Lambert Creek – and 25 miles south of the border, near Curlew Lake) had several rapidly developing claims. The **Oversight** owned by that Mining and Milling Company of Seattle reported iron-gold-copper yield of many tons shipped by wagon. This became too expensive and they closed down for two years awaiting arrival of the railroad. It came into the Curlew-Republic area in 1903-04. At **Copper Key, Anonymous, Bortle,** and **Belcher** large bodies of iron ore was said developed and some copper shipped. A 1907 report said 3,249 tons, and prior to 1940 there were 7,000 tons shipped. Other claims mentioned in the district were **Hidden Treasure, Pin Money, Granite** and **Wander.**

Belcher Mine had three tunnels. A ledge 80 foot wide of solid metal had been exposed. That was said to be the largest body of solid ore known of in Washington State. It was mostly iron ore.

A small town of homes and offices, and a power plant to furnish lights for the town and mines were said at Belcher, along with a store and post office. Some development was begun by G. Waver Loper and Associates on the **Churchill** Group, and the **Winnipeg** was mentioned. Several smelters were within a 50 to 100 mile radius of these mines.

Mining in South Ferry County–Covada Camp

The south part of the reservation was opened to mining on June 30, 1898 and activity was great for awhile but with minimal showings. Some mining in a haphazard way commenced at Covada Camp in 1901. There were towns at Covada and at Meteor (south of what is now Inchelium). From the Meteor Mining Company, operators of the **Meteor,** gold and silver in various forms was reported along with calaverite, silvanite, ruby, and native silver. Wires of gold and silver picked specimens tested $143,000 in gold (at $20 gold price), and another at $1,300 in silver (remember silver bottomed out in 1893 and was only 50 cents in 1920). Run of the mill returns showed $97 a ton. Two small shipments were made assaying 75 to 90 ounces to the ton, of native silver. One assorted shipment gave five to 66 ounces to the ton, of silver and 0.3 to 0.6 in gold.

The **New York** and **Montana** owned by Mr.'s Pea, and Fish; **Stray Dog** owned by Stray Dog Mining Company with headquarters in Seattle, **Keystone** owned by Hurley, Davis, Lodgson, and Matthews are worthy of mention. The **Advance** company had spent $25,000 on development, more than any other in the district. Frank Johnson and A.M. Campbell of Spokane were officers. (Campbell of Coeur d'Alene, and Canadian Mining.)

The district could be reached from the Stevens County side by a 30 mile trip including wagon trail from Blue Creek, then ferry across the Columbia River (now called Lake Roosevelt). The alternative was across the Huckleberry Mountains, either way making hauling very expensive. Some small shipments were made from **Longstreet,** and from **R.E. Lee.** (This prospect could be one of many made around the country by the Colville merchant by that name; and his partner, George Sizemore of Covada.) **Longstreet** was owned by Henry Garrett and both showed a little iron and traces of gold and silver.

In Columbia Camp, **Buster** was owned by Southern Cross Mining Company, financed by Howard H. Lewis of Seattle; **Orion** Mining Company's claim by the same name was owned by Robert Bell, of Portland. A 200 ton smelter was expected to be built down the river as these claims showed increased value with development.

San Poil District

When the south end opened up Keller area probably got the most attention. Mines receiving mention were the **Manila Copper Mine**, the **Last Chance**, and **Walla Walla Copper**, the last one belonging to people from Walla Walla and the development done by Jasper King. The Meadow Creek Copper and Molybdenite Company of Bellingham had Frank Hatton and J.E. Jannot as officials. Values were in silver-copper-lead-gold- and "moly."(is this early day mining man, Joe Jeannot?)

The Byrne Property named for Patrick Byrne of Spokane showed values in silver and gold. H.R. Alexander managed the **Summit** Group, H.A. Johnson owned **Silver King Group**, and J.A. Cody was manager of **Silver Hill Group**. Many other prospects and claims were in this region then booming with mining fever.

The Keller and Indiana Consolidated Smelting Company was built in 1905-06 by Allis-Chalmer Company, with R.L. Boyle as promoter. It was built to smelt the ores from **Manila Mine**, seven miles west of Keller. That mine showed 0.5 to as much as four ounces of lead-silver and some gold quartz. Sadly the 150 ton smelter stood idle for want of coke in a district that thought it was on the verge of prosperity. Even the town itself had come into being, due to the large number employed by mining activity, or service connected. Tragically, railroads were tied up in litigation and had not yet been built into the district by 1909, creating serious problems. Because of that, 1500 tons of ore had been produced, but not smelted by 1909. Lee Farr held title to the **Manila**. Sometime after July 1910 the smelter was blown in to treat ore from his mine. Mrs. James LeFavre held controlling interest in **Manila** in 1942, and Bert Summerlin of Keller bought the adjoining **Last Chance**.

Keller was formerly called Harlinda. There were a few houses, a general store, stable, post office and lodging house; and a stage tri-weekly from Wilbur. It may have been Samuel gray who located the first hotel and grocery store in Keller. He was an old time miner who had started in the Boise Basin, Idaho camps in 1863. He has been credited with discovery of **Ida May** on Granite Creek in the San Poil vicinity. Gray claimed to be a good Republican and told when he voted for Lincoln.

The **Congress** prospect on the north side of Bridge Creek showed nickel ore but the miles distant to either Republic, or to Wilbur in Lincoln County cost more than shipment was worth. **Polepick** was another named prospect that awaited railway service.

As a whole, the San Poil district was said to be the most inaccessible in the whole county and up to the summer of 1910 had been less developed than any other well recognized mining district in northeast Washington except the Metaline district. Wilbur, 12 miles away, and Republic 24 miles away both provided rail service by that time. In April 1911 a route had been surveyed down the San Poil River but construction had not started.

In a Division of Mines report of 1956, five claims in the San Poil district listed nickel and cobalt content. From the entire Ferry County there were 80 occurrences of copper listed; 164 lode gold, (only 30 in 1975); 35 placer gold, (11 in 1975); 12 iron; 140 lead; 10 tungsten and 188 silver, even platinum was mentioned again in the **Walla Walla** and the **Rogers Bar Placer**. Many of these claims are a result of the old prospects located around the 1900's. Ferry County's gold production including 1903 to 1952 was $20,442,521, according to a 1955 bulletin. The output of gold is no longer reported after 1956 because confidential data would be disclosed.

Back to Republic and KNOB HILL

Changing times and ways in the mining industry was best expressed in the Republic newspaper in August, 1937. It told that if the old miners were still around Republic they would see two huge bulldozers, two power shovels and power trucks moving ore in large quantities to the largest mill the district had ever known. They would be surprised at "What we used to call waste," as they would watch the stream of rock flowing to the **Knob Hill** 400-ton plant. The article explained that formerly this was just dumped but that it carries values and there is so much of it removable that decades may be occupied in just that effort.

Developments in the district include the **Mountain Lion** production of 3,000 tons monthly by report Jan. 30, 1937. C.M. Trevitt and A.B. Thompson of Republic operated the mine under lease from the Mountain Lion Mining Company. 25 men were employed. In March 1938 high-grade ore was reported shipped from the mine to be treated in the **Knob Hill** plant. A daily production of $1,000 was reported with an average 15 men employed by then.

Expected to increase production from 75 to 100 tons daily in September 1940, the **Mountain Lion** and **Rebate** mines were controlled by Knob Hill Mines, Incorporated. L.A. Fitzgerald, Republic, was mill superintendent. Walter Lyman Brown of San Francisco was general manager of both parent and subsidiary concerns.

The Mining Journal tells March 15, 1941 that capacity production from the **Knob Hill** Mines, Incorporated, was 400 tons of gold-silver ore being mined and milled daily. In November the company completed installation of a 30-ton flotation plant. This was necessary because the 400-ton cyanide plant could not treat the sulphide ore. Frank H. Mitchell was general superintendent and Louis Fitzgerald was mill superintendent with Allan Patterson mill foreman. The company earlier reported employing 50 men at the mine and mill.

Part of the huge cost of operation is material breakdown; material purchased with newer methods, and delays which keep the miners off the job. In the spring of 1939 **Knob Hill** had a shut-down when gears in the ball mill were stripped. The **Mountain Lion** was also idle. Besides the cost of the new flotation plant; the power house, containing mine hoist, compressor and a Diesel engine were destroyed by a gasoline explosion at the company's **Rebate Mine**, which was being operated by lesses.

Formerly having mined by open-pit method from the **Knob Hill** and **Mud Lake** pits, the **Knob Hill** put down shafts. In November, 1941 they planned to extend the 250 foot inclined shaft an additional 150 feet. Open pit was still in use at the **Mountain Lion** at that time, where 300 to 325 tons was taken daily.

Other Area Changes

There were geographical things happening that had an affect on all of the area. In 1915 a huge Magnesite plant was being located in Chewelah on the Stevens County side. This plant and the associated mining drew many of the miner-prospectors that were finding things slow.

Probably the greatest force for change was the construction of Coulee Dam on the Columbia River starting in 1933. Lands were cleared, towns were moved and Lake Roosevelt evolved around 1940. For all of the years since the beginning, the Indians had fished the Columbia and one of the favorite spots was at the Kettles in the Columbia River dividing Stevens and Ferry County. The white man fished this area with boat, hook and line, while the early natives used nets. It was a sad day when the last salmon run was seen in the Kettle Falls area in 1938.

Off the Gold Standard

One great piece of news to the mining men was on Jan. 3, 1934 when the country was taken off the gold standard and gold went to $35 an ounce. It had been held at $20.67 for many years. Gold coins were withdrawn from circulation then, but by Dec. 31, 1974 all statutory restrictions on gold ceased and citizens can now own gold. Gold price was released in March, 1968 but it stayed near $35 until it went to $45 sometime in 1972, and upward since then. (The gold standard had been effective at different times from 1896,-1900,-1917 and until 1934).

Reports on Activity

A geological report of 1962 stated only two Ferry County mines were active; **Knob Hill** and **Hillside.** A 1965 report in Bureau of Mines and Minerals Yearbook said "most of the older inactive mines in Republic's Eureka Gulch are owned by Day Mines, Incorporated – however **Gold Dollar** is operated by **Knob Hill** from that shaft. Ore is extracted from the **J.O. Mine** Number three vein." Some exploration was being done by Bear Creek Mining Company and Consolidated Mining and Smelting Company in the Keller area in 1965-66. More search was going on around the Danville, Curlew districts in late 1900's. Mining employment in Ferry County for 1971 was reported to be 70 persons, having been paid $736,279 in wages, according to Trico economic print-out for March, 1973.

Of slight interest to some is how much the boss is paid. It is where part of the earnings go from mining companies. Securities and Exchange commission requires a list of salaries for the three highest paid officers of a company... those making over $40,000 per year. The figure includes bonuses. This list printed April 20, 1975 was generally for 1974 salaries.

President Charles T. Snead, Jr. of Callahan Mining Corporation was paid $69,379 in 1974; William H. Love, president of Hecla Mining Company was paid $103,333, more than $18,000 above the 1973 salary of $85,000; and Gulf Resources Chemical Corporation president Robert H. Allen (parents of the then **Bunker Hill Mine**), was paid $205,500 as president. He had only received $123,900 in 1973.

Day Mines president William M. Calhoun was paid $40,658, while Sunshine Mining Company paid their board chairman Irwin P. Underweiser $55,000 for each 1973, and 1974. That was to be "upped" to $100,000 by the new contract. Clarence E. Nelson, their president received $75,000 and a contract for the same the next year.

(By comparison, it might be of interest to know that Wendell J. Satre, president and board chairman then of the Washington Water Power Company received $74,067, a raise of over $6,000 from the previous year– and maybe your banker – Harry S. Goodfellow, board chairman for Old National Bank of Washington received $86,057.)

Proxy statements for all these companies indicate most of those in the executive suite were "keeping pace with inflation." These salaries hardly compare with the 1988 Forbes report but as pointed out "big industry is costly!"

Hecla Take-Over

Some early information in what ended in a "take-over" of the Day Mines properties first came with a notice of registration statement filed with the securities and Exchange Commission in connection with Hecla Mining Company's proposed sale of it's Day shares. That news release was on Oct. 9, 1972.

The **Knob Hill** lode gold mine was said to be the only gold mine operating in 1975. Then came bits and pieces of news and in May, 1977 Day Mines, owners of property near the **Knob Hill** were reported negotiating for a possible purchase of the Knob Hill property. At the stockholders meeting Keith J. Droste, Day Mines general manager announced "termination of mining" at the end of the year in the **Knob Hill** gold mine. The company had earlier reported they were negotiating the purchase of the **Knob Hill** mine with it's San Francisco owners; the announcement in May confirmed that action.

April 13, 1978 the Wallace Miner news reported that during the year, Day Mines participated in certain revenues from the **Knob Hill** operation. Their share of production from the **Gold Dollar** was 7,611 tons with an average grade of 0.84 ounces gold and 2.55 ounces silver per ton. From the Number three joint operation Day's share was 9,271 tons with an average grade of 0.48 gold, and 2.98 ounces silver per ton.

The Knob Hill Mines, Inc., terminated mining operation Feb. 28, 1978 after 44 years of almost continuous operation of that mine. Day Mines acquired the remainder of the mineral rights and all of the property and surface plants of Knob Hill Mines, Inc. A drilling program was under way and the mine and mill was being rehabilitated. Gold price hanging around $200 an ounce made development work and limited resumption of production during 1978 appear feasible, according to the Wallace paper.

For years the **Knob Hill** gold mine had been expected to close down for lack of ore but by July 31, 1978 some 50 men were working under the Day Mines operation. As Thomas E. Sellers said, "a hole may be a gold mine again." In April, 1981 Day's first quarter income results were announced as $1,175,664 or 27 cents per share as compared with the same period in 1980 of $4,028,339 or 93 cents per share. Silver had dropped in 1981 to $13.37 as compared with $32.49 in 1980 and first quarter gold had gone down to $518.64 in 1981 against $632.74 in 1980.

The KNOB HILL MINE at Republic was owned and operated by Day Mines at the Dec. 7, 1978 time of photo. (S-R).

The news of that day told that the merger proposal made to Day Mines Board of Directors several weeks before by Hecla Mining Company, had expired April 9, 1981. No action had been taken and now Hecla was filing a Registration Statement stating it's intentions to solicit the exchange of stock on the basis of $1.65 Hecla shares for each Day share. Talk of "take-over" was heavy in local conversation.

As most "take-overs" are, this was to be a hostile one but did come to a head Oct. 21, 1981. Day Mines was merged into Hecla Mining Company that day including the holdings in Republic and the **Hercules** mine in Coeur d'Alene. The total figure accepted by Day Mines Company was $105.8 million payment for Day stock.

Henry L. Day, the longtime mining leader died March 22, 1985. After the take-over Mr. Day maintained an office in Wallace. The Day mining history had commenced when Henry's father, Harry L. Day and partner Fred H. Harper staked the **Hercules Mine.** At Henry Day's death, Elmer Bierly, Hecla manager of investor and public affairs said that "Henry Day was a very real part of the mining industry in this area. He was a leader in the community for many years, and he'll be sorely missed." Judge Richard G. Magnuson declared Day a friend of the Coeur d'Alenes. "Of all the people that amassed wealth in the Coeur d'Alene district, Mr. Day took the greatest pride in staying here and in fulfilling the obligations he felt to the people of his community."

Henry L. Day, respected mining leader died Mar. 22, 1985. Associated with HERCULES and TAMARACK in the Coeur d'Alenes; over 92 properties in the Republic area, the Northport smelter of the late 1900's, and other properties. (S-R).

The day Hecla Mining took over Day mining holdings. William A. Griffith, president of Hecla Mining Co. stating "It's a great day for Hecla." S-R).

Hecla Back to Work

Gold price had been going up so fast that one broker stated, "it's like a balloon that you keep inflating – eventually it has to burst." Gold had gone to $850 an ounce and on Jan. 23, 1980 it took a drop of $175 in Zurich and was down to $675 in New York. Hecla announced they were not going to have to close the **Knob Hill** mine due to additional reserves that would allow it to be mined awhile longer. 85 to 90 men were working. "If current prices for gold and silver hold, the mine will continue at least two more years," one company man said. Their gold production at **Knob Hill** had, totaled 39,192 ounces in 1985, an increase of 72 percent over the previous year, according to Hecla Mines, who forecast 50,000 ounces a year.

On Nov. 24, 1985, Norman Thorpe had an article in the Spokesman-Review. His sub-title pretty well told the story about **Knob Hill**: "A page in history or a real Phoenix." Prior to the June announcement it looked like the end had finally come for the mine. The town was notified of the closure but even they picked up their hopes. This was the third time the mine was "at closure" stage. At that time Thorpe said the mine had produced 1.9 million troy ounces of gold and 11 million ounces of silver over it's lifetime.

Saddest possibly is the family of miners who wonder "what, and where?" A waitress says her dad mines, and "we thought we were going to have to move away." There is some logging and lumber-mill work in Republic, but for the 950 people still here after the boom days, jobs are very important. They know the mine will run out one day but is this the time —? It seems to be another Phoenix rising from the dead.

A Rogers High School graduate Ronald R. Short, originally from Spokane, was given the go-ahead to search for more ore. After very great effort, Manager Short was able to announce his crew located a new deposit the **Bailey.** "It's a small zone, but we're able to produce 5,000 tons a month of ore containing 0.67 ounces of gold per ton," Short said. This would prolong the life of the mine another couple years, but a real effort had to be put forth. Short set the goal and in a whimsical moment he made a promise to reach that goal.

Another extreme effort and two veins were located, the **Golden Promise** No.'s one and two. Named for Short's promise to the Hecla chairman, the **Golden Promise** No. two vein was hoped to be producing in mid-1986. As Short commented, the young miners would like to know – can I buy a house? "The question as to how long their jobs were good for; where to go if the end was in sight, and the old "at my age," came up as Short kept his crews still searching. By sheer luck the **Bailey** deposit was a near miss of some 200 feet by an earlier mining company. He knew that there may be more veins but "it takes a lot of money to find them," but he adds, "there's a lot of people who believe another **Knob Hill** exists."

Miners are gamblers and the company gambled $500,000 on a geologist's hunch. The **Golden Promise** became a reality and along with its 1,300 foot, $2.2 million shaft, it was dedicated in late December, 1986. The mine was said fit for at least another decade.

Hecla chairman Bill Griffith says the company didn't think much of the property it acquired from the Day Mines merger. Later (Jan. 11, 1987), with prices depressed for Hecla's main commodity silver, **Knob Hill's** gold is the company's chief source of cash. (This was during the time Hecla's **Lucky Friday** at Mullan was closed down.)

He explained to David Bond, the writer for the Spokesman-Review, the **Knob Hill** mine is highly profitable with extraction costs of $160 per ounce and a gold price of nearly $400 per ounce. He pointed out that production in 1986 will be about 40,000 ounces, a 72 percent increase over 1984 output. Griffith had become president of Hecla Mining company in 1978. He explained also that Hecla plans to convert more of it's ore production into dore, bars of unseparated gold and silver. At present, 40 percent of the mined gold and silver is melted into bars. The remaining 60 percent is recovered as pyrite concentrates in **Knob Hill's** flotation mill. (Gold production at **Knob Hill** for 1985 was reported as 39,192 ounces with 50,000 predicted by some for 1986.)

Now called the **Republic Unit** Hecla spokesman Elmer Bierly said, "we have definitely proven about 300,000 tons of ore." With current annual production of 60,000 tons of ore, he said proven reserves could give a five year supply. He added the new ore bodies are rich, yielding 0.8 ounces of gold per ton.

The Statesman-Examiner news release said Bierly pointed out the Republic mine at it's present rate of recovery is probably the lowest cost gold producer in the United States. He explained that the new circular, steel-lined shaft will have an inside diameter of six feet and reach the 1,300 foot level. It will be used to transport miners and materials to ore bodies, and to improve ventilation. There were 120 employees; also two shafts, the **Golden Promise,** and the **Knob Hill.**

Republic and Tourism

At the time the **Knob Hill** was about to close for the third time, Mayor David Brown, of Republic, along with community promoters, decided the town must no longer depend entirely on the mine. The residents and businesses went about giving the gown an Old West image with new fronts on the stores. A new water system was also installed. They are promoting tourism and along with fishing, beautiful scenery, and winter activity, they have located right in city center, fossils on which they plan to capitalize. The fossil digs, right across from the town hall, contain some of the oldest roses and trees ever discovered. They have attracted the attention of universities. Brown envisions pilgrimages by colleges, with an eventual museum.

Dividing Stevens and Ferry County is Lake Roosevelt, one of the Pacific Northwest's most beautiful boating and fishing lakes . . . made from the Columbia River backup. To reach the south part of Ferry County one can go by highway, or by free ferries across the Columbia, one at Keller, and one reaching Inchelium from Gifford. The highest pass in the state is Sherman Pass, 5,575 feet; and an excellent highway goes through the beautiful Colville National Forest. Ferry County has some lumber industry. It's "Prospector Days," July 1st, 2nd, and 3rd features gold rush days including the cancan girls; and also reminds the visitor of the many abandoned mines, and ghost towns.

Knob Hill Gets More Years

Joseph Suveg, unit manager for the **Republic Unit** mine **(Knob Hill),** spoke recently for the Kiwanis Club in Colville. A very learned mining man, originally from Hungary, he has been with Hecla Mining Company for many years. Looking forward for Hecla to have at least five more years at Republic due to the recent dedication of the **Golden Promise** claim, he said exploration continues with hopes for more reserves.

Suveg said the Republic district had produced to date 2.3 million ounces of gold at a current market price of over $1 billion. He told the listeners that at

Joseph Suveg, unit manager of the REPUBLIC UNIT MINE (the KNOB HILL) recently explained the operation of the mine for the Colville Kiwanis Club. Suveg was project manager for Consolidated Silver in Oct. 28, 1984. He retired May 1, 1990. (Courtesy S-R)

least five other very large companies were involved in the area at the time. He also stressed the great amount of capital it takes to put a mine into operation, and that the profitability depends on the ore grade and metal prices.

Republic Unit Labor Trouble at KNOB HILL

Of the 29 major gold mines in Ferry County, listed in the geological report from the state as of 1975, the **Knob Hill** or Republic Unit, was the only one still working (and is still so in 1988). Perfect harmony cannot always exist and on April 27, 1987, David Bond of the Spokesman-Review wrote of underground unrest at the **Knob Hill**. Employees complained of working a month or longer without time off, around dangerous equipment. They contended that underground safety was being compromised to meet unrealistic new production quotes.

Two veteran contract miners quit 10 days earlier saying they were given a choice of working unsafely, or "tramping out." (Looking for another job.) Unit manager Fred Stahlbush resigned on March 2, and superintendent William Hamilton was said fired the next day. Nine-year contract miner Jim Mink said "we're not paying any attention to what we're doing." He contended, "somebody's going to get hurt."

Pete Brandon suggested the **Knob Hill** shaft should be given to the Smithsonian; and Dennis Hilderbrandt, one of the miners who quit, said "this used to be the best little mine in the west. Now it's just a cesspool that keeps getting deeper." He had worked at the mine for 14 years. He claimed he "tramped out" after being ordered to dynamite a new round of ore from a tunnel that he believed needed to be cleared of loose, and potentially dangerous overhead rock. He left a $95-a-day plus production bonus job.

Art Brown, chairman of the Hecla Company told Bond that he had been made aware of problems but he claimed safety was not being sacrificed and that increased production was not unreasonable. A 30 percent boost in tons of ore mined, was being required. Manager of Mines, Mike Gross blamed Hamilton for "extremely poor results in the fourth quarter of 1986, and poor performance in 1987." From the current 268 tons per day, the 90 miners were told seven-day work weeks would become mandatory and that production will be increased to 350 tons per day.

Dave Nichols, a **Knob Hill** shaft hoistman said, "if we mess up, we can kill somebody. Day after day you tend to get rummy." Ron Krause, hoistman in the new **Golden Promise** shaft said he hasn't had a day off in two months and is scheduled to work straight through Memorial Day weekend. He added, "you get rummy after awhile." There was even talk of going to the Steelworkers' Union, not because they wanted a union, but to get management's attention.

Others said the union isn't the answer no matter how bad things get. Brandon said "the unions are too big," and asked "but why does Hecla want to be the white-devil slavemaster?" Dave Hamilton said, "It's Hecla's way, or the highway." Charlie Jorden contended "they're just trying to justify the big wages in the Coeur d'Alene." Steelworkers staff representative Steve Brown in another organizing effort to unionize, charged Hecla with over-working Republic's miners to strengthen it's bargaining position with union workers at the then idle **Lucky Friday** silver mine in Mullan.

Hecla president Brown said the firm is shooting for 70,000 tons of ore annually from Republic and that production targets were set a year ago. He explained the increased tonnage will keep the **Knob Hill** mill working seven days a week but the mine should soon settle back to five days a week. His explanation towards appeasement included telling of an apology to Hilderbrant and Clancy Heideman by the mine manager, Jim Graham.

April 27, 1987 – Spokesman-Review staff writer David Bond told about Hecla miners at the REPUBLIC UNIT being mad over month-long work weeks. Pictured from the left is ex-superintendent William Hamilton and miners Charlie Jordon, Ron Krausse, and Pete Brandon. (photo courtesy S-R).

Heideman by the mine manager, Jim Graham.

Bond's expose apparently helped the miners because the Statesman-Examiner reported May 6, 1987 the **Knob Hill** mine has returned to five-day-week production. This was said to give many miners their first weekend off in months. A Hecla spokesman confirmed that Mike Gross, manager of Metal Mines, resigned from the corporation Friday "over differences in philosophy of management." His departure from Hecla was due in part to conditions at Republic, said Vice President Elmer L. Bierly. Miners interviewed blamed Gross for some of their problems, chiefly a lack of communication between the miners and the front office, according to that publication.

What of the Future

Because there is some of the most serious exploratory work going on now (mid '88's), we will tell you a bit of what is happening. Barnard J. Guarnera, president of the Northwest Mining Association told the group in session in Spokane Dec. 3, 1987, that over 20 companies were exploring in the Republic area alone, with several new gold deposits under development.

He explained the outlook for the Silver Valley (Coeur d'Alene mining district) is less certain even though the **Lucky Friday** and the **Sunshine** mines are back working. "While it's the largest store house of silver in the world, it's also one of the highest cost mining districts," he said.

Guarnera told the members of the 93rd annual convention that the lag time between exploration and the opening of a working mine can be as much as seven years, so the economic benefits of today's exploration efforts still are several years away.

David S. Bolin, geologist for Pincock, Allen and Holt, Incorporated said, "in the United States alone, gold production has more than doubled since 1982."

Echo Bay and Others

Activity is so great in Ferry, Okanogan, and Chelan Counties that a new "boom" has to be predicted. Echo Bay Mining Ltd. announced it will spend $2.5 million in the next year exploring gold properties in Ferry County. The Edmonton, Alberta, based company will study feasibility of

Arthur Brown, chairman of Hecla Mining Co., claimed safety was not being sacrificed. (S-R).

bringing the properties into production. Vice-President Paddy Broughton said a decision won't come before mid-or-late 1988. So far the company has found four deposits with estimated reserves of four million tons of ore containing .141 troy ounces of gold per ton, or 560,000 troy ounces total.

Two deposits, the **Key East** and **Key West** could be developed by open pit, as also could **Overlook,** according to the July 25, 1987 report. **Granny** is being tunneled into to further explore size. Echo Bay and it's partners, Crown Resources Corporation and Gold Texas Resources Ltd., own claims on a total of 3,500 acres about eight miles south of the Canadian border. The **Granny** property is eight miles north of the other three. Together, they are called the Kettle River Project.

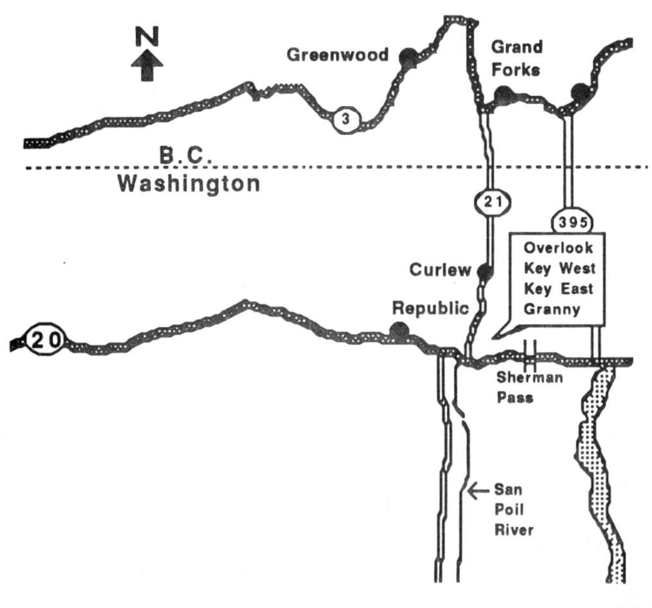

Echo Bay President John Zigarlick said, "We are very encouraged by the potential." He explained the geology at the **Granny** is complex but similar to other properties in the Republic district. "Eventual production is likely to include open-pit and underground operations and a 1,000 ton a day processing facility to employ 100," he added.

Echo Bay is committed to spend $10 million on development work on the properties by 1992 in return for 60 percent interest. Crown and Gold Texas will each have 20 percent. Echo Bay has already spent $1.5 million on drilling.

In August, 1987, Bert Caldwell of the Spokesman-Review did a report of activity by Echo Bay. Drilling rigs were working at the **Overlook** on Cooke Mountain across the mountain from **Key West** and **Key East.** Development was going on at those three claims and at **Granny,** closer to Curlew. The 23-year old company has become one of the world's leading gold producers in the last five years, with production in 1987 estimated to be 480,000 ounces, according to Caldwell.

Officials recently explained to Ferry County residents what was ahead and Mayor of Republic Bert Chadwick said speculation was that thousands of new miners and their families might be coming into the area, overloading available services. "There's an awful lot of geologists wandering around in the woods," Chadwick said. He knows there are also other companies exploring the area. (Could this be a repetition of the 1896-1900's??) It is indicated that these finds 100 years later may be even greater than before.

Nancy Joseph, a geologist for the state Department of Natural Resources said about the Republic district, "it's one of the heaviest prospected areas in Washington, and probably one of the most heavily prospected areas in the country." The search is heaviest in Republic and Toroda Creek. Other firms reportedly prospecting are Asarco Inc., U.S. Borax and Chemicals Inc., Newmont Exploration Ltd., Westmont Mining Inc., and Houston Oil and Minerals Inc. George Brown, geologist for the Bureau of Land Management's Spokane office said independent prospectors are poking around too. Some of the prospects are on bureau land, or partially so, such as **Overlook.**

Zigarlick of Echo Bay, told the audience that a mine probably is two years away. "There is no reason for us to rush into something. There's no shortage of gold."

Crown Resources President Paul Jones said his firm has teamed up with Glamis Gold Ltd., at a small heap-leach mine north of **Knob Hill.** Two weeks ago he announced a joint venture in Okanogan County with Westmont Mining.

On Oct. 16, 1987 Crown reported assays from the **Granny** property to be as high as 0.66 ounces per ton from the newly discovered **Lucy Vein.** It's companion **Brutus** vein, ranges in width from seven to 41 feet and the lowest assay is 0.117 ounces of gold per ton, according to Paul Jones.

The Statesman-Examiner of Nov. 4, 1987 printed a letter to the editor from Fred J. Richardson of Tonasket. His worry was the making of a Wilderness Area out of the Kettle Range and what it would do to placer gold mining. He said that in 1890 the Deadman Creek stream produced 40 cent per yard gold. By his figures at $460 today per ounce of gold, that placer area would produce $9.20 per yard gold. He adds the area could produce a tremendous yardage and that there is placer gold on the Canadian side at Skeff Creek.

The search expands - Inland Gold and Silver corporation and N.A. Degerstrom Inc., announced in December 1987 they had a joint venture on an option to lease 5,000 acres of mining property in the Republic area. The Inland-Degerstrom joint venture involves fee simple land and unpatented claims adjacent to developments by Crown Resources and Echo Bay Mining Company. That combine estimate their reserves at 500,000 ounces of gold. The Inland-Degerstrom venture is doing exploration on property optioned from Roy and Virginia Leland. In May, 1988, James Etter announced drilling to start within 60 days.

Development by Echo Bay has revealed two of their properties show 700,000 troy ounces of gold at an average grade of 0.164 ounces per ton of ore. Echo Bay believes that 521,500 ounces grading 0.199 per ton can be mined. The shaft at the **Granny** area was down 1,900 feet at this time.

By May of 1988, the partnerships was being changed. Crown Resources Corporation agreed to sell a portion of it's interests to Echo Bay. Crown Chairman Mark Jones III said the $5 million in proceeds from a sale of one-quarter of it's 20 percent interest in the **Kettle** and **Key** properties will enable the company to pay it's share of development costs without further dilution of shareholders equity in the project. Gold Texas Resources U.S. Inc., will sell an equal interest to Echo. Echo Bay estimates the **Key** and **Kettle** properties hold about 4.3 million tons of gold-bearing ore.

In June of 1988, Echo Bay Mines had applied for a wastewater permit for it's mine still anticipated to go ahead that fall. The Department of Ecology spokesman Jim Prudente said mines at two sites and a milling facility would employ as many as 150 during construction and 130 later. Crown Resources Cooperation estimates reserves as high as 700,000 ounces of gold. Crown and Gold Texas Resources, Ltd., also own half interest in eight other properties encompassing 15,000 acres in the area. Crown has other properties in Okanogan County to include around the old **Ruby** camp; in Ferry, and eight in four other states.

On July 20, 1988 a news release told that Crown Resource Corporation has begun surface and underground exploration at it's **Seattle-Flag Hill** property west of Hecla's **Knob Hill Mine.** President Paul Jones, of the Crown Corporation told that the **Seattle Mine** had produced 5,300 tons of ore bearing .685 ounces of gold per ton in 1983-84.

Work to be done by Crown will be financed by Texas Star Resources, Ltd., Vancouver, B.C. He goes on to explain that Texas Star is committed to spending $2 million over five years at **Seattle-Flag** to acquire a 50 percent interest in the property. Denver-based Crown; and Sutton Resources Ltd., also of Vancouver, will each own 25 percent when Texas Star fulfills it's commitment.

Crown also told the latest on the **Kettle** project north of Republic where Echo Bay Mines, Ltd., is working. Results confirm Echo Bay estimates the Kettle and Key properties hold 891,000 ounces of gold. Mining was hoped to commence September, 1988.

Crown Resources Corporation, Gold Texas Resources Ltd., and Gold Capital Corporation, said the week of Aug. 5, 1988 they had signed a letter of intent to combine ownership of the three companies into a new corporation to be called Crown Resources Corporation. The new company based in Denver will have 22 properties including a 30 percent interest in the Kettle River Gold Project in northern Ferry County. Approval of shareholders was expected to be completed by early 1989.

A Hecla Request

Now comes the question of right to mine under downtown Republic. On May 17, 1988 Hecla Mining Company tried to buy the rights to burrow under the downtown. The county is negotiating a lease and royalties for as many as 25 acres inside the city limits. Owners have title to the land sold by the county after foreclosures, but Prosecuting Attorney Allen Nielsen said the county retained mineral rights when it sells land for non-payment of taxes. As a result, the county owns the mineral rights to virtually all of the public land and much private land in Republic.

The deals involve some 50 landowners. Hecla has made an offer because they are mining only one mile away. Wondering whether it makes sense to acquire mineral rights in the area, Bill Booth said, "we presently have no plans to mine or explore under the town." Resident Wanda Berg wanted to know who would be responsible for liability insurance, "if they start blasting underneath us we're going to feel it." She also wanted to know how much the county would earn from the deal. The commissioners had 30 days to consider the proposal.

On Oct. 26, 1988 a news release tells that Republic Gold Corporation, a subsidiary of the Hecla Mining Company had signed leases with Ferry County that would allow the company to explore for minerals beneath the town of Republic. Spokesman Bill Booth said there are no immediate plans to begin exploration under the town. "We do expect that eventually we will do it," he said. "Our geologists believe there could be an extension of the ore body."

Republic Gold will pay Ferry County $4,000 in fees for it's land within Republic an amount that will climb $2,000 a year to a maximum of $12,000. If mining occurs, the county will receive a share of the net profits. Terms can be renegotiated every 10 years during the lease's 35-year term.

South Ferry and TOLMAN Mine

There are other areas of interest in Ferry county, and take into accounting the earliest citizens. In general, this reservation area of Ferry and Okanogan Counties still holds to the beliefs of their forefathers, the early Indian residents. One case in point are the tree picture men off the San Poil Highway, some 15 miles northeast of Coulee Dam. Figures of men some four or five feet tall have been carved into the trees. Rough outlines, the figures show a man offering something with his outstretched hand; another with raised hands palms-forward to prove the "greeter" had no weapons and came in peace. It is thought there once were eight or more trees, probably carved well over 150 years ago. There are only three left.

Ben McClung explained the trees marked the first post established 1807 by Hudson Bay Company. Others thought they indicated when trappers would arrive to barter. Coordinator of the Colville Indian history project, Adeline Fredin says another tree is along Silver Creek. She believes the trees are trail markings made by Chinese miners, explorers or trappers of early days. The residents are trying very hard to preserve this part of their history by fencing the remaining trees but nature is treating them cruelly. Insects and woodpeckers are causing one of the trees to crumble, many are entirely gone.

Possibly those same strong attachments to nature helped to bring the **Mount Tolman** mine to a stand-still. Spokesman-Review writer Larry Young explains it by saying that every time you dig a mine you have to expect some environmental object – But when the mine's neighbors happened to be Indians with such high regard for nature that worship of natural forces was the foundation of their old time religion, you must be prepared for an outcry.

Elders of the Colville Confederated Tribes anticipated some dissent when they agreed to let Amax, Incorporated spend nearly $30 million exploring an ore body on Mount Tolman. The huge copper-molybdenum mine is near Keller in Ferry County. Tribal chairman Al Aubertin said they held 75 meetings about the mine on the reservation, more than 50 of those were held near Keller.

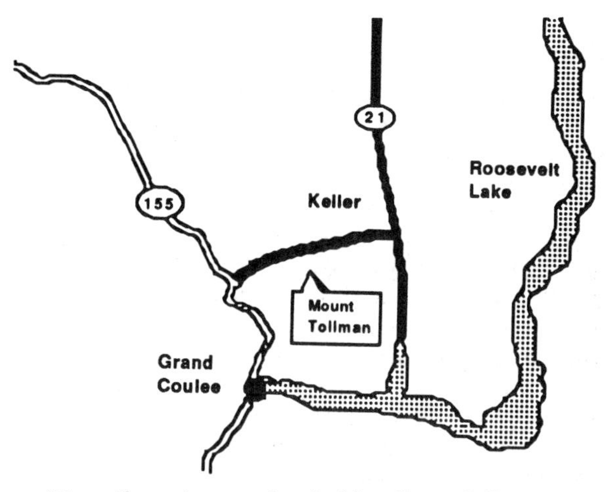

More than two and a half million dollars was budgeted for environmental and social planning studies which were done before the mammoth project started. In December, 1980, Aubertin said they had worked on the statement for over two years. The tribe and AMAX hired a Portland, Oregon firm, Beak Consultants, to take an inventory of plants, wildlife, fish, air and water quality before even a shovel full of ore was dug.

As careful as he said they had been, a group of dissenting Indians proceeded to block development. That "Preservation of Mount Tolman Alliance," was headed by Alice Stewart, of Inchelium, and Yvonne Wanrow Swan.

AMAX paid the Colville tribe $8.5 million for permission to explore the mine. That was paid as per-capita with a full tribal member getting $1,000. Stewart and Swan both accepted their share.

Charlie Stott, former manager of **Twin Buttes** copper mine in Tucson, Ariz., was named project manager. He said that peak employment at the mine would be 1,500 workers but would fall off after the first few months. A concentrator was to be built; buildings to house the giant trucks, and power shovels; administration buildings, and others as needed along with a parking lot.

Nearby towns could house the extra workers as they did at the building of Grand coulee Dam and it's 3rd powerhouse; such as at Elmer City, Electric City, Coulee Dam and Grand Coulee. No one wanted the mine to bring a boom town with drunken Saturday nights, fights, camp followers etc.; but there wasn't to be that many outsiders.

Most comments heard at the many meetings were positive, said Stott. Creating more jobs on the reservation was hoped to keep the Indians near home to work instead of them having to go out to find a job.

Called a closed system, no processed water was to be released to the environment and every precaution was to be taken. Topsoil over the mine area was to be removed and stored for use in reclamation. Because it was expected to take 43 years to extract the ore body, there was plenty of time to develop reclamation plans covering 4,000 acres.

That was to cost some $1,000 per acre to re-seed. The waste rock dump was to be reforested and the top of the tailing area to be grassland, suitable for grazing, according to Les Darling, environmental manager for Amax.

Dust from the vehicles hauling ore out over dirt roads was to be kept down by sprinkling. Including all of the latest in emissions controls, Darling said that some wheat fields in Eastern Washington could be generating six times as much as was to be allowable at the mine.

The mine area contains no wildlife which do not exist in ample numbers elsewhere, Darling said. There were no endangered species; some bald eagles use the Sanpoil Valley but none roost on Mount Tolman. "We've studied the question of vision quests in detail," he said. Vision quests were the Indians solitary periods when undergoing the passage of youth to manhood. He added, "we've found no unique areas which the aboriginal people used more than any mountain in the area."

Saying that "I suspected Mount Tolman was used quite heavily because of it's proximity to the Sanpoil Valley where there was habitation, you needed an area that was remote." He continued that, "custom called for the youth to be secluded for the number of days it took for this vision quest to occur. This is the source of talk that Mount Tolman was considered a sacred mountain. That's true but so were all the other mountains in the region," he explained.

Archaeologists studied the peak for other signs. One site was a talus slope, a ridge by the Sanpoil River which would not be disturbed at all. Another would be next to the administration building thought to have been an open camping site used by the Indians for many years and long ago. These sites were to be fenced off ;and a stagecoach stop in the valley would not be disturbed by any of the mining activity.

Yvonne Wanrow Swan protested that the Indians were concerned about disturbing sacred grounds. "They say that every mountain has a spirit," she said. "There are burial sites and alters there that have probably already been destroyed."

Amax said the mountain contains 900 million tons of mineralized material, about half molybdenum, and half copper. Actually more copper but as Stott said then, the "moly" is more valuable. He explained that as the mine was yielding it's ore, the top of the mountain would be eaten off until it would be about half it's present height. The ore was to be shaped like a tablespoon. The "moly" is used for energy devices, auto and jet engine parts, and is added to steel to improve it's strength and hardness. Stott added, "It is also used to make an excellent lubricant."

At the time of the news report, miners had been working 20 hours day for the past several years at **Mount Tolman** boring 260 holes in the mountain. After that drilling, thousands of the core samples have been stored at a temporary headquarters. More than $6 million had been spent on this drilling process.

After the most intense preparation for careful mining the Mount Tolman operation was closed down. In December of 1981 Amax, Inc., announced it was putting it's half-billion dollar **Mount Tolman** project on hold due to "adverse metals market conditions." Stott used the remark, "postpone but not pull out." Amax's plan to spent $700 million to put the mine in production would have meant sizeable royalty payments to tribal members.

Bitterly disappointed, the 120 employees were to be laid off in February and caretaker status be completed by mid-1982.

Okanogan County History

In earlier pages we learned briefly about the mines on the Chief Moses Reservation of Okanogan County. With the immense amount of exploration going at some of the old mining camps in 1988 we should review and add to our knowledge.

According to Father De Rouge who came there in 1885, the name should be spelled Okanakan. He said there is no "g" letter or sound in the Indian dialect. Having been spelled a number of ways, it was named by the Indians as meaning "head-of-nothing" because the river starts at the lake. In English the name means rendezvous. Father De Smet first came into this area from Fort Colvile in 1839 and through his efforts a chapel was built at Omak in 1889. It was at the head of the lake east of the Okanogan River on what was the south half of the Colville Indian Reservation. The honesty of the Indian people was mentioned frequently by the people of the robe.

Chief Aenas earlier occupied this area that eventually came under the care of Chief Moses. Being a "non-treaty" Indian, after the reservation was changed several times Moses laid claim to 600,000 acres in consideration for preventing an Indian outbreak. He had refused any federal authority but the matter was settled March 6, 1880 when the land west of the Okanogan River to the Cascades was set aside for reservation, by executive order. There were said 14,300 men, women and children and each was to receive 323 acres (far in excess of possible). In 1886 this Moses Reservation was thrown open to mineral search. Residents of the area said the reservation was first thrown open to the prospector to file his claim, then the Indians were located on allotments. Due to this, many mining claims were located on the finest farms to keep the Indians off. Many claims were located for the purpose of establishing a townsite such as the **Reno, Chesaw, Trenton,** and **Trenton #2,** where Chesaw is located.

Richard F. Steele in his 1904 book "The State of Washington" gives more detail than most publications. He tells of Hiram F. "Okanogan" Smith coming into this part of Washington Territory in

1858. He made mining locations in the Palmer Mountain district and a number along the Similkameen River. He had settled in the Osoyoos part of the district. When the government established the reservation, Smith held tough and refused $250,000 for his locations. Instead of payment, the government drew a line excluding a strip 15 miles south from the border and across the reservation, allowing Smith to keep all his claims. This fact was kept from the prospectors until the Moses Reservation was thrown open in 1886. The area saved by Smith was found to be among the richest in the United States.

Smith became a member of the Territorial Legislature and is said to have interested members of Congress in the preservation, and eventual purchase by William Seward, of Alaska from the Russians. Smith was elected to the state legislature and died in Olympia in 1894.

Camp McKinney into Okanogan

Our many prospector-miners had moved into the Camp McKinney region of British Columbia. Former owners of the **Dead Medicine,** or **Silver Trail** in Stevens County were with that movement and discovered the **B.C. Mine** at Summit Camp. Those three Keough Brothers worked the mine for some time and sold out around Christmas 1899 for $60,000 to $67,000 – and returned to Colville. One of the brothers built a beautiful home east of Colville that became a famous landmark until it burned to the ground in May 1966. (One publication said the Keough's paid $60,000 for the B.C. Mine and sold it later for $300,000.

What had begun as a rush to find wealth in this vast northwest area was not reaching the panic stage. Prospectors were heard from in each direction and for a time Colville could better have been known as the "hub" of mining activity. Claims were being recorded at the nearest possible site – that being Colville in many instances. Of the several hundred in Stevens County, we have knowledge of 286 major claims that were worked; some became mines and we will learn about those – many more of several thousand only made through prospect stage.

For the moment though, we are far behind our prospectors and must follow into new camps. Hobos were noticeably absent, for this strenuous existence took sheer manhood and womanhood. By now whole families often moved as their men found "pay dirt."

The move into Ruby-Conconully

Veins of silver had been located in the Ruby-Conconully district in 1886 and by 1907 some $200,000 in silver ore had been produced. Many of the 1,000 to 3,000 who were in the camps by 1893 had come down from Camp McKinney, and that field of exploration. Many found employment along with prospecting as the market dropped but soon the entire operation was stopped. At that time Ruby boasted a brewery, saloons, and the usual supply stores. Today there remains only the brush covered foundations. Among those who "papered" claims was Joe Moris. But before writing this camp off, let's return to this Okanogan area and learn some of it's mining history.

Okanogan becomes a county

Up until 1888, Okanogan was still a part of Stevens County. Upon organizing, it's first county seat was at Ruby, but on Feb. 9, 1889 it was moved to Conconully. (That town was formerly called Salmon City.) On May 31, 1891 permanent headquarters was provided at a cost of $2,495 including all the material and labor. The Indian residents were not too happy with the selection, sometimes out of fear. They declared the area had evil spirits due to a sea monster in the Conconully lake. White residents not being of the same thoughts used this as a tourist attraction between mining activity.

Loomis Camps

Word spread rapidly and miners and prospectors came in droves, the first coming from the Fraser River and Cariboo camps to the Similkameen as early as 1859. Smith's first claim was said to be in the early 1870's. North of Loomis, it was the **Julia,** since called the **King Solomon,** later owned by Spokane people.

L.K. Hodges, a noted historian said, "the first flock of investors were doomed to failure being without experience; and mostly from the cities, especially Seattle." Stamp mills were built without concentrators; ore changed from free milling to base, and most of the value was lost in the tailings. Then the price of silver began to fall. This camp was predominantly silver.

The population of the new county reached 2,500 in 1892, and 4,689 by 1900. Natural hazards had played a large part in slowing development including the blizzard of 1898. (That had also done much damage in Ferry County.) On Jan. 11, snow and heavy wind came and stayed for three days and nights, the temperature dropped to 35 degrees below zero. It was even a worse storm than the one in 1892-93. Travel was next to impossible and such stage drivers who had to get through were nearly frozen; their horses were coated with ice. Then in the spring the rains came. In 48 hours there was 3.3 inches of rain and many bridges and buildings were carried away; along with many deaths by drowning.

Mines of Okanogan

By 1904, there were 14 mining districts in the county. Palmer Mountain was called Wanicutt; Chesaw was called Meyers Creek, and Squaw Creek was Methow.

1–Early day view of TRIUNE MINE shaft #1 (Golden, WA). 2–Shaft of #2 TRIUNE with horsepower hoist. (S-R 1899).

The **Triune** and the **Jessie** had been located at Palmer Mountain near Loomis. Opened up by the Gold Mining and Tunnel Company, organized in 1895, they had the **Expert, Jumbo, Helen Belle, Wisconsin Central,** and **Dolly.** A tunnel was complete on Sept. 10, 1897 with assay running $185.20 in gold and $2.50 in silver. In 1890 a nugget brought $1,000. John Boyd was president and general manager and lived in Loomis.

(In 1954 the **Triune**, or **Crescent** was owned by Dell Hart of Oroville. The five-claim property included **Jessie** and **Occident.** It produced more than $300,000 prior to 1938 in gold-silver-lead and copper.)

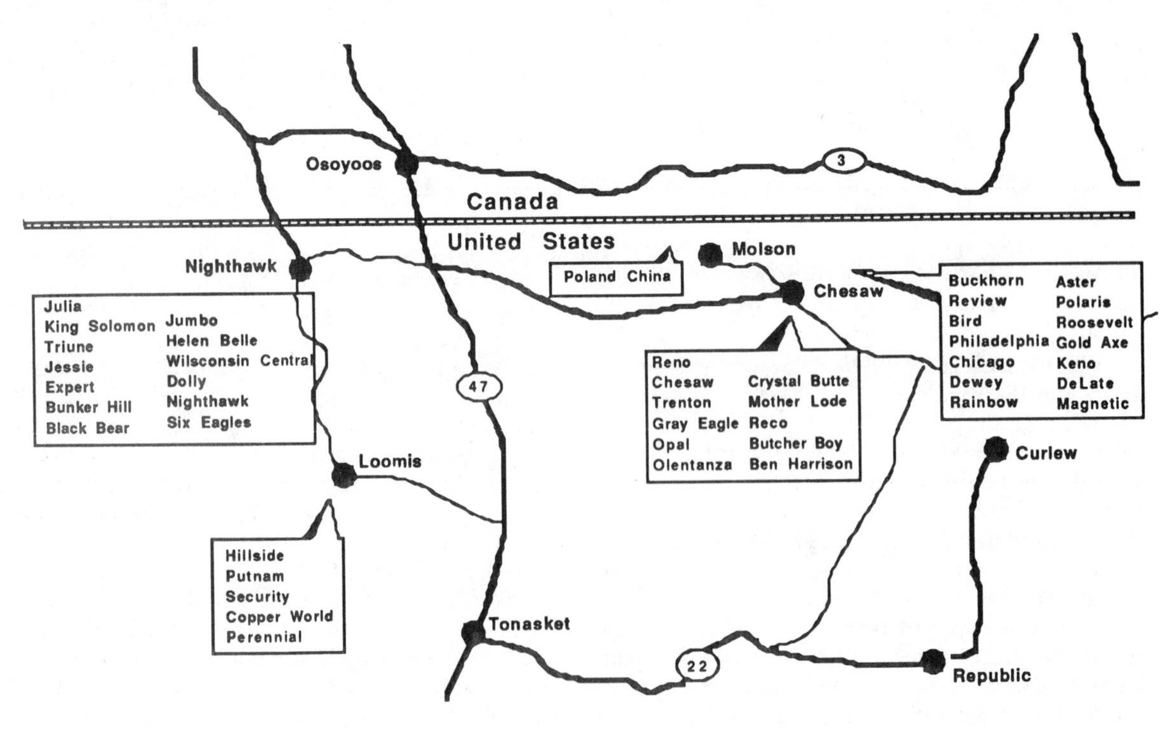

Top: BLACK BEAR MILL near Loomis. Bottom: Compressor for Palmer Mt. Tunnel Co. (S-R).

In the vicinity of Loomis was the **Hillside Mine, the Putnam Group, Security, Copper World, Perennial Group,** and north of Loomis was **Nighthawk** and **Six Eagles. Pinnacle** was a short distance from Loomis and was discovered in 1888. The Pinnacle Gold Mining Company from Renton, Wash., also owned the **Bunker Hill, Nevada, Bonanza King, Julia Franction, Telephone,** and **Bullion.**

A serious mystery surrounds the **Pinnacle.** A considerable amount of gold was taken shortly after discovery but those who located the claim went "outside" to dispose of that gold and were never to be heard from again. The claim was left in charge of James O'Connel, who relocated it and later bonded it to a Canadian investor. Refusing to rebond the claim due to gossip about a big "hit" at the nearby **Bunker Hill,** O'Connel met a tragic death and the property went to relatives where it was in litigation for years.

General J.B. Metcalf bought the claim and it was in the hands of the Pinnacle Gold Mining Company by 1904. Mostly Seattle men, they own a number of claims and had a lease on **Black Bear** five-stamp mill in Loomis which was working night and day.

A 1955 report said **Pinnacle** has six claims showing gold-copper-lead-zinc and silver. It produced $200,000 prior to 1910, mostly in gold. **Black Bear** produced some $150,000 prior to 1902 and 77 tons were taken out in 1947.

Chesaw Camps

On Meyers Creek at Chesaw, on the north half of the Colville Reservation there were more than 500 claims – many have eventually passed from their original owners. The first location was at **Crystal Butte,** and was owned by Interstate Mining and Developing Company. In 1904 they had 1,500 feet of work completed and a $100,000 concentrating plant. Under the name **Mother Lode,** that claim shipped some ore in 1937 and was under lease.

The second location was the Monterary Mining Company who owned the **Buckhorn Group.** Their ore showed gold and copper. The Review Gold Mining Company had both the **Review** and the **Bird** and had 1,000 feet of work with hundreds of tons of ore ready to ship in early 1900's. Besides these two, there was **Chicago, Dewey,** and **Philadelphia,** some located by Robert Allison and John Mulholland and sold to the Review Company for $35,000. We learn more about Buckhorn later.

The Yakima Gold Mining Company adjoining Chesaw had a 500 foot development. The Wyandotte Mining Company, owners of the **Oregon Group** of eight claims was located by John Gilihan. One ledge showed $60-600 in gold, another $11 to $114. The Opal Mining Company west of Chesaw has the **Opal Group.** J.P. Blaine was involved. Ore ran an average $30 to $500 to the ton and they were to have a smelter.

The town of Chesaw was named for an old Chinese miner who was married to an Indian lady. He also farmed and furnished produce to the advancing army of prospectors and was respected by those who knew John Chee Saw (some historians called him Joe). This was the first instance of an American town named for a Chinese.

Pell-mell, disorder and tent town was customary as new mining districts opened and around Chesaw and Bolster this was no exception. The first mine located on Copper Mountain, (since named Buckhorn Mountain by the Forest Service) is credited to Jim Grant. As in the location of the great **Silver King,** near Nelson, B.C., the hunting of grouse played a part. Step-son Louie found ore samples while out hunting, and which led to the location of the **Copper Queen** in the spring of 1896.

Copper was the only ore found on Buckhorn Mountain in early days but iron took precedence especially during World War I and II. Other claims in the area were **Aster, Polaris,** and the **Roosevelt.** John Gaffert and John Berquist are said to have shipped several carloads of gold ore from the **Gold Axe.**

The Grant holdings were said to be the earliest on Buckhorn Mountain, (also listed as on Copper Mountain), to ship ore. Several carloads of copper ore had been hauled by freight wagon to Republic and shipped on from there. One news item of 1907 tells of $50,000 transaction by the Grant people in disposing of some of their claims. Others went as high as $10,000 each. It was around that time that the assessment work was not done on the unpatented claims and they were lost. The manager Jim McLain is said to have jumped the claims and renamed them to be the **Roosevelt Mine.** This has since been relocated as **The Gay.** Early assays ran $40 to the ton of gold-iron and copper.

Three miles west of Chesaw was the **Keno Group** with gold and copper running $10 to $15 to the ton. **Delate,** one of the more valuable properties in the district was owned by Interstate, a Columbus, Ohio company. They had a tunnel, shafts, and were making ore shipments to Everett smelter that assayed $45 to the ton; one ounce gold and 24 percent lead. Manager was Henry Thompson.

Molson Mines

The **Poland China Mine** was staked by Neal Undem and J.M. Henkins, on the north fork of Mary Ann Creek on May 22, 1896. This was some five miles east of Molson. Finding rich gold ore near the surface led to the staking of the mine sometimes said named after the owner of a Poland China hog – and other times said named after interested property owners who originated from Poland, and China. How much of the high-grade ore was sent to the Cariboo mill at Camp McKinney is not known.

Development work continued from 1896 to 1906 during which time in 1898 George Mechem, part owner of the **Poland China** developed the town of Molson. In 1906 the property was said purchased by D.W. Dart, of Dayton, Ohio; and six associates who incorporated under the name of Molson Gold Mining Company. A number of men were employed, houses, assay office and an office was built and in 1907 a mill was being built. Machinery was installed to recover the ore values by amalgamation and gravity concentration. This very process led to temptation later and one high up in the company was discovered as having scraped amalgam from plates and selling some of the gold in Wenatchee. This mill later proved unsatisfactory.

In 1908 D.W. Dart sold part of his interest to Eastern parties and the property continued to be worked. In 1909 six 1300-pound Nissen stamps and two Wifley tables were installed. Development work continued and around 1914 the company reorganized as the Mary Ann Creek Mining Company. The eleven claims were patented and approximately $200,000 had been spent on the property.

Failure to show a profit up to 1924 precluded further development and in 1927 F.W. and C.F. Rankin were reported sole owners. The mine lay idle until 1931 and at the death of C.F. Rankin, a long term lease and option was negotiated with F.W. Rankin by D.W. and R.D. Dart.

Activity commenced and former families moved back to the property by then operated under the name **Overtop Mining Company.** Cold winter, and the closing of the Wenatchee bank with all the operating funds, slowed work. New processes were installed and prior to 1937 a number of carloads of ore was shipped to Trail smelter, but which proved unprofitable. D.W. Dart died in 1937 and the last work at the mine was in 1939. The property was left to Chester Rankin in 1950 and the final sale was made to R.D. Dart in 1955. The mine had produced more than $100,000 by 1936.

The **Buckeye** Mine located some four miles east of Molson was staked by William Warshow, and J.C. McNeal in July, 1898 and comprises three claims. In 1904 a stamp mill was built a mile to the north but which operated only a short time. P.C. Dunning of Wenatchee became interested in the property about 1925 and around 1933-1934 along with William McMurray and others, a company was formed called the M&D Company. In 1952 the property was purchased by R.D. Dart who is said to have deeded the mineral rights to Ralph Dunning, of Okanogan.

Placer gold first discovered on Mary Ann Creek in 1888 continues to offer a challenge to the summer vacationer. According to the Washington Geological Survey made in 1909, about $40,000 in gold had been recovered from the creek. That figure has greatly increased since then. During the depression years of the 1930's quite a few men were said able to make a living placer mining that creek.

Chesaw

In the Chesaw district including Strawberry Mountain and Porphery Peak, were the **Reco, Gray Eagle,** and **Butcher Boy.** Much gold ore was said produced from the **Gray Eagle** and some copper ore from others. Along with the **Opal;** the **Ben Harrison,** and the **Olentanza** were in the vicinity. On Buckhorn Mountain was the **Rainbow Group,** and the **Buckhorn Group.**

BUCKHORN or "MAGNETIC"

Among early prospectors said to arrive by 1896 were James and John Grant and William Fahsbender. Location of the **Buckhorn** was credited to Grant and Fahsbender, and was earlier said to contain as many as 60 prospects. (Five original claims were held in 1964 by the late Delbert Scoles and the late John Citkovitch, both of Colville.) In 1956 the State inventory showed those five unpatented claims; **Neutral, Copper Queen, Polaris, Crystal Springs,** and **Number 9,** under the mine name **Magnetic.** John Citkovich had explained that from 1918 the property was at times leased out and for the past year and a half (in 1964) it was operated on priority by Zontelli Brothers, of Ironton, Minn.

Citkovich claimed the mine had the biggest deposit of iron in the state and that copper indications were good. The late Archie E. Wilson, also of Colville, and Citkovich bought the **Buckhorn** from Mrs. Sadie (Grant) Woodard in 1936. As high as 20 men worked the mine and two and three cars a week were shipped. Up to 1945 John said approximately 50,000 tons were shipped to the Northwest Magnesite Company at Chewelah, and to the Tacoma shipyards where it was used there as ballast. At Wilson's death, his son-in-law Delbert Scoles, also of Colville, became "Citko's" partner in the operation. Some $50,000 in development work was being done and in 1958-60 the Keokik Electro-Metals, of Keokik, Iowa (with company men in Wenatchee), worked the mine. The mine is some five to six miles from Chesaw.

Colville mining man John Citkovich owned and mined the MAGNETIC for a number of years, along with partner, Archie Wilson. (Courtesy son Jack Citkovich).

The Spokane Falls and Northern had built through Colville by D.C. Corbin in 1889 and it later changed hands twice. As the Great Northern line was built to Curlew and Republic by 1903-04, then into Oroville by 1907. In 1920 the mines at Phoenix closed and dismantled. Smelters were closing at Grand Forks and Greenwood. At that time Phoenix had a population of about 4,000. In 1931 the railroad was abandoned between Molson and Oroville and the tracks pulled in 1932. Service was discontinued from Molson to Curlew and the tracks there were pulled by 1936. The last train out of Molson was Feb. 28, 1935.

John "Citko" recalled sardonically that he had just completed the purchase of the **Buckhorn** Mine and had returned to Curlew to find the Great Northern Railroad pulling the tracks. This required his building 10 miles of roadway up Toroda creek to haul out the ore.

He recalled gold being taken at the **Bodie** Mine under management of Jim Perkins, and that the mine quit operation around 1937-1938. According

Archie E. Wilson mined the BUCKHORN, or MATNETIC in Okanogan county, with John Citkovich. Wilson also operated the BIG IRON, near Pierre Lake, and other properties, Wilson in center of track. (Courtesy Jack Citkovich)

to the 1940 report; some 302 tons were produced valued at $9,770. Citko also told of there being what to him seemed thousands of prospect holes in the vicinity of his **Buckhorn.** He strongly discouraged hikers and riders in the area unless they are well aware of the possibility of suddenly falling into some old prospectors disenchanted diggings. As early as 1897-98, claims innumerable were being filed in Okanogan County. Gold and silver yields were very high in some.

The Buckhorn claims went to Mrs. John Citkovich and to Mrs. (Scoles) Holland who later sold the **Buckhorn** to Marion Bumgarner of Sunshine Valley Mineral, in Wenatchee. (He and Scoles had earlier been partners in the Grand Forks slag dumps.) Son Jack Citkovich said it was ironic that payments for the Buckhorn went for John's three-year stay in a nursing home, prior to his death. In a sad way, one could use his favorite remark and say "his ship came in."

The Bolster Boom

At Bolster, established by J.W. McBride in the spring of 1899, James McEachem, John McNeil, P.H. Pingston, George Tindall, and John Schafer staked claims. A boom occurred and Bolster boasted greater population than Chesaw. By 1916 there was said to be seven old bachelors living out a lonely existence doing their assessment work and dreaming of a big strike they hoped to make someday. In 1939 L.C. Pickering was reported buying the **Commonwealth Placer Mine** which included the ghost town of Bolster and two mining claims, **The Bird,** and **The Review,** from the Patrick Welch estate. The town of Molson also went into declined and died by the early 1920's.

Other Camps

Earlier J.M. Burns had made the first mineral discovery on Polepick Mountain near Silver. The **Red Shirt** mine reported a yield of $20 to the ton in gold-silver-iron and copper. By 1896 a 20-stamp mill was reducing ore from the dump. A lady prospector, Mrs. M. Leiser discovered a good property near **Red Shirt** and sold it to J.S. Crockett. Crockett also bought **Black Warrior,** a later nearby discovery.

In 1952 the Red Shirt Mining Company leased their property to John Russell and George Gibson of Winthrop. The mine had produced intermittently for 50 years, up to 1938 and was credited with over $100,000.

Southeast towards Squaw Creek, J.W. Dra and Nels Johnson found abroad belt of mineral in 1892 which drew a great deal of attention. The **Highland Light,** was a promising property in this section. On the left bank of the Methow River was the **Friday Group** of five claims developed by that gold mining company. The **Diamond Queen** group of two claims was west of the **Friday** ledge. It assayed $10.80 in gold, 61 cents silver and later went to $3.65 and on up to $32.70 in gold.

On that same side of the river was the **Emerald** group of three claims; the ledge crops five and one half feet wide between granite walls with surface ore assaying $25 in gold, silver and copper according to an early report. An assay ran as high as $122 to $157. **Friday** later called **Tom Hall** was owned by J.J. Sullivan of Pateros in 1942. Over $5,000 was produced prior to 1897 from five claims, and carload in 1940. **Gray Eagle,** and **Last Chance** were well developed early and the later ore shipped ore to Everett smelter that netted $30 in gold and silver. **Gray Eagle** was being leased by men from Loomis in 1941, the mine being part of the Patrick Welch estate from Spokane. It produced from 1916-1939 over $8,000 in gold ore.

On Johnson Mountain the **Hunter** was the original location with values from $16 to $20 in gold and eight to 12 percent copper. The **Washington group** included seven claims owned by Methow Mining Company. E.W. Lockwood from Wenatchee, H.M. Cooper, and Edward Shakleford made the discovery in 1884. After locating the **Washington Group,** they abandoned them as being too remote in the Twisp Pass region. **Hunter,** along with **Grubstake, Okanogan,** the **Doris Barbara,** the **Bay Horse, Ace Of Diamonds** and **Esther** veins all became part of the Holden-Campbell properties; the owners from Chelan. Very little activity went on but considerable gold ore was reported by 1902 with limited report to 1940.

The **Black Jack,** The **Spokane,** The **Paymaster, The Ramsey, The Oregonian Group;** all were names of claims that showed great promise along with dozens more. The districts were filled with prospectors filing new claims. The **Derby** was bonded to Frank Rosenhaupt, of Spokane, for $10,000. All of the claims mentioned were showing promising gold content which was the major mineral being searched for then – and is the major one searched for in 1988.

The **Mountain Goat** on Gilbert Mountain was the property of O. Gilbert, Nelson Clark, A. Raub, George Witte, Henry Plummer, and Frank Thompson. Surface assay ran $95 to $387 gold and up to $100 in the "hanging wall."

The **Portland Group** owned by the Consolidated Twisp Mining and Milling Company reported their ore carried $13 free gold throughout the two assays made of the drillings from the tunnel ran $1,500 to $1,900. Nelson Clark and R.J. Danson owned the **Washington** nearby.

The rich Ruby Hill

Upon opening the Moses Reservation for mining, the first discovery in 1886 was on Ruby Hill, just northwest of Conconully. John Clunan, Thomas Donan, William Milliken, and Thomas Fuller made that original location. The **Ruby** had a ledge 18 feet wide, wall to wall in $14 gold and silver. It proved to be the lowest-grade mine on the hill. The **First**

RUBY MILL between Loomis and Nighthawk. "Ruby Silver" was found here and by 1902, rich deposits were said located. (Byron Larson)

CONCONULLY HISTORICAL PLAQUE - Conconully, Washington sprang alive in 1886, a rip-roaring mining camp known as Salmon City. Fire wiped out most of town in 1892. In May 1894, flash flood roared down North Fork of Salmon Creek, wiping out business district. It is believed Indians called this entire area including the lake "Conconully" meaning "Beautiful Land of Bunch Grass Flats." Ore Car, right, is typical of those once used to haul rock from mine tunnels.

Miners drifting at the United Copper Mine, early 1900's (Washington Geological Survey).

Concentrating Mill and mine openings at the Cleveland in Cedar Canyon, early 1900's. (Washington Geological Survey).

Thought was located by Patrick McGreel, Richard Bilderback, and John Clydostey. The ledge 30-40 feet wide on the surface was running $28 gold and silver it's whole width. The **Fourth Of July** showed the richest ledge; followed in discovery by the **Arlington. Peacock** on Peacock Hill became the big discovery at the time, located by John Pecar, James Robinson, and James Gilmore.

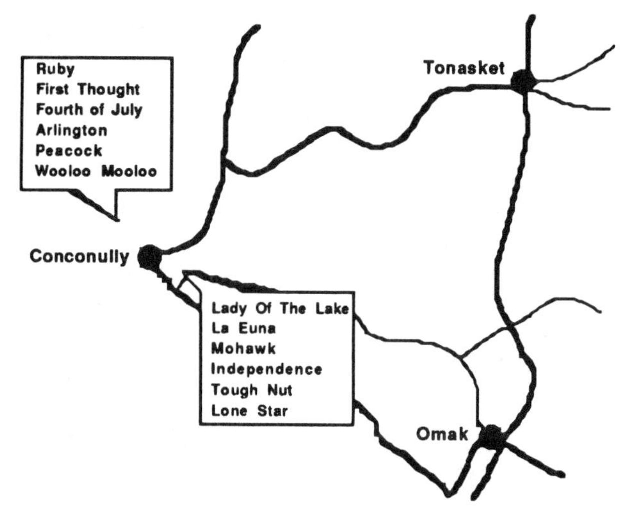

The men who discovered the **Ruby,** sold it to a Senator Jonathan Bourne, Jr., from Portland. (He also later bought the Billy Kell holdings of Eureka Mining Company, and became a stock owner in 1908 of Knob Hill and it's associated claims. Mrs. Bourne became an owner in the **Mountain Lion** Claim, also in Republic.)

A 50 foot shaft had been sunk, a 100 foot tunnel driven, and $1,000 in ore taken from a rich pocket of this **Ruby.** Suddenly many Portland people became interested. Mr. Bourne incorporated **Ruby** and **First Thought** separately and organized the Washington Reduction Company. He put in a concentrating plant to treat the ore and acquired other claims. In 1888 the Arlington Company of which he was president, purchased the **Arlington Mine** for $45,000 cash. After spending $130,000 for development it was discovered that no water could be obtained for their leaching plant, although there was water 200 feet below. Work was suspended and the ore was concentrated at the Washington mill. Some production from **Ruby** was reported from 1890 to 1937.

Mr. Bourne turned his efforts toward the **First Thought.** Assays ran from six to 10 ounces of silver and three dollars in gold, but there were said to be rich streaks. The Washington Reduction Company had erected a concentrator at **Ruby** and a cable bucket tramway a mile long from the **First Thought.** When silver went down to 70 cents the mill was stopped. It had produced $40,000 in concentrates.

The **Fourth Of July** was purchased by a syndicate. From the richest ledge on the hill with a pay streak said four feet wide, one shipment of 20 tons was reported to pay $480 a ton in gold and silver. Another outstanding claim was the **Wooloo Mooloo** located on Ruby Hill by Hugh McCool and others. From an eight foot ledge the first assays ran 3,000 to 5,000 ounces of silver, according to company reports.

In 1945 the **Arlington** was reported to have nine patented claims and it's total production from 1901 to 1940 was $144,650. **First Thought** was under lease to Benton Bailey, formerly of Colville, in 1949. It was said to have produced more than $66,000 between 1922 and 1926 and some work was carried on in later years. The **Fourth Of July** was also under lease to Bailey in 1949, according to the 1955 report. It was said to have produced $36,000 in silver-copper and gold. Bailey also leased **Last Chance.** Some ore had been shipped in 1920 and there was some production into 1924 of silver-lead-copper and zinc.

Conconully camps

George "Tenas" Runnels and J.C. Boone discovered **Lady Of The Lake** near the foot of Conconully Lake, the day the Moses Reservation was thrown open. They bonded it to O.B. Peck for $40,000 (He may be related to Oroin O. Peck of **Bunker Hill** history), but the bond was soon forfeited. A number of other claims were filed in that vicinity. On Mineral Hill the Bridgeport Mining and Milling Company bought five claims. Over $30,000 in work was done, with a ton of ore showing $300; $20 of that in gold, the balance in silver. T.L. Nixon of Tacoma, paid $10,000 for the **La Euna** on this hill, and H.C. Lawrence refused $30,000 for his **Mohawk.** John Stech, of Seattle paid $4,000 for the **Independence.**

Seeing such a rush on the Salmon river district just south of Conconully, (of Okanogan County), ex-Lt. Governor Charles E. Laughton organized a company and erected a building between the **Tough Nut** and the **Lone Star Mines.** He had a plant consisting of rock crusher, a set of rollers to pulverize the rocks, drum screens to size the material and also wooden jigs. So much mineral escaped with the tailings which was even richer than the yield, and less than half was saved. Miners refused to furnish ore when the assay values did not show up after a two-week run and the machinery was stopped forever.

About 1939 the **Tough Nut** stopped production but from three claims they had reported 500 tons of ore valued at an estimated $9,000 before 1916. The mine is silver-lead-copper and zinc.

Ruby Mine to the north

Some half mile northwest of Palmer Lake and some four miles south of the British Columbia line is another **Ruby Mine.** Discovered by A.M. and George Reist, it was owned by the Ruby Mining Company of Mansfield, Ohio in the early 1900's. J.M. Hagerty was president and Monroe Herman managed the mine. It was reported to contain from $2.50 to $3.00 in gold per ton but some of the ore ran $200 to $1,000. Taken as it comes, the shipping ore was assorted to run over $100 a ton. 30 tons have been shipped to the Hall Mines at Nelson, with net returns of $2,742. The ore was being sent down the hill by an aerial tramway to a small bin at the base of the mountain. A new tunnel was to be dug at a lower level.

Nighthawk Mines Incorporated owned this **Ruby** in 1941. It had produced $25,000 between 1915-1920. Some work was being done in 1935 and up to 1940.

Gold is discovered in Methow Valley–or is it?

Gold discovery in the Methow Valley was around 1885 when Colonel F.S. Sherwood, of Colville, was sent on a government expedition. "Captain Joe", an elderly Methow Indian was employed to guide him. As they camped on the headwaters of War Creek, and while hunting for the horses, Joe stumbled across a big gold ledge cropping out of the mountain. He broke off a piece and put it in his pocket. Nothing was said until returning over the same route, he told Sherwood. They looked but could not locate the ledge but Sherwood developed some real excitement among Portland friends with the sample. The ledge has never been found. (It was this same Mr. Sherwood who claimed to have located the **Old Dominion** after it's discovery seven miles east of Colville but who had failed to record his claim.)

Near loss of Conconully

Some of the state's richest mines are in Okanogan County, where gold quartz has assayed as high as $43,000 to the ton. Much wealth has been expended but the absence of transportation slowed developing various districts. By 1893-1898 many of the earliest claims, and mines were shut down. A severe fire at Conconully on Aug. 30, 1892 destroyed all businesses there. Some 500-600 people lived there at the time. That was followed by a destructive flood, and shortly after– the drop in silver price brought very hard times that defeated Conconully almost entirely.

The town had tried to rebuild after the fire but a cloud burst swept water 12 feet high. It was said to have carried away a stone cellar under the Moore and Irish Company store; and their heavy iron safe was never found. Beigle's Saloon was crushed in and the mirror behind the bar was forced up against the ceiling – but not even cracked! – by all the rocks and logs that jammed the room. The town was slow in it's attempt to rebuild.

Loomis grows

Samuel I. Silverman organized the Oro Fino Placer Mining Company and platted what he called Loomistown. It became Loomis after J.A. Loomis. The town was extremely well supplied with businesses; three general stores, eight saloons, two dance halls and other establishments.

Money was plentiful but much mining development was said woefully mismanaged. Hundreds of men were employed. An expensive mining machine was rushed into the district . . . some of it is still lying on the hillside, never used, never even having been erected. Much later desolation was said due to the financial depression of 1893. Mills and concentrators were erected before the owners knew what was needed. The district remained dormant from that time until 1898. (Among the residents of Loomis was Father Sherman, son of General William Tecumseh Sherman.)

Then came the new "boom" in Loomis when conservative mining men with plenty of capital came into the district. The "old-timers" who had held onto their claims benefitted mightily. By fall of 1899 there wasn't a bed left in town. Volumes of freight was pouring into town, and valuation of properties went up. And – as in too many mining camps, death walked the streets. Well-known to many, "Pinnacle Jim", James O'Connel was shot down when he brought it on himself and was killed out of "self defense" by another old prospector.

Among those drawing interest today is the **Ida May**, formerly mentioned in Ferry County. Sundance Mining-Development Inc., said June 27, 1985 that they had acquired the gold claim and five others. Said similar to the Bodie, it had been a former producer. Sundance is a Kellogg, Ida., based company.

Towns of Okanogan

In earlier pages we learned some of the town of Ruby and one that will pay to remember. Among the ghost towns, it is one the ghost is moving about in a restless pattern in 1988. It had been one of the liveliest and best-known mining camps in the northwest in 1885. The town was well built up on each side of the quarter of a mile long street. Many men were employed and an expensive concentrator was built nearby. A wire tram was built from the mines to the mill, but in time the wires were cut and the parts stolen.

The town in the vicinity of today's Conconully is remembered for such silver mines as **First Thought, Ruby,** and **Fourth of July.** When the price of silver bottomed out in 1892 the town collapsed, people moved away, and soon vandals stripped everything of value. It went to a town of nothing – a ghost town in a very short few years.

In July, 1983, there was very good news about the other old **Ruby** silver mine located at the base of Chopaka Mountain. John Wishart, president of Vancouver, B.C., based Kaaba Resources, said the company wants to reopen the mine. The company was said to have purchased a small locomotive and was installing tracks to haul ore from two silver veins known to exist but had not been mined.

Wishart explained there were three parallel silver veins in the northern **Ruby.** He said sampling ranged from 31.3 to 190 ounces of silver per ton with the average reported to be 91.9. Old timers had reported 1,200 ounces of silver to the ton. Exploration at the time of this news article indicated an ore body of about 9,000 tons of silver ore. He said that if a mill with the capacity of 500 tons-per-day was installed, as many as 80 people would be employed. (The name **Ruby** was given to the mine because the silver turns red when scratched.) Since 1904 there has only been intermittent activity until mid 1980's.

Chesaw, a "real" mining camp

There are so many great old mining camps in Okanogan County, most of them ghost towns now. Chesaw, with nearly 500 claims in gold, silver and copper became one of the more rousing, rowdy centers. Hundreds of miners, storekeepers, and ladies of the night came. The first ones came in 1896 and saloons, restaurants, and hotels were soon built. The area was best known for the **Poland-China,** the **Copper Queen,** and **Buckeye.** Chesaw was east of Oroville and connected by stage to Republic.

Many well-known men came into this area to mine. The best 1896 claim was the **Reco** located by Robert Allison and John Mullhollend. It was rich in gold. The town was located on this claim and in the spring more prospectors came. U.L. McCurdy opened the first store which was enlarged several times over the next few years. (150 tons of ore was shipped from **Reco** to Northport smelter in 1916. Very little other production was reported up as far as 1924.)

The town of Chesaw was platted Nov. 14, 1900 by the Jim Hill Consolidated Gold Mining Company by it's vice-president, J.C. McCurdy. With other additions, hotels, and new buildings, Chesaw was designated the metropolis of eastern Okanogan County with a population of around 250 in early 1902 and they had a miners union. Even today they are one of the busiest Okanogan towns on the 4th of July when they hold rodeo for several thousand who come for that celebration.

Bolster

Bolster almost became a rival of Chesaw. Located on the Colville Indian Reservation side of the county, prospectors were barred until 1896. By 1898 there was quite a settlement and in 1899 the "boom" came. J.W. McBride had bought up several claims and had platted the town; and the Bolster Drill newspaper had opened it's office. The new – expensive! – ($3,000 to $4,000) hotel, 30 some houses and a population of some 200-300 people were there when the boom ended in 1900.

Molson

Molson was less than a mile from the border and on the state route between Chesaw and Oroville. With the opening of the north half of the reservation to homestead entry in 1900, George B. Mechem saw the need of a farming area near the rich mineral deposits. He interested Montreal, Canada capital and they formed the Colville Reservation Syndicate, with John W. Molson president. Mr. Mechem became general manager and he set about building a city. He spent $50,000 of the company money including $8,100 for a hotel supposed to be a credit to a town of several thousand residents.

At 300 population, the town had an attorney, a post office, general store, drug store, and The Molson Magnet newspaper. Many other businesses were being established including four saloons, and of course the assay office – usually the first business in a town. In less than a year, in February, 1901 friction arose between Mechem

Main Avenue of Old Molson 1905 showing the site of the present museum. (Courtesy Okanogan Highland Echoes and Statesman-Examiner). (Some history of the Buckeye was written by Harry A. Sherling of Oroville)

and the company and cost of improvements to the city were stopped. This company went into receivership and it's affairs were looked after by trustees. Mechem was supposed to be the heaviest loser but it was said that he went to Texas during the oil excitement, made a large fortune and went on as a broker on Wall Street.

The first passenger train had arrived by 1906 and many supplies came from the Canadian side. The town of Molson claimed to be short of oblivion by 1963 with few of it's buildings standing. A museum there is visited by many each year.

— Earlier spelled Loop Loop

Loup Loup's history is that of money makers. More mining activity, and more millionaires made in minutes was part of this mining camp's claim. Richard Steele put it this way, "A man with a piece of rock and an assay certificate was in the mining swim, and a man who had a ten-foot shaft with good showing of ore could talk of nothing smaller than millions." That was during 1889 and the early 1900's.

It was the first town in Okanogan to be platted Aug. 14, 1888 by W.P. Keady and S.F. Chadwick. There was a mercantile store, plenty of saloons, and places to eat. The mines were silver and the downfall of the town was great. As people left and the lights were turned out, the fear was of vandalism and theft.

Ore from the **First Thought** in better times was sent from Loup Loup to south Ruby by gravity tram to a reduction plant. Upon closure, those wire cables were cut up; taken away, and the tram equipment at the ore bins was torn to pieces for the bolts.

Nighthawk

The town of Nighthawk was noted mainly for the mines of the area, the **Nighthawk, Six Eagles, Golden Zone:** and by a telephone system that connected those offices with the **Ruby** mine and with other portions of the state. Production history was short on these mines.

Bodie, Toroda, Mazama

Bodie is on the Colville reservation side. By 1903 a new town came about due to the **Bodie** mine. A $20,000 mill for treatment of ore was erected. A few miles south of Bodie is the remains of the mining camp Toroda. Shipments of ore to Granby showed $500 to the ton. There had been a stampeded to the camp due to the rich ore being reported. Failure to prove this left a town of log buildings – totally deserted. From the area of Mazama we learn of the **T&B; Oriental,** and **Crown Point** properties. This was a gold camp with many thousands of dollars worth of development around 1902.

Golden

The town of Golden boasted the **Triune** and the **Spokane;** said to be rich in gold and silver in the years 1892-1894. The same story existed with large capital being spent for mills that could not treat the ores and everything was closed down. The **Spokane,** later called the **Gold Crown** produced a small amount in 1939 and 1941. The five claims of the **Triune,** later called **Crescent** belonged to Del Hart of Oroville 1947-1954. The mine produced more than $300,000 prior to 1938, gold and silver.

Park City district

By 1911 Park City district near the boundary of Ferry and Okanogan Counties had the Castle Creek Mining Company. The Ramore Mining Company and the Wasco Group were all very active. **Mountain Boy** of Safe Deposit Mining and Milling Company was located by A.S. Soule and Richard Purcell. One surface sample assayed $340 in gold, another ledge assayed $227 in gold-silver-lead and copper. The area still awaited the projected railroad down the Sanpoil and was using a wagon road to Republic for the 28 mile trip. This is also within the radius of **Paul & Brimstone** held by Robert Neil of Coeur d'Alene Mine history.

At Castle Creek Mining Company's prospects the Colville Mining and Smelting Company was starting to erect a reduction works consisting of two smelting furnaces and the necessary power equipment and crushing machinery. That was never completed and no large shipment has been reported. **Copper King, Snowshoe,** and **Summit** were some of the claims. Most development was in the early 1890's when $8,100 was said taken from **Mountain Boy and Summit Claims** mostly in lead-silver. Ramore Mining Company's claims include **Hercules.** No great production was ever reported.

A state geological report of 1956 stated that $200,000 in silver ore had been removed from the Ruby-Conconully district before 1907. Other important claims in the county included **Alder,** southwest of Twisp, with three patented claims and having produced from 1939 to 1951, over 32 thousand tons of ore. Shipments were valued over $1 million and were in gold-silver and copper. Kaaba Silver-Lead Mines, Incorporated owner of 27 claims south of Nighthawk has shown production from 1915 to 1950. Between 1943 and 1949 95,045 tons of ore was taken from their mines said to be the leading producers of the area in silver-copper-lead and zinc in the 1940's.

Good "down-home" reading is the book "Okanogan Highlands Echoes" written by the Community Development Committee of Molson-Chesaw. It points out the real purpose of reading history by saying "the present nor the future can be complete without the past," an excellent thought to remember!

He He Stone

Before leaving the history of the Okanogan, some space should be allocated to the legend of the He

He Stone. The stone has been worshipped in the traditional manner of leaving keepsakes for as far back as the Indian people can recall - and as far back as the white man can remember, he has seen this reverence paid to the stone. This had gone on for years before the white man ventured into any portion of this western world. It was through the **He He** pass that early white men traveled and today only a "blasted" mass of rocks identify the spot of worship. Often a thing of value is still placed at the site by an Indian who believes in some unexplained power.

The story most often told - and believed is the version of the Shuswap and Okanogans waging war for many days. The Chief of the Okanogans was said captured and the daughter of the Chief of those on Meyers Creek was said to have nursed him through a dangerous illness. They fell in love and the chief dreamed of an escape whereby the maiden must not look back - but when they reached apparent safety she did look back and laughed, and was turned into stone. This version is said only repeated by white men and Chinese.

A second version dates back to the beginning of time and at a time the Indian suffered an affliction not unlike leprosy, which was thought destroying the nation. The priest of the "Siwashes" (a white man term), allegedly talked with the Great Spirit asking them to meet at the stone so reverenced by many. On the appointed day an angel like form appeared and alighted on the stone. With words of encouragement she handed around Camas seed and urged the planting everywhere. The roots when eaten were to prevent the return of the malady. Queen Camas, as she was called, floated away with a promise to return at some future date and many still believe that she will and place gifts on the stone when passing. It is visible to all who travel that route.

Into Chelan County

Chelan County came into being on March 13, 1899 having first been part of Stevens County, and then still part of Okanogan when that portion was separated. The Chinese placer-miners came first by following the banks of the Columbia River. There are hundreds of holes indicating where they made a fairly good living and formed a Chinese village at what is now Douglas. Theirs was the first business enterprise in the area. A Chinese merchant had a string of 40 horses and he brought in English, American and Chinese merchandise to customers who always gave a welcome when dealing with him.

The native Indians resented this intrusion and attacked a number of the Chinese by coming upon them from the bank side of the Columbia and conducting a horrible slaughter, around 1875. By 1890 there were no Chinese left, all had moved away or were killed. The Indians in the Mission Valley were under the command of Chelan Jim in 1883 but the powerful Chief Wenatchee headed up this tribe when President Cleveland isolated them to the Columbia Indian Reservation, three square miles in the vicinity of Lake Chelan.

Lake Chelan

Lake Chelan is one of the most scenic and beautiful lakes in the world. It is 68 miles long and one to three miles wide. Said to be one of the deepest in the United States, soundings show 1,642 feet. Towering granite mountains almost shut out the sun on each side; and in and around the mountain area are some great mines. The lake, and city of Chelan are some 40 miles northeast of Wenatchee where the temperature and soil of the area is conducive to making one of the better fruit raising centers, and also good for wheat fields.

When the **Champion Group** was discovered, they found evidence of White men having been there as early as 1866. The first quartz location has been said made in 1875 by a Mr. McKee. We had learned earlier that John Shafer was credited with the first major discovery in 1874 at Culver Gulch.

Prospectors were spreading into what is now Chelan County and by 1898 had located five camps well worth mentioning. These were the **Klondike, Meadow Creek, Railroad Creek, Bridge Creek, and Horseshoe Basin.** Copper and gold values were reported at **Klondike,** some 35 miles from Chelan, and eight miles from the lake on the north side. **Meadow Creek,** said some 50 miles from the foot of the lake was first located in 1892 and showed copper and gold resembling the Rossland ore.

The Holden Mine

The mine showing the greatest promise was located by J. Harry Holden, of Seattle, formerly a Colorado man. The **Holden Mine** was about 10 miles up Railroad Creek, it's ledge exposure said 3,000 feet and 700 feet in depth. It was stated in

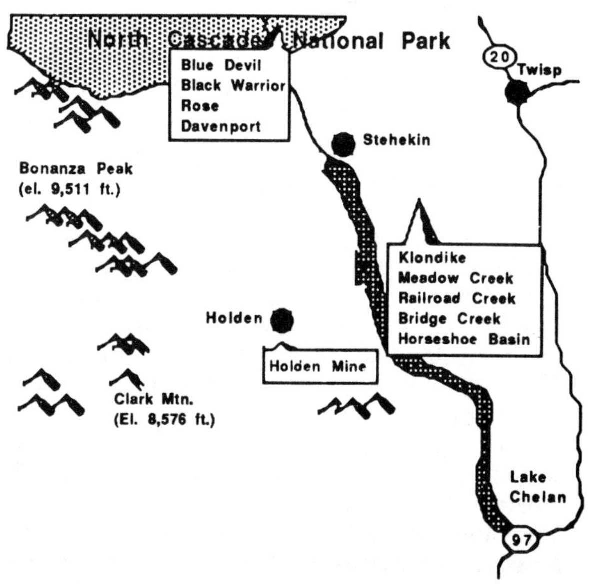

1901 to be the most monumental body of ore in the Pacific Northwest. An early assay was $18.75 in copper with an average of $12 to the ton. Ore in sight was estimated to amount to $120,000,000 at that valuation. The mine owned by the Holden Gold and Copper Mining Company with J.H. Holden president, has been said that no other mine could reach this one in magnitude and wealth.

Holden held the mine (according to reports) from 1892 to 1896. (Another report states Holden discovered the mine July 20, 1896.) The Howe Sound Company are reported to be the owners as of 1937. According to geological reports it contains 13 patented claims and 78 unpatented. A 1956 report said to reach the mine one took a boat from Chelan to Lucerne and then twelve miles by car over a good road. The amount of tons of ore produced has been put many ways because the mine was said one of the largest in the state. Copper-gold-zinc and silver were produced and from 1938 to 1957 one state geological report put it at 10.6 million tons of ore valued at $66.5 million.

The company had a 2,000 ton flotation mill, a modern camp for 450 men, roads, docks, tugs, barges etc. The **Holden Mine** was finally worked out in 1957 and the grounds has been turned to a church camp.

In the state's 1898 report the **Horseshoe Basin** was the best known camp. Near the headwaters of the Stehekin River - 25 miles or so from Chelan Lake, was the famous **Blue Devil** and **Black Warrior** claims purchased several years earlier by a syndicate at a reported price of $30,000. Assays ran 75 percent lead, 80-90 ounces silver and some gold. In that same vicinity in 1962 the major exploration was on the **Rose.** This was at that time the old **Davenport** and along with other claims was being worked by Valumines Inc., with President Max McKinley from Bayview, Idaho. He renamed the claim **Rose** prior to 1962. It is a silver-copper-lead and zinc claim.

T.S. Burgoyne as president of the Horseshoe Basin Mining and Development Company, was reported to own **Davenport 1, 2,** and **3.** The **Black Warrior** was owned by a company of the same name, from Spokane in 1946.

A 1962 report states current exploration in Nigger Creek areas brought about a strike claim renamed the **Caldo.** The **Pole Pick Mine** was under current exploration by lessees on the north side of Blewett Pass; and the Lovitt Mining Co. Inc., was then in current production of the **Gold King Mine.** The latter is in the Squillchuck area and has become one of the state's largest producers in gold and silver, of which we will learn more.

In the 1898 period the **Idaho,** and **Wolverine** claims in the Meadow Creek district, on the north side of the lake and 50 miles from the foot, were shipping copper and gold ore by 1892. The **Sunday Morning** was also mentioned at that time. **North**

The great HOLDEN MINE was snuggled up near the base of Copper Peak, Chelan County. Worked out by 1957 the mine produced more copper than any other mine in the state, along with substantial amounts of gold and silver. (Dept. Natural Resources).

Bridge Creek camp was mostly silver.

Placer mining had been the major source of gold in the 1860-1868 period. It continued as a source from 1900 through 1957 with a reported yield of $935,252. In that same period lode gold yield was $52.1 million.

Mines in the Entiat valley included **Rex** and **Ethel** owned by Mr. and Mrs. Crum in 1902-1903 and showing a yield then of $10 to $1,700 to the ton in gold. By 1930 the **Rex** was reported to have yielded $170,000, mostly in gold; and a small amount in 1940. It was leased in 1943 to Bert Rogers of Entiat.

Tip Top Mine in the Blewett district reported a yield of $10,000 by 1901 and a smaller amount in 1940. The earlier ore had been treated in an arrastre. Owner in 1911 was the Tip Top Mining Company.

Blewett, the Culver and others

The town of Blewett was the oldest in Chelan County. It is 18 miles south of Leavenworth and where the first quartz ledge in Washington was discovered at the **Culver.** A tide of miners had come into this district from the Cariboo but by 1904 there were said only 40 people.

Mines of the area included the **Peshastin** which by 1902 had produced $60,000 and by 1940 another 22 tons, mostly gold. **Black Jack** produced some 3,000 tons of $10-to-the-ton gold and between 1900 and 1940 reported a yield of $4,000. Along with **Blewett** and **LaRica,** mines of this area were mostly owned by Gold Bond Mining Company, of Spokane by 1952.

Culver

Known as "Culver" country relating back to the earliest find at the **Culver,** most of the gold with best assays was located at the top of the hills and value decreased to lower grade ore as one worked down the hills. One report is of $1,700,000 taken from the Blewett district between 1870 and 1910 with only $200,000 of that being reported from 1901 to 1910. Did the gold run out as thought? – much will be known as new explorations take place.

At Doubtful Lake camp was a molybdenite mine called the **Quien Sabe.** The mineral was very new to the mining world and used mainly at the time for medical purposes. The Cascade Consolidated Mining and Smelting Company had 17 claims at this camp near the headwaters of the Stehekin River. Capitalized at $850,000 there was $20,000 expended. Their mine reported gold-silver returns of from $22.60 to $57.50 to the ton. George L. Rowse was president of the mine that had been discovered by George and John Rowse – said not related — in 1886. (*They have been credited in some publications as having discovered the "moly" mine also.)

Copper Queen Group was on Railroad Creek and was said to yield $102.70 to the ton in copper and gold. A.L. Cool was the owner. **Tiger Group** on Bridge Creek 23 1/2 miles from Stehekin was owned by H.H. Hollenbeck, Van Smith, Professor Piper, George Young and H. Willis Carr. The claim assayed at 103-107 ounces of silver and $24 gold.

In 1891 the **Blue Jay** was discovered by Captain Charles Johnson near Lakeside. It was developed by Chelan Gold Mining Company. (Historians claim this area was prospected by row boats and the reading of markings along the mountain walls.)

Among the hundreds of claims located in the Chelan area the **Gem,** the **Winnipeg,** and the **Iron Cross** were just a few that showed early potential. The **Humming Bird** in the Blewett area is reported as having produced up to 1975. The **Bobtail,** owned by James Lockwood; the **Fraction** by John Olden and Peter Wilder; the **Pole Pick,** and the **Culver** claims were all bought by Lockwood and son E.W. Lockwood, and by H.M. Cooper.

Pole Pick #2, now called **Alta Vista,** was earlier owned by Dexter, Shoudy and Company. The **Peshastin,** earlier owned by Donahue, Dore and Cross was bonded to George W. Martin of Minneapolis in 1894. He also leased the **Blewett Mill** and ran a chute down the hill to get his ore to that mill. The ore was poorly sorted and he gave up and sold out to Dexter, Shoudy and Company. Thomas Johnson later bought this mill.

In earlier days there was a report of $2,100 earned in a cleanup the first nine days from a six-stamp mill. Then a claim of arrastre capacity of 1,000 pounds a day produced $70 earnings a day. The **Blewett** mill had run eight years before it was sold to Thomas Johnson in 1896. Johnson shut down after a short run and an ownership dispute arose at which time Johnson reportedly killed William Donahue over the dispute; but even that did not prevent the sale to Culver Gold Mining Company in 1891.

That company erected a 10-stamp mill with four Woodbury concentrators and stretched a bucket cable tramway from the mill to the **Culver Mine,** one fifth of a mile away. Some ore was shipped and one lot reported a yield of $800 to the ton. In 1892 Culver Company sold out to Blewett Gold Mining Company, with Seattle capital. New development included a 20-stamp mill at the mouth of Culver draw, with space for 20 more stamps. They moved the tramway to a new site; added every labor saving appliance and self-feeders to the stamps, and built a steam sawmill. This development continued until a system of leasing sections to the miners became a practice. Taking pride in part ownership produced more work.

One historian wrote that the Blewett Company produced $600,000 in bullion in 1896. The **Culver** was known to produce some $300,000 by 1902. When that mine was owned by Washington Meteor Mining Company the production from the mine was listed with their other operations. By 1951 it was owned by Gold Bond Mining.

The **Phoenix** owned by D.T. Cross and John F. Dore of Seattle; and the late William Donahue (killed by Johnson) assayed $20 in gold to the ton and was active from 1895 to 1897, having produced 1,000 tons. The **Olympic Group** of five claims was sold to the Cascade Mining Company. At the death of Marshall Blinn the organizer of the company, the mill was stopped and never resumed work. The property was bonded to Edward Blewett who ran a 200 foot tunnel. The mill had been creating huge losses as did other mills. Much gold in the early days was lost by milling ore in arrastres.

Pole Pick owned by Gold Bond Mining Company of Spokane, in 1943 was an early producer. In 1936 it listed seven patented claims and 17 unpatented, and was being leased to Carlton Mining Company from 1949-'52. It was said leased by Harold C. Lewis and Graeme Thorne in 1962 but the mine was no longer listed in 1965-'66. It was reported to have produced $70,000 by 1901 and still some production to 1951, mostly gold.

Activity in Chelan County started as early as

1867, 32 years before becoming a county, when Ingraham and McBride opened a trading post at what is now Rock Island. Their trade was with the Indians and they later moved their enterprise to the mouth of the Wenatchee River where the town is now. Their chief stock in trade was whiskey. Because it was against the law to sell, or barter liquor to an Indian, they had to leave in haste in 1872 to elude the law. Their property was bought by Samuel Miller and Freer Brothers and an original business building was put up in 1872 with the use of logs, built by Miller. It became a trading post and post office.

Wenatchee

The name chosen for the settlement was Wenatchee, after the Indian Chief. It meant "boiling water" named due to the commotion in the river caused by the Wenatchee River flowing into the Columbia. It also has two more translations; "good place", and "Robe of the Rainbow" taken from an Indian legend.

In 1888, Mr. McPherson built a store above the present townsite and brought mail twice a week; and merchandise, from Ellensburg. It was over terrible roads. That town was platted on Aug. 28, 1888 and named by the founder Don Carlos Corbett. In 1892, when the Great Northern Railroad was completed the Wenatchee Development Company surveyed and platted the present site Aug. 5, 1892. James J. Hill was president of the company.

W.H. Merriam had built the first modern store and Mrs. Arzilla Tripp was the first woman to make a permanent home in Wenatchee, in 1883. By May, 1891 there were said 108 people and by January, 1892 that figure was up to 300. The townspeople were organized by then and on March, 1892 they voted to exclude all Chinamen by "honorable, legal, and lawful means"; or another mass meeting would be held to adopt other methods!

On May, 1892 the townsite was thrown on the market. Within five days there was $100,000 worth of property sold. The Wenatchee Development Company was ordering and going through a change of plans. Fire struck this new settlement on Jan. 18, 1893 and damages were set at $10,000, with no insurance. 22 horses in a livery stable were burned to death. On Sept. 2, 1893 they had another fire with less damage.

As a railroad town, Wenatchee drew hobos and disreputable citizens. There were thefts, murders and riots reported as high as 10 to 12 a week – no citizen was safe. The 13 dance halls, numerous saloons, and other less desirable resorts were there until the railroad was completed and that element moved on.

Like the rest of the country, and the mining districts in particular, bad times existed from 1893 until the early 1898's. By 1903 times were looking up and new steamers were being built in the Wenatchee shipyards to haul ore, and make other shipments. There was a $31,000 outlay at the yards, and $50,000 improvements to the city that had become county seat. Fruit was becoming a big thing and in 1903 when the population was put at 1690, the orchards were shipping by express 162,743 boxes and by freight 267,743 boxes. The Columbia and Okanogan Steamboat Line had seven boats in their fleet on the Columbia and Okanogan Rivers.

Chelan

The city of Chelan lays at the base of the lake by that name. Previous to 1886 all the land north of the Chelan to the Methow River was Indian Reservation open only to homestead entry. This portion was still in Okanogan County but President Cleveland permitted the Indians to take allotments first. A townsite was laid out July 1889 by Judge Ballard and United States Surveyor Henry Carr who by error filed as a preemption. As a result, there was no valid title to 300 to 400 shacks and homes until 1892. The town incorporated in 1902. Early activity on Railroad Creek, an especially the **Holden Mine** gave greater assist to this town.

Lakeside

Lakeside came about on June 12, 1891. It had been called Lake Park for a time. The town one mile above the Chelan River counted 300 population in 1888 when Captain Charles Johnson, Benjamin F. Smith, and Tunis Hardenburg came with their families and settled. The mill became the Chelan Lumber Company.

There were docks at Lakeside. The Belle of Chelan, built in 1888-89 and captained by Charles Trow plied the rivers and lakes. The engineer was R.J. Wilkins. Then came the Omaha, the Clipper, and the Queen and Dragon which was later wrecked on Chelan Lake. It was a mail steamer and made two trips a week from Chelan to Stehekin. Superintendent (then) Trow; Captain Fred R. Burch, and Engineer R.J. Wilkins manned the "Queen" loaded with freight and cord wood. The load shifted and water began pouring over the side into the hold. It sank in only 16 feet of water and the crew got off safely but not the mail sacks. By 1900 the Stehekin, the Swan, and Lady of the Lake were busy and by 1903 Flyer and Chechachko were added to the fleet.

Chelan Falls

Chelan Falls came into being in 1891. It is on the south side of the Chelan River and had a flour mill, one general store, a brewing company and wheat warehouses that handled 350,000 bushels of wheat in 1903. The townsite was homesteaded by Joseph Snow a former state senator from Douglas County and later a Spokane County surveyor. It was platted by Sarah J. Snow on Feb. 10, 1891.

In the summer of 1891, L. McLean was going to build a metropolis there. He secured control of the townsite and spent thousands of dollars on the finest hotel, the best newspaper and other establishments. It had been a peach orchard up to May of 1890 but the trees were trampled and destroyed. McLean thought the railroad was coming through and when it did not, the boom died. The city never did materialize.

Leavenworth

Leavenworth is another major city in Chelan County, 16 miles west of Wenatchee on the Great Northern Line. It was a shipping point and supply station for the Blewett Mining District. In the spring of 1892 the first town built was a mile from the present one and was called Icicle. (There are Ice Caves northeast of Chelan but not near Leavenworth) Captain Leavenworth of Olympia; J.P. Graves, Alonzo M. Murphy and S.T. Arthur all of Spokane, platted the townsite. (This is the J.P. Graves that we met in the mining business in Grand Forks.)

Businesses from Icicle moved to Leavenworth and by February, 1893 there were 700 people; restaurants, hotels and saloons. In 1896 a bad fire consumed seven buildings. Not a building or any of the contents were insured. Loss was estimated at between $25,000 and $30,000. On Sunday, Dec. 28, 1902 they had another fire with over $20,000 loss; none of the buildings were insured.

Lamb-Davis Lumber Company Inc., with paid up capital of $250,000 in 1903, bought up all the vacant lots; 30 acres of land owned by Mary Ralston; William Douglas's homestead, an 40 acres belonging to John Holden (of the **Holden** Mine). To be a mill site, the company built a sawmill, boarding house, and hospital; purchased the city water works and constructed flume two miles up the Wenatchee River. The water works and electric light plant was incorporated as also was the Tumwater Savings Bank. A fire Sunday, Jan. 24, 1904 leveled six buildings at a loss of $25,000 but this time there was $14,000 insurance.

The town was reported down to 500 people in 1904 but there were still some active mines including the **Emerald,** and the **Esmeralda.** The **Red Cap** and the **Bryan Group** of 20 claims reported gold running from $3 to $180. A $50 to the ton average was in gold-silver and copper.

Chelan's RED MOUNTAIN MINE

According to a Spokesman-Review article written by Cecil M. Ouellette for February issue in 1972, one copper mine 39 miles northwest of Leavenworth did not process an ounce of ore. The **Red Mountain Mine** gave up without even trying when the price of copper went down and other costs went up. They were said to have a camp town at Trinity of some 200 men, around 1930. Begun in 1916 the town had 40 buildings; street lights and fire hydrants; offices, commissary, boarding house, assay and other necessary mine buildings. They had a two mile long shaft electrified, crushing plants and concentrator ready for operation. What a waste – and how representative of many of the disappointments in mining.

GOLD KING MINE

Then again – here is a story of great victory – the second outstanding mine in Chelan County is the **Gold (or Golden) King.** The location on the west side of Squillchuck Creek is credited to V. Carkeek in 1885. By 1894 it was the property of the Golden King Mining Company and produced 240 tons with a yield of $1,600. The mine changed hands often with Wenatchee Mining Company; J.J. Keegan; American Smelting and Refining Company, and Knob Hill Mines Incorporated, all working the claims until Lovitt Mining Company, Incorporated of Wenatchee took it over in 1949. A five and a half year production report through 1954 told of value received of $4,432,884. Another geological report from 1949 to 1967 was total value $14,962,305. Still another report gave $16 million for the nearly 17 years operation by Ed Lovitt.

The L-D Mines of Wenatchee had explored for gold on it's property the **Gold King** with financial assistance from the Office of Minerals Exploration, U.S. Department of Interior. Cost of the work was estimated at $133,260 with government participation of $66,630, according to Mineral Industry report from 1965 yearbook.

Ore mined and milled, or direct smelted from **Gold King** in the previous three years, according to Day Mines, Inc., annual report, was 88,728 tons of ore containing 0.30 ounces of gold and 0.32 ounces silver per ton in 1963. In 1964, 94,002 dry tons of ore contained 0.22 ounces gold and 0.70 ounces silver. Continuing to go up, thru 1965 this was the leading gold producer in 1951.

The **Gold King** suspended operation March, 1967.

Much of the returns from Gold King went into the betterment of those in Wenatchee, and improvements in Chelan County. Our reason for reviewing the mining history of this county and adding to what has been told in the past, is that a positive future in mining is being indicated here again. A heading in a Spokesman-Review article about Chelan County dated May 19, 1968 pretty well tells it like it is now, 20 years late and in 1988. "Back in the 1870's an epidemic called gold fever drew men to this rugged land and turned them into strange, burrowing creatures." As was said in Ferry, and again in Okanogan Counties, and now in Chelan, "there's a lot of geologists out in the hills."

The New Breed and the Future

That exploration and development work has brought about a new gold mine practically "downtown" at Wenatchee, called the **Cannon.** Asamera Minerals of the United States, and Breakwater Resources Ltd., Vancouver, B.C. opened their mine in December, 1985. By August, 1986, the first quarter production totaled 32,197 ounces, making the $90 million project potentially the largest Northwest gold producer at 120,000 ounces a year.

David Bond of the Spokesman-Review explained in the article in August that **Cannon,** like the **Knob Hill** in Republic, is an underground-shaft operation. The mine has added 180 workers to the Wenatchee area payroll, he said.

Asamera's Cannon

Judy Mills of the Spokesman-Review wrote the first story September 18, 1983 telling of Asamera Mineral's location and the first sinking of a mine shaft west of the Wenatchee city limits. Tests then showed the site held up to five million tons of ore containing an average of 0.3 ounces gold to the ton. Stockbroker Keith Howard at that time described the find as a "sleeping giant of gold."

On the first drilling, Asamera hit a "hot spot" with a 50-foot section that yielded three ounces of gold per ton. Dan Meschter, a U.S. Forest Service mining engineer said "in the mining business that's unbelievable." What followed was a land rush for claims jumping to as many as 4,456 in a year that ordinarily recorded 100 new claims. Meschter described the corporate prospectors as "front-end people – walking cardiac cases." Auditor Ken Housden said these were not the disheveled pick-and shovel variety prospectors, and one resident added "more like the carpetbagger type."

Some representatives of rich Canadian companies were said to be gold seekers who went as far as dropping claim markers from helicopters. These "reps" were described as "wearing Calvin Klein jeans and carrying briefcases." Bill Meyer as exploration manager for Teck Resources Inc., a Vancouver, B.C. firm was exploring 12,000 acres in and around Asamera property. He said the American companies tend to explore after a major mineral find; the Canadians on the other hand buy, or lease every piece of property in sight – then explore. Dillon Howard said, "those could have been American companies if it weren't for our over-regulated venture-capital market." He explained that the Canadians get financing much quicker.

The **Cannon** is said to be a rediscovery because it lies within the site of the now defunct Lovitt Mining Company's **B-Reef Mine** which that company operated along with **Gold King** until 1967. The properties are said in this article to have once produced and estimated one million tons of ore averaging 0.4 ounces of gold per ton. Despite the rich deposits, gold was only $35 an ounce and Lovitt was forced to halt it's operation.

The **Cannon** earlier looked like a rocky yellow hillside dissected by crude roads. The "Wenatchee Dome", a volcanic mound of rock is the landmark under which Asamera is burrowing 4,000 feet for gold. Securing leases on 3,016 acres was not easy; 30-40 leases and subleases had to be signed to obtain a continguous land parcel, according to Gary Bates, the mine manager, Asamera, primarily a gas and oil company has decided to go for the precious metals. It has been a Calgary based company for over 58 years and has two silver mines in Mexico. Peter Maynes, Asamera spokesman said that as long as gold stays above $100 an ounce, Wenatchee operations will be profitable.

Asamera Company anticipates from eight to more years gold production from **Cannon.** Tech Resources, Inc.'s Bill Meyer said, "I think there will be other discoveries in Wenatchee." Weaco Resources Ltd., another Canadian company has purchased 80 acres adjoining Asamera. United Mining Corporation of Virginia City, Nev., was also in the area with a total of some 200 acres near the action.

In September of 1983 when this all started, Montine and Andy Matthew's backyard was fast becoming the entrance portal to the **Cannon.** Actually living next door, they said the only thing they were scared about was when blasting started. Claims have been staked literally on the edge of town. Bates, of Asamera said they planned to be a good neighbor and if they broke a window they would fix it – and even take you to a movie while it was being replaced.

One group that seemed pleased was the Appleatchee Riding Club located across from the mine site. Activity had been stopped there but Asamera gave the horse enthusiasts a $108,000 donation for the right to drill on Appleatchee's 20 acres. Still owing a lot of money, Joe Schumacher, president of the club said, "they're kind of helping us out of the hole."

Overall, Wenatchee area residents welcome Asamera. Dave Griffiths, Chamber president in 1983 said "it's kind of an old-time gold mining boom atmosphere. It's a nice feeling to be sitting on a gold mine." The school district did turn down the offer to lease it's property for drilling even though some lessors will receive royalties as high as 10 percent. Some 50 locals were expected to have work at first with some 200 to build other facilities. 125-150 were to be hired at the mine. Real Estate brokers now have to answer questions about "mineral rights."

Robert G. Hunter, president of Breakwater Resources Ltd., which holds a 49 percent interest in the **Cannon** property as of June 15, 1984, said enough reserves have been found to keep the mine operating at least 10 years. That figure is expected to be increased with further exploration. Asamera Mineral holds the other 51 percent. The **Cannon Mine** was reported to be the largest gold discovery in Washington in recent years with proven and probably reserve estimated at more than six million tons of ore containing 0.25 to 0.33 ounces of recoverable gold per ton. Those reserves would yield 1.7 to 2 million troy ounces of gold, according to the new release.

Less than 30 acres of the more than 4,000 acres under lease had even been explored at the time, Hunter said. He added, there was going to be a lot more gold found there somewhere. Hunter was reported as saying the mining and milling costs less than $30 per ton of ore. At that rate, mining would remain profitable even if the price of gold fell as low as $125 an ounce. (Gold at this writing is around $417 an ounce). Down only 900 feet from the surface kept cost down; along with being right on the edge of Wenatchee giving it paved access to utility, and to home commuters. And, he added, the full-time workers were expected to be non-union. "We don't want a union if we can help it," Hunter said, "and I don't think we'll get one."

Because it is so close to the city, the mine is underground. Hunter said it would have been an open-pit had it been five miles out. The mill being built at the time will be capable of 2,000 tons of ore daily. He said the project would pay about $5 million a year in wages and another $10 million into the community annually through taxes, purchase of materials and supplies. Getting the mine and mill into operation was expected to cost $45 million; half from each corporation. (That figure went up to $60 million by July 17, 1985.)

Now that milling has begun in mid July, 1985, the companies have gone back to more exploration. Asamera is based in Calgary, but this project is under it's American incorporation. They hold the rights to more than 5,000 acres in the area, less than five percent of which had been explored at the time.

On April 7, 1988, two firms had signed 40-year lease agreements with Asamera Mineral Inc., on claims near the **Cannon.** Althouse Placers, Inc., leased three parcels totaling 640 acres, and half interest in another 240, Summit Silver Inc., owner of the other half interest of that parcel, signed a lease. Peter Laczay is president of both of the companies, Coeur d'Alene based firms. In return for the leases Asamera agreed to do $400,000 in exploration and development within two miles of the Althouse properties. There is also a royalties and percentage agreement. Asamera will extend a 6,000 foot exploration tunnel towards the other leases.

Tenneco Mineral of Denver exercised an option to buy as much as 50 percent interest in a lease on property in the Wenatchee gold belt, said lease holder Silver Strike Resources Ltd., of Vancouver, B.C. Tenneco said in August, 1984 that they would spend a minimum of $12 million in exploration and development on these properties, according to A.J. MacDonald, president of Silver Strike. That company recently released assays on the **B-Reef,** the former Lovitt Mining Company's claim. They read 0.11 to 1.35 ounces of gold to the short ton from seven drill holes. The **B-Reef** is near **Cannon.**

Total Washington State gold production from 1860 through 1956 was 2,844,204 ounces valued at $78,306,908 (at $20 and $35 gold), which ranked Washington as eleventh in the nation in gold production. In 1950, Washington produced 92,117 ounces gold valued at $3,224,095, their best year. 97 percent of that production came from three mines; the **Holden, Gold King,** and **Knob Hill. Knob Hill** is in Ferry County, the other two in Chelan County.

For the years 1903 through 1956, Chelan and Ferry Counties are at the top of the list with about $25 million each, followed in order of output by Whatcom, Stevens, Okanogan, Snohomish, and Kittitas Counties. Since 1957, almost all gold produced in the state came from Ferry County's Knob Hill. Now the Chelan mines of 1988 again challenge Ferry for leadership.

(Gold is sold by the troy ounce which converted is: 1 ounce troy = 1,097 ounces, or 20 pennyweight or 480 grains. 1 pound troy = 12 ounces troy, or 13.164 ounces avoirdupois. 1 pound = 14.58 ounces troy.) (24 "carat" is pure gold; 18 carat is 18 parts gold and 6 parts other metal, etc.) Remember that most of the gold in this story was only sold at $20.60, and few mines received over the $35.00.

Artifacts Found

Along with the yield in gold, there is also great findings in artifacts in the orchards of Wenatchee. In April and May of 1987 and 1988 different archaeologists have visited the area since the finding of 27 stone objects. Called Clovis points, named after a New Mexico site where such stone tools were previously discovered, "these are the largest Clovis points found anyplace," said Richard Daugherty, a retired Washington State University professor.

First found in the R&R orchard by employees installing underground sprinklers, the points were once used to hunt mammals, including elephants, horses and bison, said Peter Mehringer of Washington State University. The latest find had not been dated but archaeologists on the scene said all previous Clovis finds have been scientifically dated between 11,000 and 12,000 years of age.

After the initial investigation the site was covered with concrete blocks which were finally removed in mid-April of 1988. Researchers included officials from the University of Wyoming, and the Smithsonian Institute. Co-owner of the orchard, Rich Roberts, said the orchard owners own the artifacts and have not determined what to do with them. Some archaeologists believe people who used the pointed tools were the first inhabitants of the Western Hemisphere. Researchers are trying to find if this was a cache, or signs that paleo-Indian people who used Clovis tools actually inhabited the area. First Clovis artifacts were found stuck between the ribs of mammoths and extinct bisons in New Mexico, in 1927.

As more and more "finds" of one value or another are told in northern Washington State one recalls a statement made about Idaho years ago that will not fit in Washington now, concerning Congress believing that "all there was left out here was bear and deserted gold mines." New locations every day prove that not so!

There are deserted prospects, and there are Bear–but–

In Whatcom County a subsidiary of Homestake Mining Company has obtained shares in Spokane-based Steelhead Resources Ltd. Proceeds to that company will be used to continue exploration at the Company's **Excelsior Gold and Silver** deposit in Whatcom County, according to Steelhead President Wallace McGregor. The report July 7, 1988 indicated a mining plan would be ready by the end of the year with mining starting in 1990. The location is near Mount Baker.

Azurite, owned by a company of that name, from Winthrop, produced $972,000 from 1920-1941. Reaching the mine is by truck road over Harts Pass from Methow Valley. **Boundary Red Mountain** is located two miles south of the border in the Mount Baker district. It produced from 1916 to 1942, $375,496 principally from 1916 to 1922. It is owned by Gold Basin Mining Company of Bellingham, Wash.

Other gold properties in Whatcom include: **Great Excelsior** with claims, (Lincoln, and President) listed as owned by Great Excelsior Mining Company of Bellingham. It produced some $20,276 to include some 13,000 tons of ore around 1915. (It is not known whether this is the same Excelsior being explored today by Steelhead Resources.

Lone Jack near the head of Silesia Creek and owned by R.J. Cole, of Seattle, the lessee in 1951, has produced from 1902 to 1915 over $360,000. **Mammoth** and **New Light** were good producers in the early 1900's. Most of the north-end counties of Washington State produced some good mines. Since Snohomish is one of the early ones with it's **Monte Cristo,** we mention a few more and expand on that county a bit further. Monte Cristo has 14 claims and is owned by H.S. Ofstie, of Everett, Wash. as of 1949. Of gold-silver-lead-zinc and copper, it produced 300,000 tons from the **Mystery,** and **Pride** claims. Some production was reported from 1897 to 1926.

Because gold is the mineral most searched for in the 1980's period, aside from Ferry County's **Hecla,** other major producers in the state are all closed down at this time. The **Apex** in King County produced $80,000 prior to 1901 and another $300,000 from 1903 to 1934. The mine belonged to Apex Gold Mines, Inc. in 1943. It was near Money Creek (well named). **Camp Creek** Mine on the creek by that name in Skamania County produced $75,000 prior to 1934.

Kittitas County

Kittitas County had several leading producers including **Clarence Jordin** that produced $35,000 from the **Ace of Diamonds** claim in 1952. In the Swauk district, the mine belonged to W.L. Palmer of Electric City in 1941. In the same vicinity was the **Ollie Jordin** which produced $20,000 in a two-year period prior to 1934; "wire gold" occurs here. The mine was owned by Ollie Jordin and Miss Ollie Blissett in 1938. The **Golden Fleece,** also in the Swauk district produced $30,000 in the late 1930's. Orin F. Fry, of Mukilteo, Wash. was the owner in 1941. **Wall Street,** up Cougar Gulch from Liberty produced $50,000 prior to and from 1935 to 1938. William Newstrum of Ellensburg and R.J. Jordin of Liberty had three patented claims in 1934.

Back to Stevens County

We have covered most of the major mines in the State of Washington and have learned some of the future potential. We must return now to Stevens County, one of the heaviest producers of several minerals, and learn of their mines.

So as not to lose ties with the past, or the future, it has been necessary at times to go far ahead of our prospectors. In earlier pages we did learn of some of the major mines in this county, but now we follow along to Chewelah where the earliest discoveries were non-metallics.

About Stevens County

Stevens County's Embry Camp is the oldest mining district in the state. It is blessed with mineral wealth and is also the largest producer in the state, of non-metallics. In 1899 a geological report stated the major need was much more prospecting and exploration, and a great deal more capital to make development of the mines possible. It's many appendages had been separated from Stevens County except Pend Oreille which would become a separate county in 1911. That county's mines had already commenced production at that date but will be discussed in a section apart from Stevens.

The Chewelah mining district was one of the earliest in discovery and production. In the area was a limited population but one well preserved memory is the mill stones from Yantis mill built at Arden in 1859. They are mounted on the Stevens County courthouse lawn at Colville, courtesy of Colville's late Judge Lon Johnson, and Dr. S.P. McPherson, of Chewelah.

Cord wood cutting was a necessity for fuel, and income. In a list of Stevens County bills was an approved payment Dec. 24, 1892 for three dollars to James Steele for a cord of wood. (In 1988 that

Ad from the 1899 CRYSTAL MARBLE COMPANY MINE, southwest of Colville. It was once the property of Colville business men, R.E. Lee, C.W. Winter, F.H. Chase, and C.A. Mantz. (Pat Graham)

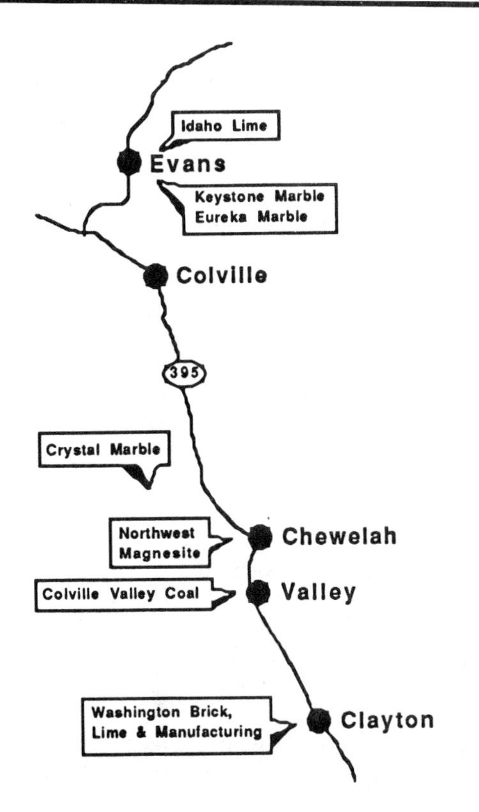

cord would cost around $60.) As a timber, and sawmill area it may not be widely known that a coal mine did exist around 1918. A news item of Sept. 6, 1968 tells that 50 years prior, development work was underway at the **Colville Valley Coal Mine** near Valley. The Chewelah Independent reported Oct. 25, 1907 that coal from the recently discovered mine near Valley was being used by people in the vicinity and said to give good results. The 320 acres of deeded land was owned by Frank Weatherwax, the coal said to be sub-bituminous. A little over 1,000 tons were mined from the property but it did not reach the production, nor the notoriety of coal in the west side. A news item from west side Washington Aug. 3, 1893 tells that a lump of coal from the state was shipped to the World's Fair and weighed 16,860 pounds.

CRYSTAL and KEYSTONE MARBLE COMPANIES

The discovery of marble so beautiful and of a quality seldom seen outside of Italy, brought about a flurry of locations. West of Valley, green onyx and black marble was discovered in 1899. A carload was reported worth $1,000. Crystal Marble Company, located June 1899, nine miles southwest of

Colville contained over 1,340 acres. The company was owned by Robert E. Lee, president (a man we learned about earlier in Colville history); C.W. Winter was treasurer, F.H. Chase manager, and C.A. Mantz secretary. Many tons of the marble was shipped east and Larsen and Greenough paid $25,000 for just one quarter interest in the property. There were five quarries. (The two men are known in the Coeur d'Alene and Pend Oreille counties).

Richard Steele tells us that Charles F. Conrady mined several properties and was vice-president of Crystal Marble Company (he said in 1894). Conrady had been in the Cavalry and settled in the Blue Creek area in 1902. He had also carried mail from Chewelah to Colville. In 1903, Conrady is supposed to have homesteaded near Addy and also near the **Alice Mine** (probably was vice-president for Crystal at this time).

Keystone Marble Company, 16 miles north of Colville, was another of the many marble quarries opened up and doing a good business at one time. The white marble brought $12 a cubic foot and in this vast deposit it was claimed to produce a superior quality to Italian marble, for statues. W.L. Sax, of Colville was secretary-treasurer along with J.F. Lavigne, George Bell, W.R. Baker and T.F. O'Leary as trustees. E.M. Heifer of Spokane was president and William E. Richardson, who served as a judge for Spokane and Stevens Counties, was vice-president.

The company also included the **Eureka Marble Quarries** and they produced 30 different varieties of textures and all shades, said equal to any product of it's kind in the United States. A shipment to Greenmark, N.Y. brought $21,000; another to New York for $10,000, and one to the west coast for $11,000. A product from one of the quarries in Stevens County was shipped to Olympia for facing of the new capitol building when it was rebuilt, and much later another product was used at the World's Fair in Seattle for one of their buildings.

Other Non-metallics

Clays at Clayton supplied the Washington Brick, Lime and Manufacturing Company, who produced terra cotta, sewer pipe, and brick. It was the largest brick manufacturing plant in the state, and was destroyed by fire in late July, 1897. The plant had been built in 1893 at a cost of about $50,000. It employed 25 and at the time of the fire announced plans to rebuild, which it did and later it employed up to 60 at times. Most of these operations have ceased, but marble is still taken in the county and there is much other production in 1988 of stone, silica, terrazo chips,and other non-metallics.

"The Idaho Lime company plant has been sold to the United Gypsum", according to a news release. That price in February of 1936 was $250,000. The Company's kilns at Evans provided work for over 50 men. In 1953 after more than 50 years in operation they had become unionized and soon went on strike asking pay raise of 26 cents an hour above the $1.72 an hour they received. It was not long afterward that the company closed the plant down. It had returned some $50,000 each month into the economy of the county, when in operation.

Stevens County is noted for being a great farming, dairy, and fruit growing district, much earlier production meeting the demands of the miners. By 1914, eggs were 30 cents a dozen; butterfat 49 cents; alfalfa yield was only three to six tons to the acre, and timothy hay ran from two to 25 tons to the acre. Things improved and by 1920 eggs were 50 cents a dozen and butterfat ran 57 to 70 cents. Timothy production was reported as high as 35 tons to the acre. (One farmer suggested the Historian was high on his tonnage figures.)

Mining interest in all districts was marked with alternate prosperity and depression. By 1903 mines were assessed to include full value of each mine and it's improvements. A 1904 report states that from Boundary, north of Colville and just below the Canadian border, down 75 miles to the south and including Cedar Canyon mining district– a belt running about 10 miles east of the Columbia River to the Pend Oreille County line, has produced more silver bearing ore than any other district in the state. The need for more public study of the situation brought about the Stevens County Mining Association of which Senator David McMillan of Colville, was elected president June 17, 1936.

Stevens County produced the most gold in the State of Washington from 1905 to 1908, and again in 1922. In 1928 the county led the state in the number of producing mines. According to the Bureau of Mines 19 produced ore at a value of $89,396; in 1954 they led again with 24 operations. Most of the gold mines in the state closed October, 1942 due to the Government War Production Board Limitation Order L-208. By 1962 a geological survey stated Stevens County produced half of the minerals of the state of Washington. With metallic mines pretty well shut down, Stevens County still employed 368 miners on Dec. 30, 1984.

Very Important Non-Metallics

It is not the intent to cover all the non-metallic operations of the state but Stevens County once claimed the largest magnesite plant in the world – the Northwest Magnesite, at Chewelah. It now has operating at Addy the first magnesium and silicon plant of it's kind in the United States. To better understand, we do have to include some background history.

The town of Addy, Wash., was platted in January, 1893 by Gotlieb Fatzer. He farmed at Arden

Clayton's brick manufacturing plant was the largest brick plant in the State of Washington. (Pat Graham).

and had his grist mill at Addy. By 1899 the marble quarries were active in that vicinity and Addy became a shipping point. There were three quarries operating making regular shipments to Spokane.

Northwest Alloys

It means moving far ahead to explain this phase of what is happening in Stevens County . . . the subject being dolomite. Virtually all of it is found in this part of the state. The **Tulare Quarry** east of Colville and owned and operated by a prominent citizen, the late Eric Carlson, took out 500 tons per month. Between 75,000 and 100,000 tons have been removed, most of that used in the manufacture of paper, according to the Dolomite Resources of Washington report of 1944.

In 1943, the marble fields southwest of Colville were investigated again and there was found to be a supply of dolomite estimated to be a 700 year supply (according to a Statesman-Examiner report). Marshall Hunting, from Olympia; Everett Haugland of Republic mining acclaim, John P. Thompson from Colville, and Lee Brooks of Addy helped in the exploration and mapping for a state report. It would be many years later until the Aluminum Corporation of America would admit to knowledge of that 1944 report and provide the answers for it's use.

There are deposits of dolomite at Fort Spokane in Lincoln County, and at Riverside in Okanogan County. Northport in Stevens County has a deposit that shipped and was used by the Electro-Metallurgical Corporation in Spokane, to make magnesium during war time. That Pacific Northwest Alloys was out of business by 1952. Dolomite has long been used together with limestone for many purposes to include smelting; lime manufacture; agricultural, crushed, and building stone; whiting for putty; paint, rubber, etc. Alone it is used for certain cements.

Addy quartzite makes up most of Dunn Mountain, six miles west of Addy. Deposit #1 near Dunn Mountain was quarried about 40 years before the 1944 report was compiled (around 1904). That was when **Crystal Marble Company** took out marble for building stone. The deposits in #1, including those old quarries, belonged to W.G. Merryweather of Spokane, Essie Taylor, and the late L.E. Gotham of Colville, in 1944. There was estimated to be 200 million tons of dolomite in the deposit, two million regarded as fair grade.

Deposit #2 was located immediately northwest of Addy, with owners A.B. Lind, Emma Weatherman,

The old Pacific Northwest Alloys Mill at Northport, out of business by 1952. (Pat Graham)

and Cecil Parker. The estimated dolomite tonnage was 68 million short tons. Another 10 acres showed 13,700,000 tons. In the two deposits, 63 million tons was said high-grade, or high-purity dolomite with a possibility of there being 185 million tons of the high-grade material. And the world wondered why Alcoa chose Addy!

A minor mystery occurred in the summer and fall of 1969 when the Washington Water Power's law firm represented by Robert L. Simpson from Spokane, began taking two-year leases with option to buy, just outside the townsite of Addy. He first offered 10 percent then raised it to 20 percent; his offer to eight farmers. That option cost $50,000 for an area two miles by one half a mile valued at $300,000. These options were put in the law firm's partner's name, John R. Quinlan. The secret was finally out when it was learned the Bonneville Power was to supply power, and the Washington Water Power to supply gas for some sort of industry.

In November 1970, Alcoa announced their company was going to build a $25 million magnesium plant at Addy. This is Washington's newest, and Stevens County's largest industry. In January 1971 the company gave the "go" to build. By then they had acquired over 1,000 acres of farm and timberland and construction was to begin by April, 1973. 300-400 were to be employed.

Then unforeseen problems arose to include major environmental opposition in the middle of 1973. The Colville Valley Environmental Council under the leadership of John D. Kirkman, blocked construction of the Bonneville power line to the plant. They finally had to give in when the plant itself was built. Along with those, and other problems to include inflation with nearly two years delay, the plant actually cost well over $50 million. Fifteen percent of the investment was for environmental control equipment, according to George Hutton.

By Jan. 7, 1976 heating of the large ferro-silicon furnace took place. In March the magnesium furnaces were fired up. It was to be production all the way, according to S. Alfred Jones, president of the new Northwest Alloys. Earlier George Hutton, general manager explained that "to make magnesium we use alumina, ferro-silicon, and dolomite. We have to ship in the alumina." He explained that 15 pounds of magnesium was used for every ton of aluminum in it's making.

May 13, 1976 the magnesium and silicon plant at Addy was dedicated by Governor Dan Evans. Alcoa's president William B. Renner participated.

Stevens County's largest industry the Alcoa Plant, was dedicated May 13, 1976. This photo taken July, 1989 showing part of the plant, and the pit at Addy (Statesman-Examiner, Dan Ghramm photo).

The 24,000 ton a year operation employed 276 at the time and the plant takes up 240 acres of the 1,200 they acquired. It uses the dolomite from the local claims of which samples contain 60 percent – a very good grade. Bill Kelly, head of the raw material said the three quarries would be working. He added,"the area holds the finest supply of dolomite and quartzite in the world". The plant produces 16,000 tons of silicon a year, and 24,000 tons of magnesium. Officials have explained the plant has met the very highest of environmental standards possible.

In it's operation, coke-coal-and wood chips are used. This gave the Brauners, a group from Kettle Falls the opportunity to set up the Metallurgical Chip Co., which supplies the plant, also provides around 20-25 more jobs, bringing in the wood chips.

George Stevens, the Northwest Alloys plant manager, announced April 13, 1988 that over supply of ferro-silicon had closed that plant for awhile but it was to reopen that week. Separate from the magnesium plant, that plant had been closed since 1985. "We will start up the furnace this month and hope to be in full production by next month. We're opening for the long term, well over a year. It will raise the employment to near 500." In 1987 the plant averaged 457 employees.

He added that, "52 percent of the local production of magnesium goes into alloy aluminum (we can judge the amount needed by the amount of cans, and other things, in use). The payroll in 1987 was $15 million at the average of 467 employees. This does not count the 80 or so employees of contractors working for us in the plant." He continued, "during 1987 we spent $10 million with local suppliers; $12 million for energy; three million dollars for freight, three million for capital improvement and safety and two million dollars for taxes." All nine furnaces were said in operation through 1988.

Early Chewelah

The city of Chewelah to become the first lode mining camp in Stevens County was first visited by Solomon Pelcher in 1842. He claimed to have visited the area when he was the only white man here until the 1860's. He died and was buried in Chewelah in 1882. Tom Brown tried the Addy area first but moved to Chewelah in 1859. He had been a miner on his way from California to the Canadian fields but possibly it was his Scottish blood that saw the good in opening up this new camp.

First called Embry, the earliest location was claimed to be the **Summit Group** in 1892, the most datings go back to 1883. John N. Squire is said to have sent prospectors into the area. A lead-silver location was made in 1886 by the **Eagle Group** some three miles northeast of Chewelah. The six claims were owned by C.D. and C.W. Ide; and I.S. Kaufman, of Colville. The ore assayed at 40-70 percent lead, and 25-100 ounces silver.

Tom Brown's daughter Mary was the first school teacher. According to the Washington Almanac, that first schoolhouse in northeast Washington was built in Chewelah in 1868. They continue, "only five years previously, the county superintendent of education had asserted that the people didn't want schools; he suggested that the school funds be diverted to building a county jail."

The first store in 1882 was said owned by J.T. Lockhard whose stock was mostly whiskey. As miners increased and times passed, prohibition came and was said not to be a problem, the "Addy's product was just as good as the store boughten."

In 1871, the area had been homesteaded by James Monaghan (later known in the CDA's). He called it Fools Prairie and was located south of Arden on what is now the McClean place. In 1873 the government put the Indian agency in Chewelah and on March 28, 1884, E.J. Webster, I.S. Kaufman, and Eugene G. Miller platted the present site.

The Indian agency was moved to Nespelem by 1885 and in 1903 the building was bought by Dr. S.P. McPherson. In 1944 it was restored and is a National Landmark. Daughter Alice McPherson Hutchinson was living in the building in 1976. By 1889 the Spokane Falls and Northern railroad had reached Chewelah. Later two passenger trains ran each way daily from Spokane through Chewelah to Rossland and Nelson, B.C. Those tracks were said taken up in 1920.

The late attorney John Raftis Sr., was born near Embry camp in 1892. He told this writer that in 1883-84 there had been two stores in Chewelah, both log buildings. He tells from memory about August Krug's store of the 1890's. A miner, Krug owned a two-story building in the town; other business and a bank. There were five saloons and a brewery with beer selling at five cents a glass and whiskey from 10-25 cents a drink. Butter was 15 cents a pound; eggs 10 cents a dozen, potatoes 50 cents a hundred pounds; and good wages were $1.50 a day for ten hours of work, six days a week. A Colville attorney in later years, Raftis told these details in the 1960's and with some sense of nostalgia said "and back in those early days there was no sales or income tax, and the people supported themselves – and the government."

By 1905, Chewelah had a population of 650; by 1914 there were good mines working in the area and large quantities of copper, silver, and lead – and some gold, was being mined. By 1920 it's population was 1,600. The population and economy was to change due to the urgent need for magnesite.

NORTHWEST MAGNESITE

It was in 1914 that the need for a magnesite plant was felt and exploration for it's supply was going on. Raymond Allen has been credited with being

American Mineral Company ALLEN CAMP around 1917. A view of the Valley, Washington camp - and one of the quarry with the men loading out truck. (Courtesy of the late E.P. Wheeler of Colville).

first to locate magnesite at Brown's Lake in 1914. Prior to the war, over 70 percent of magnesite was imported from Austria (until Kaiser Wilhelm cut our supply); from Hungary, and Greece. A better quality came from California, mostly for the making of paper. We go back to the marble fields and find the material urgently needed for making steel, right in our back yard

In 1915, R.S. Talbot of the Inland Empire Paper Company of Spokane acquired samples and soon began acquiring properties of the abandoned quarries of U.S. Marble Company, west of Valley. He recognized the deposit was surrounded by commercial value magnesite. The Washington Magnesite Company was organized. In 1916 Professor F.M. Handy, from Pullman began to acquire properties and later the American Mineral Production Company took over his holdings. There was estimated to be seven million tons of magnesite in Stevens County and from all of these findings and transactions, the **Northwest Magnesite** came about at Chewelah.

In 1916, the plant was reported to be ranked as the greatest producer in the country and at full production has been said to be the largest producer in the world. Shipments of crude magnesite from Stevens County averaged 700 tons daily. A railroad was being built in 1917 from Valley to Deer Creek, near the **Red Marble** quarry. A five mile aerial tramway was built from **Finch Quarry** to Chewelah. The Valley Magnesite produced only a calcined product and that plant was said closed about 1918.

In April 1917, there was to be $100,000 spent on the Chewelah plant by eastern money and by 1918 on June 4, California capitalist W.H. Crocker said at the time, according to the Colville news, that the Chewelah plant was producing 75 percent of the magnesite used in the nation.

By 1921, all crude magnesite was being hauled to Chewelah (that is not calcined) – selling for seven to nine dollars a ton and calcined $31 to $35. Both were shipped to Pennsylvania. Magnesite is used in the manufacture of bricks and furnace linings to withstand high temperatures. In calcining of magnesite, crude rock is heated to drive off the carbon dioxide and moisture. It is then called deadburned, or sintered form. During the war and later there were said some 800 working at the plant making refractory brick.

The five-mile cable tramway that was installed in 1921 was extended an additional eight miles during World War II to facilitate getting the ore to the plant. Each day the quarry operates, they are reported to have removed 4,000 tons of ore and waste rock. One half of that ore was transported via the tramway to the rotary kilns where the magnesite was burned to produce the deadburned product.

Up to 1917, 75,000 tons crude magnesite was mined from the **Finch** quarry; 5,000 from **Keystone,** and was shipped to Pennsylvania. 75 men worked at the quarries; 100 men at the calcining plant. The cost to quarry was about 75 cents a ton and sold for $7.50 a ton. Cost of calcining was $14 a ton and was selling for $32.50. The American Mineral Production Company had four quarries: **Allen, Moss, Woodbury** and **Red Marble.** Northwest operated **Finch, Keystone,** and **Midnight** after consolidation in December of 1916.

In 1935, with orders booked ahead for months, the plant opened a fourth kiln, and in 1940 they had a 300 ton mill underway at the **Finch Quarry.** The company was producing the crude; calcined and dead-burned magnesite – as well as building board and various kinds of insulation. By 1943 they were contemplating the building of that five mile plus tramway from **Finch** to **Keystone** quarry. The company at that time also owned the **Pacific Claims** of **Detroit, Wabash,** and **Checops** with values in copper and gold. There was said to be 150 men working handling the ore and waste rock by hand until 1943 when the company obtained machines.

Up until the ending, the only serious problem they faced was being sued for a quarter million dollars March 1923, for fire and smoke damage. On April 16, 1966, Gene A. Schalock, supervisor, told the Inland Bureau of Spokane Chamber of Commerce that **Northwest** is processing more than 100,000 tons of ore a year from the **Red Marble** claim and they recently had 150 to 200 persons employed. By 1964 Howard Ziebel was general manager.

Then the blow struck! On April 16, 1968 the plant was closed by Harbison-Walker, the largest manufacturer of refractory products in the world. "The Japs could land magnesite in Pennsylvania cheaper than it could be loaded on the car in Chewelah."

Payroll loss to Chewelah was set at $750,000 plus $56,000 in county taxes. The closure was said due to technical changes in the steel industry; new processes require higher quality product than can be produced at Chewelah. The plant was closed down the summer of 1968. **Fibco Inc.** asked to buy the company and the property by Sept. 5, 1969. That company was composed of Charles Singer, former Kettle Falls; and Bill and Dick Brauner of Kettle Falls. Much equipment was to be put up for sale and the Brauner Lumber Company was to acquire the timber land. A particle board plant was to be built on the site. One report said in July 1968 the plant was owned by Dresser Industries of Dallas.

The reaction of the closure was horrifying to the towns-people but suddenly they took boot straps in hand and have done much for themselves after 52 years of a "one company" town. On Jan. 23, 1970 the Pyro-Metric Inc., of Seattle, Joseph J. Carr president, helped buy the Magnesite mills-

The late Earl P. Wheeler of Colville was operator at the Northwest Magnesite Mar. 24, 1919 when this picture was taken. (Courtesy of Mr. Wheeler).

site and tailing pond at Finch quarry. They are changing the mill circuits for processing antimoney. The sale was said to be for $104,000. $25,000 was loaned the company by Pyro-Metric and they were seeking agreement with **Fargo Mining** and **Cleveland Mining Co.** concerning operation. The Singer Enterprises with Charles D. Singer and Charles D. Singer Jr. were reported to have bought the main plant for more than one million dollars. Ninety employees had lost work and the town is doing it's best to come up with replacement jobs.

Several rock producing companies have opened up; 49 Degrees North ski area employing some 110, has drawn a huge following since opening in 1972; and a sewing factory with a payroll of $174,000 a year, making clothing for Pacific Trail Jacket Company is always needing more help. The city of some 2,030 in 1980 is "taking care of themselves". A seed potato processing plant adds $225,000 to the community, the ski hill was to add another $18 million in investment within 10 years; there is a modular home factory that employs 15-22, and Chewelah made "All-American City" – they will make it!

Back to those Chewelah district silver-lead-copper mines.

United Copper

The **United Copper Mine** has been the major metallic mine of the district. It was located by 1891 but not developed until 1906. The mine four miles northeast of Chewelah was reported to be the best silver-copper producer in the state in 1922, and was a dividend payer. There are seven full claims and five fractions – non had been patented up to 1910.

A.V. Shepler made the early discovery and in 1906 Conrad Wolfe bought his holdings for something between $20,000 and $25,000. There was said to be included 100,000 tons of tailings. By 1914 the company had spent $50,000 on an adit driven 4,220 feet and other exploratory effort. There was extensive tunneling one mile long and 150-ton capacity mill replaced an old one and is still used for retreating the tailings. Around 1919,

Northwest Magnesite plant at Chewelah – once said the greatest producer in Stevens County, and at full production has been said the largest producer of magnesite in the world. Capitalist W.H. Crocker said this plant was producing 75% of the magnesite used in the nation. The plant operated from 1916 to 1968. (Washington State Historical Society).

the company named was changed from United Silver Copper Mine, to United Copper Mining Company, with offices in Spokane.

Our old prospector-friend Al Stiles had known Wolfe when he was a compressor man at the **Le Roi** Mine in Rossland, B.C. and said Wolfle really knew his business. A vein carrying 0.01 ounces gold, 5.35 ounces silver and 2.5 percent copper produced a shipment in 1908 grossing $9.59 a ton. There were differing reports as to the output and one states that between 1906-1909 the mine produced over $50,000 copper-silver. Another report is that between 1906-1920, there was over two million dollars taken above the 1,400 foot level. Still another report said that between 1909-1925, and again in 1931 there was two and a half million dollars mined up to 1934. Judge Lon Johnson, and Judge Albert I. Kulzer – avid mining men, both recall over three million dollars taken out of **United Copper** up to 1934. They added that stock went up to as much as one dollar a share and down to nothing. They both knew Wolfe and his operation.

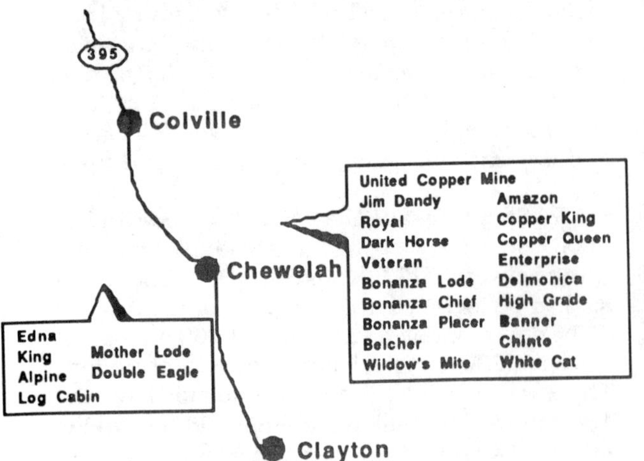

In 1914 the price of copper went down and Wolfle ceased mining in Chewelah until the end of the war. He went to other mining fields including Alaska and returned to the Spokane and Chewelah area sometime between 1940 and 1948. The mine was said to have continued from 1919 to 1920 with a high-grade wide vein yielding between $20,000 and $30,000 annually. The assays were 75-300 ounces of silver to the ton.

There were 12 new four-room cottages built for families and a modern bunkhouse for unmarried. Actual mining really picked up around 1919 but problems including bringing water from Chewelah Creek by a two-mile flume; and power clear from the Meyers Falls plant northwest of Colville. When the **Magnesite** needed an overload the **United Copper Mill and Mine** had to close down. This happened on several occasions over a two-year period. Pete Brady was the mill-man whom Stiles also told us about. Stiles had gone to school with Brady in Spokane.

In January, 1922, geologist F.M. Handy encouraged a smelting plant be built at Chewelah to care for the copper in the district. That discussion was still around in April, 1930 but copper prices were down and other metals were more inviting.

Sometime in 1942 many of the claims were relocated and **Jim Dandy, Royal, Dark Horse** and **Veteran** were recorded to Conrad Wolfe. Later the Eagle Mountain Mining Company came about.

The UNITED COPPER camp was pretty primitive in this photo taken sometime around 1914. Ore bin of the COPPER KING is in the background. (Dept. of Geological survey Bulletin 550).

UNITED COPPER gasolene locomotive hauling train of ore from the main adit, around 1919. (Washington Geological survey Bulletin N. 20).

"Wolfe and the boys" were said taking over the mining again. From the Stevens County Mineral report of 1943 – Charles Delk, of Yakima was secretary to the firm re-opened by Wolfe, and some other associates from Yakima, Wash. It was in 1948 that Wolfe was in Spokane trying to raise more capital for operating and development of the mine, when he caught a bad cold and died.

Exact periods have been left out of some of the mining history but Phil and Norman Lavigne were reported to have had an option on **United Copper**, then there follows a period of closure.

In 1955-1956 the Chewelah Copper Company was organized to reopen **United Copper, Amazon, Copper King** and **Copper Queen**, and several other unpatented claims. Yakima capital was still involved and there was stock sold. The company entered a contract with Earl Gibbs on the mill (of Bonanza Lead and Old Dominion mining companies). The arrangement went for some 18 months and legal difficulties arose between lessors and Gibbs, causing his Bonanza Lead to go to trial. End result of the trial was that Chewelah Lawyer Phil Skok eventually ended with **United Copper.** (It was during his operation of this mine that Gibbs had built a 200-ton flotation plant at Palmer Siding, north of Colville, to treat the ores from his various mines).

In 1960 the **United Copper** was up for taxes. Phil Skok bid and got the property for $6,200. The county claim was for $2,425.15 on 287.8 acres. Stevens County Treasurer Elsie Thayer reported this figure to the news.

In 1962, Colville mining men Joe Delles and the late Pat Sullivan leased the mine for two years, from Skok. There was said then to be 18 patented claims and 120 acres of deeded ground all on Eagle Mountain. The two men also took a lease on Gibbs $200,000 Palmer Siding mill and ran ore from Darrell Newland's mine through it. Phil Skok, J.W. Culverwell, Ole Alm and Frank Lehrman, all of Chewelah were mentioned at that time as owners of the **United Copper.** On Oct. 27, 1967 Spokane National Mines, Inc. with H.J. Tibbits as president, had an exploring option on **United Copper.** The state geological report was that the mine would pay dividends of $30,000 a month instead of the $10,000. A mill with a capacity of 400-ton a day was built. Due to mineral prices the mine was finally closed down but it was encouraging to note that Continental Oil Company filed claims called **"Eagles"** near the **United Copper** in September, 1976.

EDNA or KING

Axel Herman located **Edna** or **King** (or the combination of both) in 1896 and a shaft was sunk and three cars of ore shipped by 1898-1899. The mine is about nine miles west of Chewelah. It shows copper-silver and gold in ore assay. Production was reported from 1899 to 1917 to total $5,000 from four claims owned by Valley Mining Company.

In 1901 the property had been taken over by J.B. Tuttle and Sons and machinery installed to make possible active development until 1904. It was idle from 1904 until 1917 but around 1920 the company began shipping ore. From 150 tons the assay ran seven percent copper; four percent silver and $2.75 in gold from claims **Alpine, Log Cabin, Log Cabin Extension** and **Mother Lode.** In 1943 the four claims were reported by Arthur L. Hooper of Spokane with some ore shipped but no returns were available. In 1953 the property was said owned by C.R. and J.M. Carr of Valley.

DELMONICA-ENTERPRISE or BONANZA

Embry Camp – or Chewelah – was certainly the place to be, with prospectors swarming up the hills and especially Eagle Mountain. As prospects were staked it sometimes becomes difficult to follow development. In this case it is the **Bonanza Lode** located January 3, 1893. The same prospectors had earlier staked **Bonanza Chief** on Oct. 2, 1884, some four miles northeast on Eagle Mountain.

The Colville Republican of Nov. 19, 1892 tells that the **Enterprise** is owned by Jack N. Squier of Spokane (spelled some differently than other news, it is still the same man who sent the earliest prospectors into the Embry Camp). His partner was T.P. Hertzel of Chewelah. Explained as a chloride property, "assay of 350 ounces of the white metal (silver) from the 110 foot shaft; and pure black copper chunks has been hoisted, carrying 3,200 ounces copper to the ton." The property was expected at that time to develop into one of the richest in the area.

On Jan. 21, 1893 the same newspaper tells of considerable litigation likely resulting from the "jumping" of the **Bonanza** claim. A Mr. Talor relocated and changed the mine's name from **Bonanza** to **Enterprise.** All work was suspended but on March 11, 1893, 15,000 tons of ore was reported to have been concentrated in the previous year. There were said indications of a large ore body being found and that the litigation was not going to interfere with a shaft being sunk.

On Oct. 5, 1893 Consolidated Bonanza Mining and Smelting Company, with J.E. Foster as president, applied for patent for **Bonanza Lode Mining** claim. Other claims listed to include **Bonanza Placer.** No more was heard from this good silver claim until 1914, the State geological bulletin listed the claim name as being changed from **Enterprise** to **Delmonica.** The last trail we have of this prospect is in 1920 when it was called **High Grade.**

(The point we make with the previous few paragraphs is that one should not despair in tracing some old family ownership because the name could have been changed many times – and frequently a prospect is dropped entirely as was **Bonanza Chief.**)

AMAZON, CHINTO or BANNER, INDEPENDENT, KEYSTONE, COPPER KING and COPPER QUEEN

To follow all of these great claims it is necessary to lump them under one operation at times; and others to attempt to tell the individual history. Of the claims **Amazon** was likely the earliest find, having been located five miles northeast of Chewelah by T.C. Meader, M. McCrea, Edward Burr and Joshua Story, in 1898. By 1920 the claims were owned by local merchant Joseph Oppenheimer of Chewelah. The three patented claims making up the property, showed values in copper at two and a half percent; silver two to 37 ounces, and gold at 50 to 60 cents per ton. That year 550 tons of ore was reported having been mined at a value of $6,212 most of that taken out before 1914 and shipped to Trail and Grand Forks smelters. Two carloads of oxide ore was sent to Granby Mill with a yield of $841 in copper and $125 in silver. The claims were said to be an extension of **United Copper** and when it was estimated the **Amazon** would be the camp's next largest shipper they were showing from two to eight percent copper and 100 ounces of silver to the ton.

Amazon Mining Company owned the property from 1918-1926; United Mines Corporation, Oppenheimer's company, owned it in 1926 and from then until 1953 A.I. Kulzer of Valley, the Chewelah Copper Company was listed as owner. By then a quartz vein to 16 feet wide had been discovered and there was a 650 foot tunnel and 310 feet of drifts. Earl Gibbs, of **United Copper** ownership, had the property in 1954 and shipped 2,000 tons of ore from **Amazon** to the **United Copper** for treatment.

Chinto, once known as **Banner,** came about in 1904. Also on the favored Eagle Mountain, it was near **United Copper.** There are three patented and 11 unpatented claims owned by Eagle Mountain Mining Company in 1949. Chinto Mining Company, with W.A. Hall, of Spokane, had the **Chinto** from 1938 to 1944. A 1,500 foot adit; 350 foot shaft, and a total of 7,000 feet of workings on three levels were done by that company. There was a 25-ton flotation mill by 1940. Assays produced from the upper level averaged $12.59 silver, copper, an gold. Eight carloads of sorted crude ore ran 12-15 percent copper, and 80 percent silver. One geology bulletin reports that 50,000 tons of ore was mined in a two year period; another that $50,000 in value was mined - which is fact?

It is interesting to note that E.M. Spedden, the manager and also engineer in 1938 for Chinto Mining Company, encouraged systematic development and compared the ore body with **Amazin** and **United Copper,** but more extensive in depth and covers a wider area. **Belcher** became a part of the **Chinto** works by 1920. There were nine claims cut down to only seven as exploration proceeded. The Belcher Group Development Company mined the claims from 1908. In 1949 they were listed as part of Eagle Mountain Mining Company. Ore values were in copper, lead, silver and gold with one vein four foot wide and another from three inches up to six foot wide. A 420 foot audit, 135 foot shaft, and another 12 foot with short adits and shafts were reported to be part of the property by 1920. Eleven claims were combined with **Chinto** and **Belcher** becoming a company on Oct. 18, 1939. Problems occurred here and litigation put the mine on idle during 1943. Odd claim names such as **Widow's Mite,** and **White Cat** were part of the **Belcher** claims.

Copper King was also part of the **Chinto** property. It belonged to Copper King Mining Company in 1905-1907 and that name was changed to Chewelah Copper Mining Company in 1907. United Copper Company operated the **Copper King** from 1916-1918 and by 1930 it had made several changes of hands and was under Northwest Mines Corporation. The ore was said came from a low-grade 40 foot wide and 500-foot long development shaft. 4,000 tons of ore was taken from the #2 level and averaged 2.5 percent copper, 3 ounces silver and one dollar gold. **Copper King** produced from 1904 until 1917 according to a 1956 bulletin. (Daniel Harbaugh who came to Stevens County in 1894 and lived near Colville was said to have some ownership in **Copper King** and in **Granite.)**

A 1920 report stated that 6,000 tons of copper ore was shipped, the 4,000 tons from just one stope. In 1950 **Copper King** was said owned by Banner Mining Company of Sprague, and by Jan. 27, 1976 it had been acquired by an eastern syndicate who predicted they would "make the mine yield on a higher scale." By Nov. 24, 1968 trouble was occurring and **Copper King** was sold by a receiver to H.S. Wales for $74,000. Debts owing amounted to only $45,000 so considerable money was left over to distribute among the stockholders. (Unintentionally that became a "yield on a higher scale!")

Copper Queen is one of even farther north mines from Chewelah, and on Eagle Mountain. There are two claims and 80 acres of deeded land. Aurora Mining Company opened up some of the works in 1915. The claim changed hands several times and in 1938 was said to be the property of Northwest Mines Corporation along with **Copper King.** That company consists of H.E. Durkee and the Ole Alm families of Chewelah. There had been some production but the tunnel was said caved in by 1943.

Independent Keystone was owned in October

25, 1907 by A.I. Kulzer, of Valley; and Florida interests. That year he employed W.W. Hardig of Ainsworth, B.C. to do assessment work on those claims near the **Copper King.** The property was idle from June of 1904-1917. Chewelah Copper Company at one time leased the four claims from Western Guarantee Company; and Kulzer.

Kulzer also owns **Windfall** with claims **Evening Star** and **Nickel Plate,** located in 1909. Incorporated into the Windfall Mining Company, the claims are just south of **U.S. Copper Gold** and have values in gold-silver and nickel. Mike Kulzer had **Kulzer Clay; Uncle Sam, McKale** and **Double Eagle** in the Chewelah and Valley district.

Other KULZER HOLDINGS and the DOUBLE EAGLE

One of those claims, the **Double Eagle** was some 12 miles west of Valley on the eastern slopes of Huckleberry Mountain. It was located by John Kizer in 1902 and by 1922 belonged to the Double Eagle Mining Company, Inc., of Valley. It was said to have 290 acres of patented ground and had high-grade silver and lead ore. Early day tunneling through loose rock yielded boulders of such quality as to frequently pay for some work – some four feet in diameter and composed of solid ore. According to a 1920 geology report, ore mined and shipped yielded 100 ounces of silver to the ton – and some as high as 1,000 ounces to the ton. Some ore was shipped but no returns were available at the time.

Another report said that ore shipped by 1916 to Trail Smelter was running to high-grade silver at 800 to 1,000 ounces to the ton. Still another report said a shipment of 15 tons hand sorted netted $427.34 a ton with values in silver, lead and some copper.

The Kulzer families have for many years been prominent in the Valley and Chewelah area. Some early members came from Minnesota and became merchants. Albert I. Kulzer became superior court judge for Stevens County. All have been concerned in some way with mining. The **Double Eagle** property was said to belong to Mike Kulzer with the Double Eagle Mining Company at one time listing Alex Morrison, Billy Rose, Andrew Diedrich, Fred W. Dickey's father, E.E. Edwards, R.J. Davis and John Lutjen.

On Jan. 31, 1968 the Spokane National Mines, Incorporated was said to have taken a lease and purchase option on **Double Eagle** and **Jim Hill.** The announcement was made by president of Double Eagle Company W.F. Kulzer, and H.J. Tibbetts, president of Spokane National Mines. The Kulzers still hold several iron mines with production having been shipped to Tacoma; and to Northwest Magnesite in the past. From other of their claims, several thousand tons were reported shipped to Orofino Cement for flux.

Prices – Good and Bad

Price for their production was, and still is, a major item in the game of mining. It does fluctuate due to demand, world problems, and output. In 1890, just as some of the Chewelah district mines were being explored, lead was 44.5 cents and it dropped to 29.8 by 1896. In 1935 it was 38.6 cents and up to 60 cents in 1937. By 1949 it was down to 21.5 cents and then to 13 cents. In 1973 and 1975 it ranged from 14 to 19 cents and on Oct. 19, 1988 it was 38 to 40 cents a pound. Lead is cast in 100 pound pigs and some is put into one ton blocks.

Silver had been $1.01 an ounce in 1885 and bottomed out in 1893 causing severe depression and very hard times for those in the mining business. By Oct. 19, 1967 it was back to $1.75. On January 1973, silver was freed from federal price control and in August of 1972 it had passed $2.00, the highest price since 1969. Silver was up to $4.64 by 1975 and was starting to shoot upwards when by 1979 it reached $13.35 a troy ounce. The boom came when silver went to $52 in January 1980 but it fell rapidly to $8.42 by July 1981 and continued to fall to $7.30 June 3, 1988. On Oct. 19, 1988 it was $6.33 and $6.38.

In earlier days companies were not paid for zinc because the smelter often could not separate the lead from the zinc. The new smelter at Kellogg solved that for many and payment was added for the zinc. Prices did fall in 1949 from 17.5 cents to 11 cents a pound for zinc. By 1975 it was up to 38.5 cents.

Zinc is cast in 64 pound slabs and put into 2,400 pound blocks. Copper was up to 60 and 65 cents a pound by 1975 and $1.38 in 1988. After gold went to $164.45 in 1975 it boomed to $677 in Hong Kong in January, 1980, then to $875 a troy ounce. By July 1981 it plummeted to $414.50; then up to $471.50 June 30, 1988 and down to $411.45 on Oct. 19, 1988.

With prices so low, the mines in the state were in jeopardy and especially those of Stevens County. Dozens of telegrams poured into Congress from mining men of the Northwest protesting the prices. "Further cutting in prices of lead and zinc are beginning to demoralize the industry," said Paul L. Sandberg, president of the Spokane Stock exchange to Senator Warren G. Magnuson. The statement regarded the belief these metals were being stockpiled for future defense use. By May of 1949 many companies had spent money on mine improvements (with that belief in mind) and were now threatened with closure. By Nov. 14, 1952 many mines had been forced to shut down due to overproduction and declines in prices.

On Jan. 23, 1953, M.E. Volin of Spokane, chief of the Mining Division Bureau of Mines region II in an annual summary of the division's activities said the federal program had spent $5,798,000 in

Washington, Idaho, Oregon and Montana to encourage exploration for ore vital to national defense. This was a joint expenditure between mining operators and the government.

More About Chewelah's First – EAGLE, or BLUE STAR, Sometimes called REDWOOD, or CHEWELAH EAGLE

When Mr. Squire sent out prospectors into Camp Embry there were others from Colville who were equally curious. It was in 1883 that the Ide's and Mr. Kaufman were said to have discovered the **Eagle** and in October 1896 they applied for patent to their property according to a news release. It was the first mine to be developed in the Embry camp and it was said to have produced over $20,000 by 1897. The assays were running from 25-100 ounces of silver and 40-70 percent lead to the ton. It was in 1914 that this mine became known as the **Blue Star.**

The mine is about three to four miles northeast of Chewelah. By 1914 ore deposit bulletin #550 stated that $80,000 in ore had been shipped in the past. At the time in 1913-1914 the mine shaft was full of water. A few shipments of ore aggregating 315 tons were sent to Tacoma Smelter several years earlier and reports were an average 0.068 ounces of gold, 27.4 ounces of silver and 39.6 percent lead to the ton. Copper content was never determined.

By 1920, the company had three claims with offices in Chewelah. Considerable work had been done following a new company takeover by 1899 – the Eagle Consolidated Gold Mining Company. A later list included claims **Eagle, Nabob, C.O.D.,** and **Oden.** Between 1911 and 1925 there was claimed to have been up to $150,000 in ore shipped. A 1922 report said that **Blue Star** had shipped earlier, from the upper levels, of high-grade silver-copper ore but that due to a stockholders disagreement the mine was at a standstill. The claim was that they reorganized as **Blue Star** under lease to "dependable parties" and were waiting until spring to re-open the mine.

A news release of Dec. 20, 1958 said the Blue Star Mining and Survey Corporation, with offices in the Collin's building in Colville, is offering 400,000 shares of it's capital stock to the public at 25 cents per share. And at that point we run out of information concerning this, Stevens' County's earliest mine.

and Other Locations

Because there were hundreds of prospects in the Chewelah mining district; many of them developed, and some did make good mines, it is impossible to cover them all. One important point, State geological reports state there were some of the richest mines in the state in the Chewelah district. Sadly in the earlier prospecting days, many looked for "free gold" mines and were "wild catted" giving the camp a bad name at that time.

A very large number of early settlers did some mining and many of them were prospectors and miners who had followed the pack from camp to camp. One of those was James B. Tuttle Jr. who in 1897 became part owner and superintendent of King Gold and Copper Company. It was nine miles from Valley in the mining district we now cover but can no longer be traced. He had spent $35,000 on improvements. Tuttle had mined at Nelson, British Columbia, and in Colorado so he was no newcomer to the game. **"The King"** was said the best equipped mine in the area. Tuttle also owned the **Skookum** Mine, another claim no longer listed.

Josiah M. Davey or Captain Davey as he was called then, was superintendent and manager of Iron Hill Mining Company east of Valley. He had worked the mines and had mined for Hecla before coming to the Chewelah district.

JAY GOULD, or HIGH-GRADE

There are some claims that were developed that need telling about and this includes the **Jay Gould,** another mine belonging to T.P. Hertzel (owner of **Enterprise** and **Jay Dee).** The mine originated in 1892 and by 1897 Hertzel was preparing to ship and also to build a concentrator. The mine was located between 1880-1884 but not developed until 1890 and up to 1904. There were four claims. Four carloads were said shipped to Tacoma in 1903 averaging $14 gold and $10 silver.

By 1914 **Jay Gould** was said to have the richest ore being mined and it was incorporated into a company called the New Currency Mining Company. In 1915, 40 tons were mined but never shipped. In 1920 the mine closed then later came into the hands of Alex Morrison who planned to re-open in the spring of 1922. A shipment was said made and assayed $154 to the ton in silver and lead. Chewelah Silver Mining Company, with Pete A. Brady as president spent $5,000 in development between 1920-1922. Owner of the mine by 1943, O.C. Niles was said to have made a truck shipment in 1939 to **Bunker Hill** with a net return of $50 in lead and silver

Fred B. became part of the **Jay Gould** operation by 1943. The claim had been explored as early as the 1920's with ore values showing in lead, silver and copper. Earl Gibbs had operated this mine prior to his death. At the time he also had an option on **Blue Bell** and **Camel.** In 1969 the Chewelah Silver Incorporated with Maynard J. Davies of Spokane, as president; Phil Skok of Chewelah as secretary. By May 12, 1972 we pick up information that they were planning to re-open **Jay Gould** and **Fred B.**

Blue Bell claim belonged to David Barman (of Barman's Store in Colville); and John Hanley (from the **Old Dominion** and **Bonanza).** Sometime between Dec. 22, 1896 and Feb. 25, 1897 a notice of forfeiture was printed in the local paper and also included the name of Miss M.J. Kearney (of the Kearney family that discovered the **Old Dominion).**

SUMMIT GROUP

Many of the old prospects are listed either in the Chewelah district, or the Summit district and the **Summit Group** is one that somehow has become lost along the way. Alfred McKinney located the **Summit Group,** along with James Friend, some ten miles from Addy in 1897. McKinney lived in the Marcus area that year but had earlier mined the Wood River and Florence mining camps in Idaho. He had arrived in Spokane July 5, 1884 and like a great majority of the early settlers, he prospected and mined. He organized the Summit Mining Company and that year a vein eight inches wide was revealed by a 125 foot shaft showing solid gray copper, and silver and gold. One shipment assayed $90 to $1,000 in silver and the load netted $136.15.

A 1922 report said the discovery had justified packing over four miles of trail, then shipping by wagon to Spokane for shipment on to San Francisco. E.M. Spedden and C.C. Bunker were listed as operators and planned to mine and develop more in the spring. By December, 1942 a news report said high-grade galena ore was being mined by the Summit Valley Mining Company, from it's property near the Addy-Gifford road. The mine was again listed in 1967 when we come to the end of the trail.

SECURITY COPPER

Security Copper was another of A.V. Shepler's discoveries (he found United Copper). Located in 1900, S. Weese and E.A. Van Slyke were also involved. In 1913 the mine claims were incorporated as Security Copper Company, of Spokane and listed nine claims and 80 acres held under State lease. By 1943 the property was down to three claims and owned by Homer Shepler of Chewelah. Like many prospects, this one was said to need more development. There were several shipments of ore but no returns were available.

Iron King was discovered April 16, 1886 by L.H. Snyder, W.H. Mead, and Charles Kords, further south and on the Spokane Indian reservation. There was little to report from this prospect but in September, 1976, **Exxon** filed 51 claims in the area called **"Tunkey".**

JUNO-ECHO

Juno-Echo is two miles northeast of Chewelah and reported shipping 47 tons by 1916 with values in copper, molybdenite, silver and gold. By 1922 the company reported having eight unpatented claims, five taken over from Standard Chewelah Mining Company. F.L. Reinoehl was president; R.F. Chase, F.H. Alm, and H.S. Spedden were involved; and Albert I. Kulzer was listed as secretary-treasurer.

In 1941 there was a report of 300-400 tons mined and milled for the "Moly" but it didn't meet specifications of the buyer. In 1943 there was said to be three claims owned by the Juno-Echo Mining Company of Spokane, and a 50-ton flotation mill and some mined ore was on the property. A second report stated the shaft was full of water. It looked like another hoped for producer was a lost cause until suddenly in 1966 a report from the same property told of it being the sensation of the mining week as an eight foot copper ore vein was opened on the 150 foot level.

CONSOLIDATED SILVER MOUNTAIN

Consolidated Silver Mountain, a late bloomer was listed in 1922 to show a reserve of 40,000 tons of ore. Sorted, it would show 6,000 tons of ore that would yield $130,000. Values were in silver, gold and lead at $21.75 a ton. The company reported 44,000 tons of milling ore would average $15 to the ton. This company was capitalized for $150,000 with shares at 10 cents each.

Philip Creasor again

Philip Creasor who located so many great claims and mines in the Republic camp had been busy in the Stevens County camps also. He located what he called the **Silver Lode Republic,** on April 27, 1886, along with Colville druggist Frank B. Goetter, and J.H. Mumm. The prospect was near Brown's Lake south of Chewelah. Creasor was well-known in "Colville Country" mining camps and after **Admiral** was located west of Valley in 1894, Creasor acquired the property in 1896. He kept the **Admiral** until 1900 with active development. His major interest was in finding gold. In 1900 the property was taken over by T.R. Tate and T.H. Greenway. They sorted ore from the dumps and shipped three cars to Granby at Grand Forks. In 1916 the mine was taken over by J. Richard Brown and Associates of Spokane. Three carloads were said shipped to Trail, with values reported in copper, silver, and gold.

A quartz vein as much as 14 inches wide parallels the schist. Much underground workings was done with assays running 4.5 percent to 6.6 percent in copper; 2.5 to 4.5 ounces silver, and up to .04 ounces of gold. Production from 1915 through 1933 was estimated to be $10,000. In 1943 it was reported owned by Admiral Consolidated Mining Company, of Spokane, with J.R. Brown as president. There were eight claims included. **Banshee** was one and was said to have produced in 1907 and in 1910.

WELLS-FARGO

Wells-Fargo is perhaps one of the more interesting locations made in 1890 and the most development work was done after 1897. In 1920 it was owned by Springdale men the Wells-Fargo Mining Company having headquarters in Springdale. It is three claims some three miles north of the **Cleveland Mine** in the Huckleberry Mountain region. The mine is perhaps the only pure antimoney mine in Stevens County. A three-foot solid ore ledge was discovered of the material used for type casting, and alloys. In 1903 the metal was worth nine to 10 cents a pound. At that time Wells-Fargo Mining Company was capitalized for one million shares at one dollar each.

Stockholders were told on Feb. 20, 1970 that the company had been negotiating with several large firms concerned with antimony. Joseph J. Carr of Seattle, **Fargo** secretary told at that meeting, a major chemical company wanted to buy 30 million pounds of antimony over the next five years for use in plastic. The news release said that there is only one company in the United States that processes the zinkenite – a rare lead-antimony mineral ore that occurs in the **Fargo Mine,** according to Carr.

In January the company had purchased the **Finch** millsite from the then closed Northwest Magnesite plant at Chewelah. The stockholders were told that Wells Fargo did not have sufficient funds to buy the mill at the $104,000 asking price so Pyro-Metric from Seattle had stepped in to help with the down-payment of $25,000. At the time, Cominco at Trail had only offered 12 cents a pound for antimony and the company was aware that the year before in 1970, the price had shot up from 40 cents to four dollars. Roger L. Fisk, Chewelah, said the mill had the capacity of 220 tons per day for magnesite and would handle about the same with the **Fargo** ore. J. Merv Carr, of Valley reported at that time that work had been done so far that indicated 50,000 tons of ore available.

The treasurer Clair Hull, from Spokane, reported the company had spent $132,410 during 1969 to which one member, Mrs. Ann M. Logsdon said she thought an $18,000 item for administrative salaries was excessive. This prompted Joseph Carr to immediately cut all company salaries in half.

Other Locations

Other activity to mention included Tungsten Mines, Inc. of Addy Mining Company was said in December of 1932 to have reduced it's capitalization from $200,000 to only $20,000. Activity evidently did continue because in 1955 R.J. Weller, of Spokane was reported to have started construction of a 25-ton mill to treat the tungsten ores. The Hills were reported having spent $150,000 diamond drilling **Jim Hill.** There was said to be 500 tons of high-grade iron on the dump running 50 to 60 percent iron; John S. Gray who came to this country in 1875 and started mining the region in 1880 owned **Huckleberry, Thunder Mountain,** and **Buffalo Hump.** His holdings in the Seven Devils area was estimated worth the huge sum then of $40,000.

A news release of July 7, 1975 states that more than 10.5 million pounds of copper have been mined from the Chewelah-Loon Lake area with about 94 percent coming from mines around Eagle Mountain. Nearly 100 percent of the area's recorded silver production also came from these mines. Smaller deposits were mined on Embry Hill about one mile southwest of Eagle Mountain, with recorded recovery of 4,011 pounds of copper and 22 ounces of silver. The geological survey reports said that **Loon Lake Copper** yielded 622,555 pounds of copper and 532 ounces of silver from 1916 to 1919.

Loon Lake Copper Silver was the property of John P. Oschgar of Loon Lake. Formerly called the **Loon Lake Blue Bird,** smelter returns received from the first carload of ore reported April 15, 1916 went 13.6 percent copper, weighing 26.25 tons and netted $1,357.

By July 21, 1917 the company reported disbursing $14,404 in dividends to stockholders, the first one paid from the two year operation. By 1920 there was said to have been 50 carloads shipped with 12 percent copper reported. Up until 1918 the company reported $125,000 production. The company has a flotation mill and has offices in Spokane. The 1943 report of having shipped several thousand dollars worth of high-grade copper ore from this mine would reconfirm the geological report. Owners in 1954 were listed as Ernest Laukka and E.L. Tilton of Chewelah.

Just as the prospector could not cover every inch of ground, we cannot cover every prospect, claim or mine in each district but the information we have tells of some activity of the old timers and gives encouragement for newer activity. In that direction according to a Sept. 16, 1976 article Peter W. Laczay of Coeur d'Alene filed 240 claims southwest of Blue Creek on what he calls the **"M"** and **"MM"** claims. In the same news releases, Kerr McGee Corporation filed 17 claims called the "Cal" east of Chewelah on the Pend Oreille County line, and there are others to be associated with their specific districts.

Into Northport Mining Camps

Prospectors were busy in these new camps and luckily opened up ground with a number of very important mines. Many had moved into the Northport area as also did new settlers who looked to the future of mining, some having been in the California, Idaho, and Canadian mining fields.

D.C. Corbin was looking to the future when he purchased land from the government for a sum less than $250 and went about making plans for "his" town. Corbin sent E.J. Roberts, the president of this organization, and T.A. Herrick to lay out the townsite. When dedicated, Northport consisted of three log cabins; a wooded area, and no trains or even roads into the area. This was purely a speculative gesture on Corbin's part whose plans included this site offering him a jumping off point onto the Canadian side of the Columbia River, and access to those wealthy mining camps.

He did build his Spokane Falls and Northern Railroad into Northport and the final spike was said driven there on Sept. 19, 1892. The Kootenay Steamship Company was operating several of their fleet up the Columbia River from the Little Dalles to Revelstoke, B.C., where they could connect with the Canadian Pacific Railroad. Some of these early steamers were the Columbia; the Kootenay, and the Lytton. This line no longer operated after a road to Nelson was completed.

In 1896, Corbin's railroad was extended to Rossland, B.C. with the use of barges to cross the Columbia. A ferry was also in use to transport passengers across the river wanting to do business on the other side. A bridge was built across the river in 1897. It was a day of nostalgia when that bridge was dropped into the river as reported Oct. 18, 1949; it had served highway and railway. A new bridge was said built in 1951 and a second one in 1956.

Prospectors and miners were entering the region, both from the Chewelah Embry camps, and from the Pend Oreille, and Canadian camps. By Oct. 8, 1892 there were enough people in Northport that 25 got together to form a Northport Mining district. By Jan. 5, 1893 there was reported to be 1,000 men in town, some railroad men, and more were miners. To that time it was reported enviable there had been no shooting scrapes or highway robbery.

On May 8, 1893, Northport suffered it's first fire. There was little of real value left but the town did rebuild as new residents an business people were arriving; the report of silver and gold was reaching the ears of the outsiders. Another fire Aug. 10, 1893 apparently started by a tipped lamp, or a lost cigar, ended the life of prospector George Schild who had found a gold claim on Sheep Creek. He was said having come to town to celebrate; got drunk and his body was found in the ashes of the fire the next morning.

In the 1880's and 1890's the town grew; it had a post office; some 21 saloons; gambling houses and dance halls; the usual camp-followers, the "ladies of easy virtue", a hotel and a doctor. On June 3, 1894 a terrible wind and rain tore out trees and raised the river to what was said 75 feet above low water mark. It did flood a mercantile; the Northport News, and several other businesses and homes. One wag said with the limited damage, it was proof they could start building sky-scrapers and make Northport into a city.

When the Colville Indian Reservation opened for prospecting in February, 1896 just across the river and to the northwest of this border town, more prospectors poured into Northport. There were several hundred claims located within a week, especially on the northwest side of the Columbia. William P. Hughes had arrived with his printing press in 1892; had set up the Northport News, and had associated himself with the local development. In 1895 he was named Mining Recorder, and also United States Commissioner.

A fire again in March, 1896 did less damage than the last but there was still no insurance to cover the losses; the rates at 10 percent were said to be just too high because the town was declared a fire trap. On May 3, 1898 the worst fire to date took three business blocks and losses of over $100,000 were reported. An assistant to the barber was suspected to have been inebriated and the fire started with his overturned lamp, or cigar. It was very late at night and the fire had gotten too much of a start before enough help arrived.

Trouble between the operators of the mines and with Augustus Heinze at the Trail Smelter had led to some of the mine owners trying to build a city of their own on this side of the line, with prospects of another smelter in mind. D.C. Corbin was involved with this anticipated development and was said to be trying to by-pass the town of Marcus; M. Oppenheimer the local merchant, and James Monaghan. He chose a spot between Marcus and Kettle Falls and had platted the town he called Stevens. The Stevens County Land and Improvement Company was formed with mostly Spokane men who were mine owners in the Rossland camps. This included Col. I.N. Peyton, George Turner, Col. W.W.D. Turner, Chris McDonald; Custom Collector Martin J. Malong of Northport, Mark J. Shaffer of Springdale, and Eber C. Smith.

There was one building when the townsite was platted on Oct. 9, 1896. The group located a site for their intended smelter and went about trying to turn this city of Stevens into the county seat of Stevens County. After a brief time they gave up and the site disappeared.

The LeRoi Mining and Smelting Company then registered in British Columbia August, 1897 did build their smelter but in Northport instead. Built by American capital at a cost of about $250,000 (a figure of $25,000 has been quite reliable), it was owned by the LeRoi people; and on land Corbin had given the group on the banks of the river in Northport. By some transaction not entirely understood, Heinze's men James Breen, and Herman Bellinger came along with the LeRoi people, very much against Heinze's wishes. Breen faired well by receiving $300,000 by 1899, and the smelter was said never to show a profit until he was paid off.

The new smelter at Northport was some of the early part of Heinze's downfall and leaving Trail he returned to Montana where he failed totally. The smelter was blown Jan. 1, 1898 and by August serious fumes were ruining the local orchards and other growth. In 1899 the smelter was owned by other growth. In 1899 the smelter was owned by Northport Refining and Smelting Company and was running ore from the mines of Rossland, and some local ore to include that from the **Velvet Mine.** Because the smelter was about half the size of the one in Trail, Col. Topping of Trail called this one a "peanut roaster". (Geological reports said it did have at one time, a daily capacity of 1,200 tons.)

Fights and murders did take place as big stakes crossed the gambling tables and on May 14, 1897 the news told that "Red" and "Frisco Kid" were arrested Monday eve at Bossburg by Constable Griswold. They were taken to Northport by Sheriff Denny to answer to charges of holding up a bridgeman and relieving him of his shekels. "Red" was held for trial in Superior Court and in default of $2,000 bail, was committed to the county jail. The "Frisco Kid" was discharged and ordered to leave the city.

Good things of the time included a brick yard with George Hamilton and Leonard Clark doing some of the manufacturing. The town also bragged about a lime kiln and brick kiln, a saw mill, shingle and planning mill – and they were very pleased with the advent of the new smelter. The town was incorporated June 5, 1898.

Various news articles are of interest concerning the planned, and completed smelter. The brick yard became a very busy place when we are told that on July 16, 1897 a contract had been let to make 3,000,000 bricks. The Early Brothers took this to indicate the news of the coming smelter to be a "near reality".

On July 30, 1897, we are told that steps are being taken to get a new burial site. The article went on to say the "Northport Smelter site covers ground where we have been burying our dead, which was on a townsite ground and in town limits." It was noted that the new cemetery will probably be located on ground belonging to Thomas Perrott, just to the south of town.

L.L. Tower was reported Dec. 25, 1899 to have been an engineer for the Northport smelter during construction. He was later to become the United States deputy mineral surveyor for Washington State.

The smelter had started with some 200 employees and was up to 500 and 600 depending upon how much ore was on hand. In 1901 a strike came about when the smelter company objected to forming a union, the Mill and Smelterman's Union. The strike lasted for nine months. At one time the management discharged all the carpenters. They were shipping in and attempting to hire outsiders but the union was ahead and converted the newcomers faster than the employers could hold their faith.

The Union members did take their side of the argument to court because the Englishmen owned the plant and were operating on American soil. (Even though this was American finance, the company was registered in British Columbia.) In turn the management started closing down the furnaces until there was few help left that had not been laid off their job. Sixty-two men were brought in from Joplin, Mo., but after arriving, 45 of them declined to work for the smelter company. A miner gun battle brought Sheriff Ledgerwood, Prosecuting Attorney Wade Bailey, and Deputy Graham from Colville to settle the difference when one man was slightly injured.

The strike was declared ended March 12, 1902 and by 1903 the furnaces were all off due to the effects of the strike, and the fumes. Richard Steele tells that at the meeting to end the strike, some wanted to continue so the Western Federation of Miners let it be known they were cutting off the weekly allowances of the strikers. The next morning the free eating house of the Western Federation was closed. It made the strikers realize their predicament and the vote passed to abandon the charter of the Northport Mill and Smeltermen's Union. Up until closure, there had been some 1000 population in Northport, with the smelter the major employer of some 300. Wages had ranged from $2.75 to $5 per diem. At that time there were no developed mines in the vicinity but some of the richest prospects were being worked.

The Day Brothers of Wallace, Ida., purchased the smelter to run their ore from the **Hercules Mine** in Idaho but did take care of the ore from many of the area mines, and some from Canada. They planned to spend $750,000 on reconstruction at the smelter. By 1917 they were said to employ 400. Great ore bodies were being discovered, and property values went up 100 percent. But by 1921 many mines were closed and the smelter closed for the last time. Smokestacks and other momento recalled those days when Northport was "quite a city".

By the early 1916's, the value of matte scraped from the bottom of the copper kettles was too much temptation for those employees called "glommers". Some was apprehended being removed in a suitcase; some in coat pockets, and a story by Ruby El Hult still holds much fascination for the local people in Colville. Because it is her story in **"Lost Mines and Treasurers"** we can only tell you that several sacks of matte was said stolen; was brought to Colville and re-buried (at, or near what is now the golf course). Some was supposed to have been found in the 1930's but none has ever surfaced enough to make the story ending. This thievery was going on at both the trail Smelter, and the Northport smelter.

ELECTRIC POINT MINE

The most interesting story of the finding of this mine and reason for it's name, involves J.E. "Josh" Yoder who was prospecting the area in 1914, and came upon the cabin of Chris Johnson. Johnson

The Northport Smelter sometime around 1914. The largest employer of Northport. (Statesman-Examiner).

Camp of the early ELECTRIC POINT. (Pat Graham).

was said to have hauled many rocks on his back to put around the base of his cabin and when Yoder saw what they were he knew that the origin would also be the source of what he searched for. There was said to be numerous lightning strikes on one part of this mountain and Johnson took Yoder to that point where he uncovered the galena that seemed to draw the lightning – almost solid lead. They staked five claims and included Roy Young who had grubstaked Yoder. The claim was recorded by Yoder on July 2, 1915. (Young was at times referred to as "Art" Young.) (It was later learned the Indians had for some time melted the galena and made bullets from this area.)

By 1916 the Electric Point Mining Company was known in Northport. A wagon road four and a half miles long ran from Leadpoint to the mine located in a very isolated area four miles east of Leadpoint. Thirteen claims have been credited to the incorporation; **Electric Point No. 1,** and **No. 2** being of major importance. A 1,400 foot aerial tramway was built to transport the ore to bunkers at Leadpoint, and then the ore was hauled to boundary by truck. By 1915 there was reported to be an estimated value of net ore in sight of $1,300,000.

Yoder disposed of his interest to Johnson. Young also sold his and he went to Alaska where he was reported to have died in July, 1923, at Eagle, AK. A promoter was said at this time to offer Johnson $22,000 for his share of the mine. Later this same man, Walter Nichols was also reported to have paid one million dollars for the properties. A good road and automobile road to Boundary, some 14 miles away; or 25 miles the other way to Northport, was part of the development prior to 1920.

In 1916 the greatest lead discovery was told; a 40 foot vein was uncovered and two shifts were put to

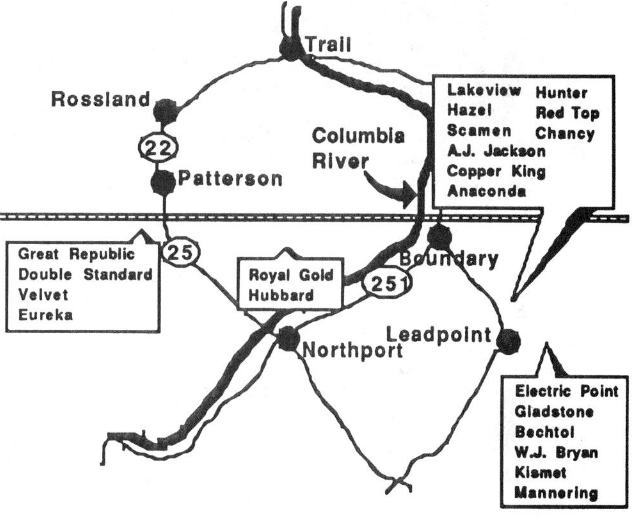

work. The mine soon consisted of 9,880 feet of workings not including surface excavations; two shafts and eight 100-foot levels from which the ore chimney were reached. A tunnel on the side of the hill on which the mine is situated, on the 300-foot level, extended to the surface on the south side of the hill.

From 1916 to 1920, there was 21,569,146 pounds of sulphide ore averaging 69 percent lead; and 30,438,010 pounds of carbonate ore averaging 25 percent lead mined; silver content was so low it was not measured.

In 1917 Walter Bower made a copper strike four miles east of the **Electric Point,** the outcrop being two and a half foot wide and running north and south for more than two miles. The nearest copper known then was in the Metaline district several miles to the east. This helped create more intense search of the area in the vicinity of **Electric Point.**

ELECTRIC POINT MINE as it developed and families were also living at the camp. (Photo Northport book).

In 1918 the State of Washington reported that a heavy shipment from **Electric Point** saved the state from a material decrease in the value of metal production the year before. The mine had produced 9,665,513 pounds of lead in 1917 valued at $2,289,285. It was reported to be the greatest lead producer in the state.

The first shipment of ore was made in July, 1916, and the last before the end of 1920. A second news report stated the mine had been forced to discontinue work by reason of additional assessment of income tax for 1917, which they were not able to pay in 1920. That report said the mine closed down in February, 1919. A more happy note was that in January when dividends of three cents per share were paid; the total payments to date was $301,530.

Yoder had gone on to prospect the **Red Iron Group.** By 1931, he was considered a wealthy man and had purchased a large ranch on the north end of Deep Lake. O.J. Bigler was renting the ranch and in March of 1931, he shot and killed Yoder. The large six foot six Yoder had come from Pennsylvania, where according to a news release of that date, he came from a family who had received a land grant from the British King, and a portion of that grant was still in the hands of his family. This colorful man was placed in the Yoder Mausoleum at the Northport cemetery.

The town of Boundary was shipping point for the mines of the region and at one time had over 1,200 active population. On February 13, 1925 we read that Joseph Klass passed away in Seattle that week. He had at one time worked at the smelter, then moved to Boundary where he ran the hotel for several years. His passing saw also the passing of the last hotel in this village.

The **Electric Point** was leased several times without anyone getting a good start at production. Among those listed "and back on a paying basis, thanks to the Finnish miners," was Isaac Martella; Eli Raisio, John Mackynan and Victor Hagan. At this period in the early 1930's the Colville paper also reported that Otto Ronka and his sons Les, and Edmond had leased the mine. The other men were apparently part of production crew.

"Art" Simonton operated the mine from 1939 to 1943 and A.G. "Kelly" Lotz had it in 1944. It was during this period of ownership that a second mine cave-in took the life of Albert Epsen, a tragedy that was all too familiar in those early mining days. Between opening and closing, it was the good news to report that in December of 1944 the mine was re-opening – and it was during that time owners declared the ore in the past had been so rich it was shipped directly to the smelter.

The mine operation continued to change and in 1953 was under lease to The State Mining Company, from Mrs. Maude Simonton (the widow of the late "Art" Simonton). At her death her will called for leaving the **Electric Point** to the Shriners Hospital for Crippled Children in Spokane. Northwest Mining Syndicate of Spokane was named owner in 1955. In 1964 Stardust Company had a lease and purchase option on nine claims at the **Electric Point,** from Mrs. Fred M. Viles of Spokane. That company had a working agreement with Shamrock Mining Company who in 1963 had uncovered boulders and shipped 35 ton to **Bunker Hill** of high-grade lead ore.

The old tramway had been taken down by 1924; better roads had been built, and by 1947 a new ore reduction plant was being installed at the mine, with a 100 ton daily capacity. (Ore off the dump was being processed when lead went to 15 cents, up from six cents.) Between 1963 and 1964, five new ore chimneys had been located by Stardust Company in addition to the eight formerly mined.

In 1964, the reportedly new company Stardust Mining Company went about sinking a 6x8 foot shaft to reach a chimney of high-grade ore. Firmer Walkley Spokane, the president of Stardust said the shaft was going down from the summit of Gladstone Mountain several hundred feet uphill from the old **Electric Point.** Howard Kruse was Stardust foreman. Shamrock also had a lease on **Gladstone Mine.** Even Bunker Hill sent a drilling team into the area in August of 1965 where they drilled on a 22-claim prospect owned by Andrew J. Anderson. His property adjoined **Gladstone.**

Total production of **Electric Point** included up to the end of 1952 (except for six years when the mine was down); 30,260,297 pounds of lead produced; 6,950 ounces of silver, and 1,143 pounds of copper, according to the Divisions of mines and geology report.

The State Mining Company owned the **Electric Point** as of Sept. 4, 1975. Fred M. Viles had died in 1959 and his daughter Marcella Grady, (Mrs. Dr. Grady) became president of the State company. Robert E. Grady became an officer; Muriel Viles of Deer Park, was vice-president, Nancy Sanders of Kettle Falls, was secretary-treasurer and Shirley Agar of Deer Park was director (along with Dr. and Mrs. Grady).

The Grady's had moved from Spokane to Colville, where he opened a new dental office. Their mining operations also include 11 claims in Montana; the **Eddy Creek,** and the **Johnny Miller**, between Plains and Thompson Falls. On Aug. 28, 1975 the Statesman-Examiner used a story about visiting **Electric Point** and that it being United States Government land. On Sept. 4, 1975 Dr. R.E. Grady came back with a response notifying the property was "private", not public since it was a patented mining claim. He stated there was a county road to the property but he was posting it because of recent thefts and vandalism . . . such as tearing lumber off the old buildings to use as paneling.

It might be pointed out here that Dr. Grady and every other mine owner have the right to keep people off their property, and doing damage. These are expensive developments and possibly even more important is that being around **any** unoccupied mine is a dangerous place to be with old mine shafts, and places to be hurt. Stay Away!!

The **Bechtol** properties were once owned by Fred M. Viles and at his death became some of the holdings of Mrs. Marcella Grady. The three claims, **W.J. Bryan, Kismet,** and **Mannering** were formerly owned by William Bechtol and Ben Hofer, of Northport. The mine had produced $16,000 in lead as reported in 1943 and had been operated rather consistently up until then. An unfortunate mine cave-in killed Louis Pointer in 1939 while further development was going on. Bunker Hill explored this property in 1964.

Gladstone

Adjoining **Electric Point** on the northwest side is **Gladstone Mine.** These two were stated in Washington geological reports to be the greatest lead producers in the state. The main chimneys of the **Gladstone** were located by Raul and Jim Hodson in 1913. Some 18 miles via the Deep Creek road from Northport, and 35 miles north of Colville, the mines were at an elevation of 4,200 feet. Leadpoint was the nearest post office and Boundary the shipping point.

Old timers gossip was something that could not be verified but they tell that the **Gladstone** was once sold for whiskey; or better put by some – that the proceeds from the sale went that way. In the history of mining, it would not be the first to go that way.

Dan Dodd helped bring the mine into production. Of the ten claims; production reports usually list from 1901 to 1953 but we find no history earlier than 1913. The huge chimneys of near solid lead made up the ore mined. 125 tons had been taken by 1917, the ore hauled by team or a sled pulled by caterpillar to where it could be trucked to Boundary and shipped via the Great Northern Railway. A 1920 report described a "small" chimney that was being worked. It was 12 to 15 feet in diameter and ran between the 100 and 200 foot levels.

There was considerable low-grade material, much of which was never mined, and that material was not shipped. Until 1924 the high-grade material was sent out on the Electric Point tramway; 20 tons daily from each mine was taken to Leadpoint.

By April 3, 1925, the Gladstone Mountain Mining Company (organized in 1916), was paying one half cent per share. On Sept. 16, 1926, they declared a dividend of 11 cents a share on it's stock with a par value of 10 cents a share. The large dividend was reported due to especially large shipments from the mine. By November of that year, they reported paying the ninth dividend and by

June, 1927 had paid out $200,000 in dividends. All was going well at the mine excepting for a couple incidents – Jason Clemons, one of the miners fell 100 feet to his death in the mine shaft. A $200,000 loss was suffered later when on Oct. 28, 1956 the mine had a fire. It was suspected to be arson but difficult to prove.

A 1943 report stated the **Gladstone** was owned by Walter J. Nicholls, Spokane, and leased to Swen Anderson and Associates, of Leadpoint. The reports to then were that the mine had produced over one and a half million dollars from 11 chimneys, and more development work was being done.

Well-known mining man, "Kelly" Lotze had a ten year lease on the mine and during that regime it was reported several tons of lead for shipment to the smelter paid results of $3,000 per truck-load. Production of $11,507.83 was reported for 1917 followed by the mine being inactive until 1920 when production was said $19,217.37 and for 1922 was $17,203.41. From then production boomed to over three times the tonnage of other years. The mine product assayed in lead, silver, zinc and copper with the mine best known for lead output. By 1943 there was less demand and only two men worked at the mine. In August of 1944, two carloads of lead carbonate was shipped to **Bunker Hill** by the lessees.

In the winter of 1963, the mine work was suspended but by May 28, 1964 William E. Cullen, secretary-treasurer for Shamrock Mining Company announced that company had a lease and had Clifton C. Hill of Spokane, as operator. That Shamrock lease was for about two years according to Ira C. Tester, vice-president of Shamrock. They also had a lease on **Electric Point.**

On Jan. 26, 1964, Firmer R. Walkley, president of Shamrock, announced new chimney discovery had been made, of high-grade lead. Stardust Mining Company was doing the work under lease from Shamrock. A late 1964 report was that **Gladstone** production was small and sporadic in recent years. That geological report gave 1901 to 1953 total production was 14,793,421 pounds of lead; 9,274 ounces of silver; 41,357 pounds of zinc, and 835 pounds of copper.

A 10 year lease was reported by Norman C. Smith, Feb. 21, 1966, to Stardust Mining Company, of Spokane. By June 16, 1970 Gladstone Mining Company let it be known they were spreading out and had acquired a purchase contract on seven mining properties in South-Central Idaho, and in the **Virginia City Group** in Montana. They had also four leases in the same local in Idaho, and on **Bodie Claim** in Pend Oreille County.

But – strange problems were evolving – not enough stockholders reported to meetings to conduct proper business. Those officers of the Gladstone Mining Company in 1964 were listed as: Board Members Jerry T. O'Brien, Spokane and George M. Grismer, Wallace, Ida. James P. Swan was president; Henry C. Smith vice-president and Mrs. Edna E. Nicholls was secretary-treasurer. By 1966, some changes included Board Directors Quin Hallbarg, Chicago; Ray Moore, Seattle; Dr. Douglas M. Bush, William B. Bantz of Spokane, and Norman M. Smith of Kellogg, Idaho.

Attempting to do any conclusive business at their June 14, 1966 meeting was nearly impossible when only 322,000 shares were represented out of 1,499,986 outstanding. G.C. George a mining broker had purchased 257,000 shares for Eastern interests and could represent those but that left less than one third of the total represented. One officer exclaimed **"There must be hundreds of thousands of shares of Gladstone Mountain Mining Company hidden away in dusty trunks and forgotten."** (Are you holding any??)

The problem had gotten worse and in October, 1967 John F. Campbell was elected president. He

A book of stock certificates was found by Dennis Blair of Colville of the DOUBLE STANDARD GOLD MINING Company incorporated in 1906. This old mine was one of the holdings of men who worked with Al Stiles around 1898. It was near his ROYAL GOLD or HUBBARD and Stiles was one of the registered stockholders including Parshall, Schaffer, Almstrom, and others. (Dennis Blair and the S-E).

wanted it known they "were not bankrupt on March 8, 1968", but from the standpoint of stockholders, Gladstone reorganized under provision of the Federal Bankruptcy Act. Only about one third of the shareholders were present for the meeting. By June 6, 1968 the company anticipated a court decision early the next month on it's petition for reorganization. Campbell explained this was the only way reorganization could be accomplished without a two-thirds majority. He asked to amend the structure from one and one half million shares of non-assessable 10 cent par value, to five million shares no par value.

A Feb. 4, 1970 news report said **Gladstone** was currently leased to A.G. "Kelly" Lotze with option to purchase for $90,000.

Al Stiles in Northport

When we last heard from Colville's senior mining man and prospector Al Stiles, he and C.W. Gerbeth were in Ferry County, heading east into Stevens County, still searching for that one great lode that would retire them from this weary life. Stiles said he located 16 claims near Paterson, just south of the Canadian border. These claims lay not far from the **Great Republic** and the **Double Standard**, both showing evidence of being good copper-gold producers.

Going on to Grouse Mountain, north of Northport, Stiles said he staked the **Royal Gold** which later operated as the **Hubbard Mine.** He told how he had three groups of mining claims and helped sink a 70-foot shaft by hand. When his outfit ran out of development money Stiles went to work at the **Lakeview Silver-lead Mine** in the Metaline district, "rawhiding" ore three miles down a mountain (according to an interview he gave to Wafford Conrad of the Spokane Chronicle).

Having to be away to earn more money left his claims wide open for a "friend?" to jump. This also happened to the **Velvet** located by several of his acquaintances and "almost" a part of his holdings. "The jumping occurs when the necessary work is not done by the first of the year", Stiles explained. He admitted to this writer in 1964 that after the **Royal Gold** incident he was tired of working in the hills and being "jumped."

After Stiles had discovered the claims, he had five and six men working. They sunk a shaft and built a cook house. Times were so bad he had to find other work and he shut down with intentions to return and keep up his assessment work, when it was jumped. (As an update – on July 28, 1977 Paul Hurn, Ron Hurn and Seth Smith, all of Northport filed 20 claims called the **"Curt"** – on Flagstaff Mt. and on Onion Creek. It is interesting to note the Flagstaff claims were listed as the **"Old Hubbard"**.) What money Stiles still had he said he pocketed it and went to school in Spokane, at Spokane Business College.

The full course at the school cost $50 and Stiles said it took him three years to keep payment ready, and to go to school. He did participate in some claims that were paying $20 to $25 to the ton and John R. Cassen, at the college offered to take stock for part of Stiles scholarship. Al did learn well and in later years published articles on mining of great interest to the industry. While in school he also worked for the Jones and Von Ausdale Laboratory, well-known Spokane chemists where he learned the industry that took him through his remaining active years.

Sometime in 1899 Stiles went to Northport, after the smelter opened and started his own chemist shop. Proving that he had learned something about ore, Stiles recalled that "one day a man came in and wanted a gold assay for a carload." He fired up the furnace. "The final test came out just 15 cents from how I figured." He advised the smelter to re-check, and his client got $400 more than another assay would have allowed. This was for a man by the name of McKenzie who later hired Stiles back into Canada to the **Arlington Mine,** near Nelson. Al remembered when the Arlington shipped $50,000 worth of gold ore a month to the Northport smelter.

Stiles worked for several different mining companies in Canada and finally returned to his little office in Northport where he worked for John Hall and the **Red Top Mine** people. He told one company that their shipments couldn't pay their freight. He said their ore was hauled by cayuse down a steep mountain over three miles to the railway at Waneta. "Rawhiding down that mountain the cayuses would throw off more ore than reached the landing," he explained. After helping many, as he did with the couple from **IXL** Mine that we read about earlier – Stiles went into the Springdale mining district, where we meet him again.

Other Mines of the Area

The **Great Republic** was mentioned earlier and was owned by C.C. Knutson of Northport. It is only a few hundred yards from the International boundary and close to the Red Mountain Railway. A good amount of work had been done as told by a 1914 report, with a picked sample of one vein containing 11.25 percent copper, $2.40 in gold, and 12 ounces of silver to the ton. A deposit a few feet below contained even better values.

Just north of the **Great Republic** is the mining camp **Velvet.** In 1943 it was held by assessment and owned by Elmer Godfrey of Northport. There was reported to be two quartz veins about 12 inches wide, with values in gold, silver and copper. **Velvet** was said located May 23, 1886 by C.H. Montgomery, Silas Johns and William B. Moore, Colville business men.

Eureka, the south extension was located June 6, 1886 by L.G. Barton, Billy Cable, Dick Chilson and big John Hanley (from the Old Dominion). **Double Standard,** also listed amongst these early time claims of importance (but have faded from existence) and was just north of the **Great Republic.** There at one time was several claims held by assessment and assays were in copper, lead and zinc. Stile's **Royal Gold** was later known as **Hubbard.** There were five claims listed at the time as owned by Hubbard Mining Corporation, with values in zinc, lead, silver, gold and copper. One carload had been shipped but returns were not available. The shafts were full of water by 1943.

RED TOP; ANACONDA, and COPPER KING MINES

Foster and **Iron Horse** claims came about around 1892, with what was called the **"Red Top"** name taking over later. By 1897 there was some information that George Foster and W.C. Taylor had located the earlier two claims and by that year were reporting ore assaying at $78 in silver and lead.

Copper King was said located by Henry and John Little in 1896, and included **Copper King No. 1** and **No. 2.** A news release of March 19, 1897 said that **Anaconda,** owned by Julius Pohle, John Lenders, Henry Lenders Sr., and Henry Lenders Jr., all of Colville, was bonded last week to British Columbia Gold Fields Company Ltd., with work to begin immediately. The two claims adjoined, one mile east of Cedar Lake and at 4,000 elevation on Red Top Mountain.

On Sept. 3, 1897 an Administrator Sale notice was filed by A.J. Chindgren, administrator for the estate of Antone Nelson, deceased. Advertised for sale in a legal notice was – undivided 2/3 interest in and to **Hunters Lode;** a 1/3 interest in **Chancy Lode,** an undivided interest in **Red Top Lode,** all on Red Top Mountain.

By early 1912 and up to 1924 these claims all became the property of Charles and Bill Flugel of Colville. They were credited with development of the claims into a producing mine although shipments were slow. In the 1920 geological report it stated that in 1902 there had been 30 tons shipped. Small shipments were reported in 1938; 1944, and again in 1953. About 200 tons were said shipped prior to 1939 and contained values in lead, silver and zinc.

Milton Flugel, of Colville, tells us today that his great-grandfather, Carl William Flugel, located these mines and that man's sons, Charles and Bill Flugel inherited them. Charles was married twice and his son Carl was Milton's father. "Milt" recalls Grandpa Charles and Grandma Golda living at the mine and that his aunt "Toots" lived there also. As a small child playing outside of the home near the mine, she saw what she called a "big kitty". When telling her parents, they were horrified that a mountain lion would approach so close to their child and home.

Great-grandfather Carl Flugel had come from Berlin, to Pennsylvania, where Charles and Bill were born. Milt tells that his grandfather, Charles came to Stevens County when he was seven years old. The only children he had to play with were those of the earliest citizens, all Indian children. He said that Charles later had a livery stable and tannery in Colville.

Milt recalls the families of Charles and Bill Flugel retiring from the mines with something like $30 to $40 thousand each (a goodly sum in early 1930). The money went into farmland but unfortunately with bank failures, over spending and "high living", it was soon gone.

The smelter was down at Northport and most shipments in these earlier days had to be made out of Boundary. This town on the railroad was at times a wild center of activity with population running to 1,200. There were dance halls, saloons, gambling houses and "resorts of immorality." Early development of mines in the area was very slow but it is easy to understand why many early gains from those mines went into wine, women and song.

In July, 1923 new rich mineral strikes were reported from **Red Top,** and ore was being shipped. There was other activity on the mountain – at **Hazel,** or **Lakeview,** H.V. Rieper was finding good silver showings in his six claims; Fred Scamen and brothers from Wenatchee had shipped some 65 tons of ore from their **Scamen,** and **A.J. Jackson Mine** reported receipt of $221.99 a ton.

By 1943 the Flugel claims were reported owned by Red Top Mining Company with J. Richard Brown of Spokane, president. There were said four patented claims; three by assessment work, and 400 acres of deeded ground. The property had 4,000 feet of tunnels, crosscuts and raises, all in good shape and were worked by three men. A good road was also said for access to the mine. **Copper King** and **Anaconda** were included in the transaction but the **Anaconda** was said to have only a 30-foot tunnel with 40 foot shaft both caved in when inspected prior to 1943.

A terrible fire had occurred on the mountain and endangered the **Red Top,** and superintendent of the mine John Colby, of Leadpoint had literally "lost his shirt." His life was saved and on Feb. 21, 1944 he was reported in town buying a new wardrobe and household items to replace his losses. Back into production, the mine was reported making an ore shipment to Fruitland (Springdale district) by Dec. 5, 1944.

From 1952 to 1954 the Pacific Northwest Mining Company, of Bremerton was listed as owner of the five patented and three unpatented claims. In 1958 ore had been opened up by A.C. Nielsen who in the process was killed by a premature dynamite explosion. Earl Gibbs was the last to operate **Red Top** and new ore shoots had been found. He shipped the ore to his mill near Colville and no report was released as to output.

By 1962 the Red Top Company leased the mine

to Rare Metals Corporation of Salt Lake, with $150,000 purchase option. This was merged into the parent firm in August, 1962, El Paso Natural Gas Company. In November, 1962 Bunker Hill Company was assigned a 50 percent interest in that lease with exploration going on at **Iroquois** and and **Red Top Mine** at that time.

Milt Flugel also had knowledge of one of the better silver mines on the Red Top Mountain. He could not remember the name of the mine but said solid silver was skidded down the mountain-side on green cowhides to keep it from splintering off and portions lost.

LUCILLE, Sometimes Called OWEN

Lucille adjoins **Copper King** and **Anaconda** and was operating from 1910 to the late 1940's. It had six claims owned by E.C. Owen, of Spokane. It was so situated that it was known as "Owen's Camp". A report of 1924 said there were piles of mined ore containing good amounts of gray copper with high silver content. The ore occurs in irregular shaped chimneys.

In 1942 **Lucille** was reported to be owned and operated by Mrs. E.C. Owen, of Boundary. 50 tons of ore had been shipped in 1926. In 1948 an ore body was estimated to be eight to 25 feet wide and contain 11,000 tons of ore. 160 tons were shipped that year, and another 44 tons in 1949.

From 1952 to 1972, a twenty-year lease was active by the owner Pacific Northwest Mining Co., from Consolidated Speculator Corporation, Spokane. Values are in zinc, lead, silver and cadmium with zinc of high quality. There is a compressor house, cabin, and new ore bins.

IROQUOIS (FLANNIGAN OR COLUMBIA), ADVANCE and SCANDIA

Iroquois Mine is eight miles southeast of Boundary and some four miles by road to Leadpoint. The three mines are in a near vicinity and most opened up in the early 1900's. **Iroquois** was once owned by Dr. Roy Wells, the one doctor in the Northport area to care for the population then consisting of the smelter workers, and the same doctor who delivered most of the babies born during that time. He was a physician in Colville in the 1930's and 1940's.

The three claims plus 120 acres under option was said later owned by Magnus Wulff, the owner of the one grand hotel in Northport. His mine was under lease to a Mr. Stewart. In 1920 there was said some ore shipped with values in galena and lead, most coming from a large "glory hole" in the mine. By 1943 there were said to be three patented claims named **Columbia, Prosperity**, and **Progress.** Production was reported in zinc, lead, silver and gold and some shipments were made from

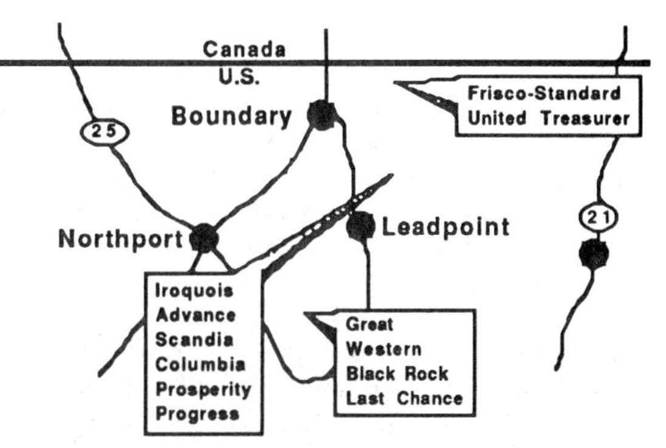

1917 through 1928, with 2,898 tons shipped in 1950.

Mines Management, Inc. of Spokane was owner of **Iroquois** and it's 17 unpatented claims and 40 acres of deeded ground; and of **Advance** and it's 28 unpatented claims and 840 acres of deeded ground, by the mid 1950's. The owner reported 640,000 tons indicated ore in 1950 showing 3.02 percent zinc at Iroquois, and 50,000 tons showing in 1952 at **Advance.** There has been a great deal of development at each mine.

The **Advance Claim** had been filed by George McCorquodale on April 18, 1941. By 1964 these mines had all shown their values and **Scandia** was under option to Grandview Mines, Inc., of Spokane. In that same year, Bunker Hill Company took leases for exploration on the three mines **Iroquois, Advance,** and **Scandia.**

In **"Tales of the Pioneers",** Mrs. Joe Garvey, Sr. tells of those early days near Boundary when she and her family took up homestead in 1908. Her remarks are more directed to timber claims of which there were many but she did recall the activity of the mines in the area; the **Electric Point, Lucille, Red Top, Frisco Standard, and Iroquois,** all who shipped ore by one means or another into Boundary to be taken out by railroad to the smelters. Sometime in their later development she remembered fifteen horse-drawn outfits hauling ore to the train, and those men losing out when three trucks took over the job. She also recalls; "One night my husband was called to haul some mining equipment to Northport for the **Iroquois** Mine. The next day we heard the remaining equipment was repossessed." – (and such tough times were part and parcel of early mining!)

Grandview Mines, Inc. who in August 1967 owned property interest in 8,000 acres of this region and into the Metaline district, contracted with Cominco American, Inc. to explore 1,080 acres adjoining **Iroquois.** Cominco was to search for lead and zinc and after five years – if they desired to exercise their option, they were to pay $540,000 for the ground. **Scandia,** and **Just Time** were to come into some of this activity.

FRISCO-STANDARD, JUST TIME, or STAR

In Richard Steele's biography of Colonel Daniel

The old mine bunker at the ADVANCE MINE seven miles southeast of Northport. Photo June 23, 1977 (S-E).

J. Zent, Zent was said to have come to Colville in 1898. Zent's company was credited with having located the Jefferson Marble Company. His company was also said to have spent $50,000 in developing various mining properties. Because he was said to have located **Frisco-Standard**, it is possible some of that money went into developing that prospect.

Because so many prospects are given up after a brief time it is difficult to be positive regarding locator, and date. In 1910 John Keough is also credited with this find, along with a total of some 20 claims. By 1914 **Frisco-Standard** was reported to belong to Buffalo Syndicate. A total of three small shipments were made, one in 1905, in 1908, and in 1926. No report of the returns were available. (Keough will be remembered at **Dead Medicine**, and later at **B.C. Mine.**)

Frisco-Standard

The mine is in the extreme northeast corner of Stevens County, just a quarter mile from the Canadian border. In 1924 it was listed in the state geology report as having seven patented claims owned by George Harrington. Lead, silver and copper ore had been shipped the 12 miles by wagon road to Boundary. In 1943 Colville Attorney John Raftis (resident born in Chewelah), was secretary of the Northport Mining and Development Company, Inc. That company was operating the mine with three men working steady and four part time. Several buildings including a cookhouse, cabin, and a mill were soon to be listed as holdings of **Frisco-Standard.** Assays were then said to be in silver, lead, copper, zinc and gold.

A 1965 report said that intermittent exploration had been carried out in recent years, by several different lessees, with most values in lead, silver and copper. On July 8, 1966 a news report stated the mine was still owned by Northport Development Company, with headquarters in Olympia. A Spokane mining engineer Cline E. Tedrow said he laid out a plan to trace downhill, a vein showing in the upper tunnels. The property has been under exploration since, with drilling anticipated.

(One assurance, in years of plenty the local pickers can tell you this is a great mountain to look for Huckleberries!)

United Treasure

United Treasure was a late bloomer on the mountain just a mile away from **Frisco-Standard.** Best value are in silver and by 1917 some 62 tons had been shipped. In 1924, Newton Hartman was said to own three claims and had produced several thousand dollars from assays running high in silver; in copper, lead and in gold. In one shipment of two tons of ore, 117 ounces of silver to the ton was reported. In 1943 four claims held by assessment work was owned by H.P. Howard, of Northport. Most of the early shipments were made on horseback, to Boundary.

Still making shipments in 1933, and up to 1953, **United Treasure** is owned by Singlejack Silver Mining Exploration Company, with William and Richard Weaver, of Spokane, listed. No returns were available.

Great Western

A Northport butcher George Thomas staked **Great Western** on May 14, 1886. W.W. Lane's name was added. The lead and zinc property is located six miles from Northport, one and one half miles off the Deep Creek road, near **Black Rock**, and **Last Chance** Mines. The Colville Republic reported on Sept. 30, 1892 that George Thomas and Tom McAuley owned the mine. The report said a tunnel was being driven to connect the main shaft, and the silver bearing galena was assaying 50 to 60 ounces of silver to the ton.

Reports of shipments include $20,000 in zinc ore shipped to Pennsylvania in 1906, and a like amount shipped in 1915, and 1916. 1956 Inventory of mines confirms that figure of $40,000 but not all shipments and returns were reported.

Sidney Norman ran the mine during World War I, and made regular shipments to Mineral Point, Wisc. From 1924 to 1931 Carl Sauvola, from Deep Creek, leased and operated the mine. He made shipments east, and to Bunker Hill and estimated total output to be around $100,000. He said the depression closed the mine in 1931.

Carl Sauvola is an outstanding citizen and is also a man of good humor. Some years back he told this writer about the time when he and son, Sam were mining the **Great Western.** A promoter whom Sauvola called Lightner came up to the mine in late 1925-early 1926 and taking these miners as "yokels" (in Carl's words) tried to talk them into a promotion deal. He stayed around for several days and found that Sauvola could not be easily duped. Knowing there would be moonshine somewhere in the district but being afraid to drive down the steepest road in the country – narrow and built out of rock, he asked Carl to go. At the time, Carl was busy cooking supper so being the sly one, he sent Sam to drive Lightner. When they came back Lightner was so shook he couldn't drink and said, "that kid drove like he'd never driven a car before." – "He hadn't" was Carl's only answer.

Sauvola did say that Lightner got Wilson to mine the **Last Chance.** Wilson was a road contractor from Walla Walla.

In 1943, **Great Western** was owned by L.J. Magney. One carload shipment was assayed at 16.4 percent lead and 28.2 percent zinc. In 1953 the mine was owned by Sylvia K. Lenma, Long Beach, Calif.

Last Chance

Since 1948, the Last Chance Consolidated Mines, Inc. has been listed as owner of **Last**

Chance. The company consists of three patented claims and a millsite. There was reported to be up to $600,000 in ore produced to 1937 with small shipments up to 1949.

Christian C. Knutson was credited with locating **Last Chance.** One production report gives him credit for shipping 30 tons per day at $20 a ton and does mention 2,000 tons. Knutson was listed as General Manager. The mine was owned by Jupitor Lead Company by 1914 and ore was shipped to Pitcher Lead Company, in Missouri. It had a tramway and bunkers located just above the level of the valley at 2,250 feet. The mine was at 2,700 feet. This mine was also shut down due to low metal price; and hard times in the industry.

ANDERSON-CALHOUN, DEEP CREEK, SIERRA ZINC, BLACK ROCK, NEW ENGLAND, BLUE RIDGE ADMIRAL, SCANDIA

All of the listed mines here have a place in the eventual **Calhoun** that became one of the state's greatest zinc producers. Many of these were little neighborhood mines, located, developed and supported by the men of the Deep Creek region. The 1956 Inventory of mines lists **Gorien Zinc** with the explanation that in 1919 John Gorien was treasurer of the Northport Mining Co., and this company owned or controlled 280 acres of mineral land on both sides of Deep Creek including **Deep Creek, New England, Black Rock,** and probably including others also.

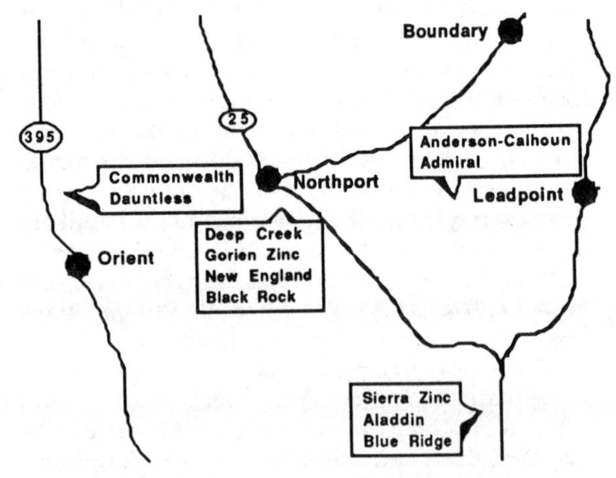

Frank W.A. Uterhardt came to Stevens County in late 1897. There is no explanation but he is said to have had charge of the **Deep Creek Gold and Copper Mines** until 1901. He lived east of Orient and was also said to own half interest in **Commonwealth Mines,** and to own **Dauntless,** both in the Pierre Lake district.

On July 30, 1897 the Statesman-Index story told of Miles Merrill, Anton Silver and Adam Goldsong having sold their Deep Creek properties to William Shelton of Rossland B.C., for $1,000 cash. On July 23, 1897 a suit was told in the same paper and said that Susan A. Fleming, versus A.J. Littlejohn et al, had brought suit for possession of plaintiff's one half of property now in corporation as Deep Creek Mining Co., a case of ownership question.

Deep Creek

From 1944 to 1947 the **Deep Creek** property was in the names of Jamieson-Higginbotham. Cy Higginbotham was perhaps the best known mining man in Stevens County and was noted for putting together mining properties and development deals, for the betterment of the owners. He was born at Rice, Wash. on Aug. 5, 1904 and worked at mines in his teens. He later managed and owned mines in Idaho and northeast Washington, and eventually the same with **Sierra Zinc, Anderson, and Deep Creek.** Higginbotham died July 14, 1982 after seeing this combine of mines come together and be the producer he always believed them to be.

A March 1943 Mining World story supplied us by Ron Nixon (owner of the property in 1989) tells that George Foster, of Northport located six claims in 1898 known then as the Foster Group. In 1906, John L. Magney, superintendent then of some mines in the Coeur d'Alene district, visited this property and certain of financing, the Aladdin Mining Company was formed. John and C.A. Magney were first officers. Much of the mining equipment was set up under the direction of John White and Oscar Nordquist from Wallace. A 4,000 foot flume to carry water from Deep Creek, still served in that capacity in 1943. Too much zinc that could not be cared for then caused the final shutdown. The property changed hands several times and by 1910 became known as the **Blue Ridge.**

It was in 1940 that it became the property of E.W. Jamieson (later to become a partner of Cy Higginbotham), and was called the **Sierra Zinc.** (sometimes spelled Jamison).

The **Deep Creek** was 100 yards from the west bank of the creek and was owned by The Goldfield Consolidated Mines of San Francisco, under management of Higginbotham. The mine had produced 350,000 tons between 1944 and 1952 and continued to produce up to 1953. Assays were running 5.7 percent zinc, and 11.4 percent lead. A very large amount of return was told from this mine. Operations were temporarily suspended on Oct. 1, 1965 due to a serious fire in the **Deep Creek Mine.**

Sierra Zinc

Sierra Zinc, sometimes called the **Aladdin,** or the **Blue Ridge** earlier included eight claims and 180 acres of deeded ground. It was on the west side of Deep Creek and some four miles north of Aladdin, or 17 miles south of Northport. In 1924 and earlier, some development work was done by T.R.

The SIERRA ZINC MILL was said one of the earliest in Washington State, and the largest plant treating metallic ores in Stevens County. Machinery had been removed by 1989 and the building was being totally dismantled. (Spokane Daily Chronicle).

Roberts. There was said from only four claims that lead and silver ore was shipped to that date. Amos E. Huseland filed claims for part of the mine on April 19, 1935.

The commercial operation for over two years saw 25 men employed; with zinc, lead, silver and gold produced. Past production was said to be 9,000 tons with present mining putting out 1,500 tons a month. No actual count was given from 1909 to 1944 but a report from 1950 to 1952 said that tons had been shipped. One account tells that $500,000 was earned between 1940 and 1955.

Higginbotham was managing the mines and a mill built in 1942 at **Sierra Zinc.** In 1948 he sold the mill to Goldfield, then in 1960 he bought it back again. By July 24, 1950 Goldfield had ordered **Deep Creek** and **Anderson** Mines unwatered and that company went back into operation. They hauled ore from the **Anderson Mine** to the **Sierra Zinc Mill.** Goldfield did considerable development work at **Anderson** and on March 17, 1952 four of the companies were taking part in securing federal funds for road improvement. It included **Goldfield, Scandia, Pioneer and Mines Management.** By 1960 workings at the **Sierra Zinc** were abandoned and the effort put into **Anderson.** A report in 1955 said that Goldfield netted a million dollars from **Sierra Zinc, Deep Creek** and **Anderson.**

Black Rock

Black Rock, another mine of this eventual combine, was located about 1918 by Gus Maki when he was road boss. He said his horse pulled the road grader and one day upturned a peculiar rock that proved to be very high-grade zinc. The seven claims and seven acres of deeded land became the property of a man named Harris. By 1921 it was sold to John Gorien for $30,000. Much exploration and development was done but around 1928-1929 the property was shut down. Lessors John Makynen, Victor Hagan, Otto Ronka, Ed Maki, and Dan Langlie put the mine into production and considerable ore was again being shipped, principally zinc.

By 1923 a new mill was built. There was a large reserve of zinc ore ready for milling from this mine, located on Gus Maki's place on Deep Creek. Production reported included six carloads to 1920; 5,280 tons between 1922-1924, and $300,000 earned prior to 1930. The mine was still producing to 1943 but no results were available. It was said at that time to be owned by Gorien and Mullen of Spokane. By 1964 **Black Rock** was under option to American Zinc.

New England

The **New England** also had a round-about history. Bill Cook had worked at the **LeRoi,** in Rossland and was known by Al Stiles. Cook is said to have grubstaked Ben Stout and Archie Robinson's prospecting Deep Creek. In 1917, the Statesman news said that platinum had been discovered on the **New England** claim. In 1920, there was said to be five claims that were being developed by two open quarries. It was reported the ore was similar to that at the **Young America Mine** but no more was said about platinum.

Carl Sauvola had close knowledge concerning

this mine and it's operators. Among other things he told us about Archie Robinson's cabin with the dirt floors; but what was more impressive was the beaver stew meals, and the poker parties with Robinson, Cook and Stout.

In 1943, Knob Hill Mining Company from Republic, took an option on these leases then under Gorien's name for Northport Mining Co. They planned to open the old workings and had eight men working. They also planned to install a 300 ton mill. This property eventually became part of the Deep Creek combine.

Anderson-Calhoun

Anderson or **Calhoun** was located in the early 1930's. Jan. 26, 1937, six claims were filed on by Andrew Anderson of Leadpoint; and Nels Anderson of Touchet (no relation), was listed as another owner. The mine, one mile north of Leadpoint was said to assay from 50 tons of ore about 10 percent lead but later acclaim was in zinc. It was said sold to Dr. Mowrey; then T.C. Higginbotham became resident manager for Goldfield at the property.

The mine was closed down in 1952 due to depressed prices on lead and zinc. Two production reports made by Goldfield about **Anderson** was 100,000 tons had been shipped by the end of 1951, and 500,000 tons by the end of 1952 prior to shutdown. By 1954, Goldfield Consolidated began a limited operation running 400 tons a day on a three-shift basis. By 1957 things worsened. The zinc could not be sold and had been produced at the rate of only 800 tons per month. This caused another close-down. In 1959 the company had Boyle Brothers drilling for a new source of ore. They went down 2,000 feet and new indications were reported. Higginbotham was mine superintendent by then for Goldfield.

By late 1960, a multi-million dollar deal was in the making. The first exploratory stage cost $175,000. Current development at the time was expected to cost $500,000. The third stage would be for a mill expected to cost $1,500,000. A tunnel 15 feet wide and 15 feet high was to enable diesel-powered trucks to haul the ore out of the mine to the mill. Workings had been open-pit to that time but would go underground and the company was to employ 60-65 men when operating. By 1961, the mine was again closed due to low prices. President of Goldfield, Willis A. Swan did announce the mine would open again when the price of zinc went up to 13 cents a pound.

American Zinc had taken all the ore that was profitable at their **Grandview** Mine in Metaline Falls and closed that operation by Sept. 5, 1964. In the meantime, they had taken a one year lease on **Anderson** from Goldfield Consolidated. By July 11, 1964, American Zinc Lead and Smelting Company had purchased **Anderson** for $750,000. Zinc had gone to 14 1/2 cents per pound. Howard I. Young, President of American Zinc, announced that the company would build a large mill. By 1965-1966 the company had built a 1,200-ton flotation mill for zinc, lead and silver. When first purchasing the mine, they had to drain a large lake to even start operation; which actually began Jan. 31, 1965.

In 1965 a news report stated that American Zinc and Smelting Company had negotiated a $21 million loan indicating that some of that money was to be used for what was then known as the **Calhoun Mine,** named for R.E. Calhoun, manager of American Zinc for many years. His son W.M. Calhoun was named Washington Manager of American Zinc and came into charge of the Anderson-Calhoun operation. Work had to be completed by 1966.

Net property purchases did increase the worth, and obligation by $516,000. By Dec. 16 a mill costing about $2 million (according to Wafford Conrad, Spokane Chronicle Mining Editor); all kinds of shops and warehouses were completed, and 40,000 tons of zinc-lead ore containing some cadmium and silver, had been stockpiled. Ore reserves were sufficient for many years to come according to W.M. Calhoun. By Oct. 27, 1966 operations were so well automated that only 65 men were needed. The entire ore crushing circuit was controlled by one man and the concentrator only required two operators per shift.

Admiral

Along with other local holdings, American Zinc took on **Admiral.** Owned by Admiral Consolidated Mining Company of Spokane, the mine produced zinc, lead, silver and cadmium. In 1947 it had produced 1,441 tons, and to 1955 that figure went to 308,750 tons mostly zinc. J. Richard Brown of Spokane, and others were listed as owning the property April 1952.

In 1947 **Admiral** had a flotation mill operating and had shipped carloads to Kellogg. John Colby and Ben Melby rebuilt the mill for better production. On May 15, 1965 American Zinc signed a lease agreement with Admiral Consolidated, of Spokane, with Karl W. Jasper as president. Exploration and development was to be done at the property across the valley from **Calhoun.** American Zinc was to mine the ore for 60% profits. At the Admirals stock-meeting of May 3, 1965 they showed assets and properties valued at $85,033. Stockholders made the capital stock non-assessable at that meeting and it's corporated life perpetual. Jasper was re-elected president and director. Admiral leased it's 500 acres of mineral rights six miles east of Fruitland to James V. and John C. LeBret of Spokane.

Things were looking very good "up Deep Creek way" by Oct. 17, 1967 when the company reported that **Calhoun Mill** produced 16,795 tons of concentrate during the fiscal year ended in June. The products were trucked nine miles to Boundary, the zinc concentrate shipped to Bunker Hill and the

ADMIRAL MILL. This is the 50-ton mill of Admiral Consolidated Mining Company, one of Stevens county's newer producers. Regular shipments were being made from the property near Leadpont to the Trail,B.C., smelter.

lead concentrate to East Helena, Mont. By Dec. 1, 1967, **Calhoun** was employing 75 men and had an average payroll of $650,000 besides spending $900,000 for raw material they bought locally such as gas, lubricants, tires, etc. etc.

The parent company Goldfield Limited of London was reported owning 60 percent of American Zinc and Smelting. The latter company was said plagued with losses which were reported occurring since Dec. 31, 1966. On Oct. 11, 1968 the company felt a great loss at the death of Ralph E. Calhoun (namesake of the **Calhoun Mine),** who died of a heart attack at the mining convention in Las Vegas.

Just prior to that, and on Sept. 25, 1968, the **Calhoun Mine** shut down. It was said the property would be placed on a standby basis and mine development would resume when conditions were more favorable. The $2 million operation that was so automated and had been producing 1,200 tons of ore daily since Oct. 1, 1966 was closed down. On Jan. 24, 1969 William M. Calhoun went with Day Mines as assistant manager.

Some time earlier the company had cut the length of it's name to only American Zinc. On Jan. 18, 1971 this company was discussing the possible sale of some, or all of it's assets. The losses in 1970 were to be $2.4 million and had been $1.9 the fiscal year before.

Produced up to 1971 were 875,000 tons of ore; 3.18 percent zinc, and 0.10 percent lead. **Calhoun Mines and Mill** was purchased by Washington Resources Inc., of Spokane from American Zinc that year. Cominco American was later said to be carrying out a drilling program at **Calhoun** and also exploration work at **Deep Creek** and **Iroquois.**

In this ever changing search for ore, and profit, Scandia Mine **came into this combine.** Six hundred feet above Deep Creek and some six miles from Northport, the mine in 1943 was owned by John, Donald, Theodore, and Effie Nasburg; and Raleigh Hellenius, of Spokane. Assaying in zinc, lead and silver, the mine shipped one carload in 1950.

On Aug. 21, 1974, Grandview and Admiral Consolidated leased their Stevens County holdings to Cominco American. The 20- year leases and mineral rights included the **Scandia Mine,** and the **Hartbauer** lease adjoining **Calhoun. Admiral** was said to have shipped 14 carloads amounting to 50-60 tons each of crude zinc ore averaging 43 percent zinc, to the United States Smelting Company of Salt Lake, many years prior. 765 tons of zinc concentrate was said produced at the property.

The future will depend on the finding of new ore bodies in this section of Stevens County, and maintaining a price that makes operation of the facilities feasible.

Goldfield and Yellowstone Park

A bit of a detour – but an interesting news article of Aug. 9, 1966 informs us that the parent company, of a mining company we have just read about, the Goldfield Corporation has bought the concessions at Yellowstone National Park, Wy., from the Yellowstone Park Company. That company had operated the concessions at the park for over 70 years. At the same time, The Yellowstone Company was granted a new 30-year franchise.

The Yellowstone firm owns and operates hotels, lodges and cabins that can accommodate about 9,000 persons, 14 restaurants and four cocktail lounges. It also operates all park transportation facilities, fishing and sightseeing boats, gift shops, ice machines and public baths. It has a half interest in park gasoline stations. The contract calls for the Yellowstone company to spend $20 million over the next 20 years for new lodging. (Those should have been complete by 1986 – well before the terrible fires in the park the summer of 1988.)

Erwin A. Bauer writes in the February, 1989 issue of **American West** that financial loss (to the investor as well as to the employee) had not been too severe because lodging within the Yellowstone housed fire fighters on rest and relaxation after all tourists were shut out. The terrible fire burned seventeen buildings and Bauer said 40 percent to 75 percent of the park's area was consumed. He relates that 1988 was the driest season in the park for it's 116-year history and that since President Ulysses S. Grant opened the park in 1872, it was the first complete closure to visitors.

Van Stone Mine

One of the "great" ones, the **Van Stone Mine** on the upper end of Onion Creek, near Northport, became the state's second (if not largest) lead-zinc mine. It's life spread over some 55 years, up to and through 1975.

George Van Stone and Harry Mailor (sometimes spelled Maylor) prospected the hills on a regular basis. Van Stone claimed the Roger's Mountain area should be a second **Bunker Hill** for lead and zinc output but "no one paid any heed." This was said about the mine as written by Mrs. (Ivy) Gus Anderson in her "History of Onion Creek."

Most histories tell that the claims were located in 1920 by Van Stone while he was hunting deer — probably true because prospecting and hunting were both a natural part of his life. (John Colby has also been credited with the location of these claims but there is no verification of that.) History holds tough to the Van Stone story and the huge mine was named for him. There were 14 and later 16 claims included in the vast operation which Van Stone kept until 1926. He died in Spokane at 88 years of age in 1956, having lived long enough to know the property was the great mine he had believed it would be.

Hecla owned the property in 1926, and Van Stone Mining Company took over in 1930. Willow Creek Mines, of Nevada operated the mine from 1939 to 1950. It has been stated that Company netted over a million dollars a year. In 1950, American Smelting and Refining of Salt Lake City, bought the claims and the surrounding land. Some 1,200 adjoining acres were purchased from Ernest Lotz, and Louis Menegas of Northport.

Good production reports were made for 1930, 1937, 1942, 1952 to 1955. Some 20 men were said employed and there were camp buildings and a 1,000-ton mill. Willow Creek had done some diamond drilling and drove two short adits through ore bodies. By 1945 diamond drilling was going on a 24 hour basis – done by U.S. Bureau of Mines. A lead-zinc area of about 3,000 feet in length was being exposed.

When AS&R took over, Isabel Construction Company of Reno, was doing the open-pit mining with 35 men employed. AS&R had 25 men working at the mill. By Nov. 14, 1952, the company was building up production at it's two million dollar project. P.A. Lewis was general superintendent. Ore was trucked to Marble, from there the zinc was shipped to Anaconda Copper Company's Black Eagle, Montana plant. Lead was sent to AS&R's East Helena, Mont. smelter.

The company was reported to have built a 1,000-ton concentrator near Northport in 1952. Plans had been for a 1,500-ton flotation mill in 1951. The company had a Defense Material Procurement Agency order to fill that had guaranteed 15 1/2 cents per pound of zinc, after which having filled that, the mine was forced to close down in July 1955 due to low price. It knocked 65 men out of work. They did re-open briefly in the summer of 1957 but then had to close until 1964. The company did report production of 9,864 tons of zinc with a small amount of lead.

Nolan Probst had been superintendent and was replaced by Walter Barlow when Probst moved to Colville. General Manager Norman Visnes, from Wallace, did say the operation was put on standby at that time and "if price is right we will go." The low-grade ore body was spread over the main mining area "big enough to hold at least three football fields end to end", according to Wafford Conrad, Spokane Chronicle Mining Editor. He did say the pit was 280 feet deep at it's uphill end.

Openings and closing became difficult to keep up with but in 1965 **Van Stone** yield was 130,000 tons of zinc and some small amounts of lead with earnings of $52,456,000. That was said to be "top year." In 1966 production of concentrates averaged 820 tons of zinc a month. By May of 1967 all shut down again but reopened awhile in 1969 with Al Kingman, mine manager. On March 16, 1970, a new ore strike was reported – but by winter of 1970-1971 AS&R closed the open-pit operation. It was a hard blow to many who had been employed.

In the winter of 1970, Simplot Company of Idaho was named as a possible purchaser of the mine to

The VAN STONE crusher and mill in 1953. (Courtesy Herb Buffan).

use the waste material for their fertilizer manufacturing. At the time they had a large Far East contract to supply. Nothing came of this.

On June 18, 1971, Callahan Mining Corporation of New York acquired the mill and mine property for $500,000. This included mineral rights on 1,224 acres. The news release also said the Company was looking at **Calhoun.** New mining problems arose when on March 13, 1975 Callahan's application to discharge water into Onion Creek at the mine, became public news. A hearing regarding this situation included wanting to pump the inactive mine of groundwater to allow under-ground exploration. The company proposed to pass the water through a holding pond before allowing the overflow to pass into Onion Creek, which flows into the Columbia River.

By April of 1975, the news told of American Smelting and Refining Company, or AS&R becoming officially known as **Asarco**. Herb Buffan says **Asarco** took out over nine million tons of ore between 1950 and 1970. Other news told that fall was of Callahan financing 49 percent of underground drifting and diamond drilling at Asarco's former mine. Callahan's partners were U.S. Borax and Chemical Corporation, and British Newfoundland Exploration Ltd. Hope is always strong for the future of this mine, but it was being offered for sale by Callahan in 1989.

A brief September 1985 news note tells about Herb Buffan who had worked for Goldfield under Higgenbotham until that operation closed. He then went to work for AS&R as a contract miner, becoming powder foreman, and then pit foreman. The Buffans had moved to the **Van Stone Mine** in 1970. When Callahan took over that operation Buffan went on as general foreman. The article states that in the fifteen years since that time, underground and experimental work has gone on. At this time, former District Supervisor Buffan says Callahan has developed another nine million tons. Operation again is dependent upon prices.

Frank Paparich Jr., now of Rossland, B.C. told of working 19 years with Van Stone.

Clayton Mining and Fertilizer Co.– Clayloon

Mining, like many other industries is sometimes operated on a financial string and once in a while bad luck steps in to end all the hopes. This is the case of the **Clayloon** operation that came about in 1955 when the company organized to mine some uranium claims.

Back in 1924 the **Lead Trust,** the **Lead King,** and the **Keystone** were operating on Gladstone Mountain. The **Lead King** was a group of three claims with the "Father of Metaline Falls", Lewis P. Larsen as president of the company. Jens Jensen, the other important developer of Metaline Falls, was secretary for the **Lead King** Company. Larsen was also president of Flusey Lead Company.

In 1918, there had been three or four carloads of lead-zinc ore shipped from the claims and by 1941

At VAN STONE MINE April 15, 1966. Steven's County's largest lead-zinc mine, reopened under Equinox Resources August, 1990, closed down again 1991. A 1992 report says prices will influence opening.

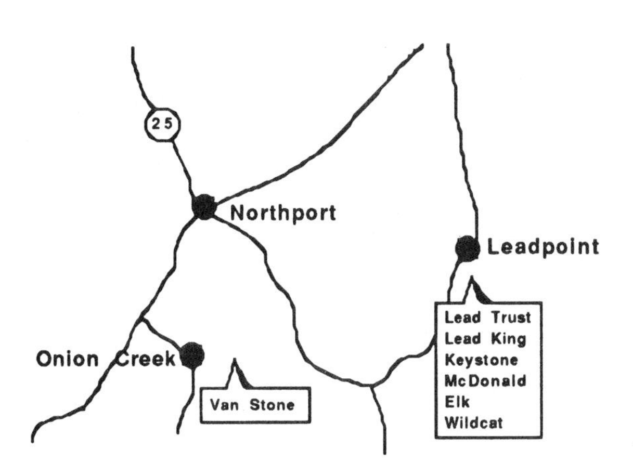

they were owned by Elmer Godfrey. William McCue was also involved in some of these claims.

The **Lead Trust** was operating from 1910 and by 1924 was reported to have four or five claims, that property being under the old aerial tram of the Electric Point Mine. It was leased by Andrew Anderson. Some 30,000 tons had been shipped prior to 1924. By 1951 the claims were said owned by Ray Cater of Marlin, Wash., and was being leased by Felix J. Cardinal of Spokane. In 1955 Goldfield was said to have purchased **Lead Trust** from A.C. Neiman of Northport.

The **Keystone** was four miles east of Leadpoint and had five unpatented claims owned by C.R. McDonald of Northport. Said to be iron, lead and silver, a few tons of high-grade lead had been shipped to Portland Cement for use in that plant from 1936 to 1940.

By 1959, Clayloon Company had acquired **Lead Trust, Lead King** and later **Keystone** by lease or purchase which was said to give Clayloon 54 1/2 claims on Gladstone Mountain. Beryl T. Goodwin became president and V. Eugene Goodwin became executive vice-president, both from Spokane. Also of Spokane was Arnold W. Barns named vice-president and Patrick H. Shelleday named secretary-treasurer. It looked like there was going to be a miner revival of the old mining town of Leadpoint, once the center of dozens of thriving mines. They had all been closed by the sinking price of lead and zinc, the major metals in the district.

In June, 1960, this company more frequently termed Clayloon Uranium Company, installed a 500-ton concentrating mill at Leadpoint, the ore was to come from their claims. The company was said to have made $40,000 in improvements and to have invested $300,000 by 1965.

In August of 1960, a test run of the new mill was made on the first of 60,000 tons of lead-ore that had been stock-piled. The mine and mill was said to employ 50 at the time. Their major claim was that of being the only mine in the west which has achieved 100 percent utilization of their ore. They were milling some 150 tons a day and the dolomite was being used for their blending plant near Mead after the lead concentrate was removed. A later report tells that they had bought the Saco Abrasives plant at Mead to process low grade uranium – too low to be accepted at the Ford Uranium plant. Hopes were to increase the output to 1,000 tons per day at the mill.

A news story in 1963, tells that the name **Clayloon** had come from Clayton-Loon Lake, the homes of the Goodwin brothers. It went on to tell about their blending, or fertilizer plant at Mead

CLAYLOON Mill was active in 1964 operating from several old claims on Gladstone Mountain. (Photo August 1964 Spokane Chronicle).

having sold fertilizer to a golf course in San Francisco who wanted tons more of it.

In 1962 they had to shut down the **Clayloon** and mill because lead was down to 9 1/2 cents a pound. Mining commenced later and 17,000 tons was stockpiled. In 1964 they resumed operation on a one shift a day basis. A great deal of work had gone on at the plant in the way of housing in the mil with aluminum siding to make better working conditions all year around.

The purchase of **Keystone** or **McDonald Group,** was made in November, 1965. The claims along with **Elk** and **Wildcat** had been located by oldtimers Raul and Jim Hodson. From the 10 claims there was said to have been a few tons shipped in earlier days showing 72-80 percent lead and two ounces of iron (some reports gave this as two ounces of silver).

Earlier in the year of 1965, the company had reported a net income of $19,000 on it's first full year of operation, that being in 1964. 20,000 tons were drilled and blasted and two carloads of lead concentrate were recovered in milling operations and shipped to Bunker Hill. 8,000 tons of dolomite had been sold on the Pacific coast for use as soil conditioner on ground ruined by too much rain. The rest was used at the Mead plant with products being distributed over five states.

In 1967, the blow was reported. A judgment of more than a quarter of a million dollars against the Clayloon Mining and Fertilizer Company was signed by Superior Court Judge John J. Lally. The property in Stevens County was to be sold at sheriff's sale Sept. 1, according to then Sheriff Albert "Dutch" Holter.

Clayloon was a defendant in a court action brought by the Continental Investment Corporation on the contention that four mortgages were negotiated between the two and had not been paid off. Continental was allowed $224,401 as principal of the debt and $31,790 as interest through July 1, 1967. Additional interest after that date was also allowed.

The Statesman-Examiner news story went on to say that George P. Davis, another defendant in the investment firm's suit, filed a cross complaint and was granted a $22,347 judgement for the principal of another Clayloon obligation plus $1,781 in interest to July 1.

Continental was given the "first and paramount lien" upon Clayloon property in the judicial decision. Sale of the property was authorized. Continental, Davis and the U.S. Borax and Chemical Company were allowed to bid and purchase the Clayloon property. No price was quoted.

Slow settlement for smelter fumes damage

The Northport district was certainly the most prolific and we have only covered a few of the more major mines. There was other activity going on – some not so good. The bad part of mining involved smelter fumes and in February, 1934, the U.S. Government was telling the Canadian government they were stalling in $350,000 payments. Those had been awarded for damages to Stevens County farmers over two years earlier – no payments had yet been made. That damage was from the huge Trail smelter. In April of 1938, Northern Stevens County property owners were awarded $78,000 in damages in addition to the $350,000. Those payments were eventually made with some of the duty falling to Judge Lon W. Johnson of Colville as official dispenser.

The importance of the Northport district explained

The good news was that from 190 mines in nine districts, the Northport district was the most important, having provided 72 percent of Stevens County's zinc and 63 percent of the lead but with only five percent of the silver as told in a 1971 report.

In December 1962, a geology team from Bunker Hill Company was in the area and stayed for a year and a half doing exploration work at the **Bonanza, Advance, Clugston Creek, Old Dominion** and on **Red Top** in the Northport district. The team consisted of Al Nugent and Elmo Thomas geologists, and Ray Horsman project engineer who were also the three men who provided us with so very much material used in this history. Nugent speaking at the Colville Chamber of Commerce at that time did say that "we had in early days many **Bonanza** type mines. These included the **Electric Point, United Copper,** and **Old Dominion.** All had large ore bodies. Then we had the large low-grade bodies such as **Anderson, Deep Creek, Bonanza** and **Van Stone."** He went on about other mines in the area; "Today good work is going on at the **Delles** and **Sullivan, Triton,** and more on **Electric Point.** (About to re-open at that date.)

Peoples of Northport

There were several residents of the Northport area that were involved in all the mining and the growth of that district. Peter Janni was certainly one. Born in Grimaldi, Italy in 1874 he joined his merchant father in the Butte, Montana area. Coming to Northport in 1893 he worked on the Great Northern rail line to Nelson, B.C. From 1904 until 1918 he was interpreter and acting inspector for the U.S. Immigration service at Northport. Aside from learning the English language, he is said to have later mastered Spanish, French and Hindu.

A hard worker all of his life, Janni had earlier worked at the **Morning Mine** in the silver-belt of Idaho. When he retired in Northport from being U.S. inspector, he took up road contracting into various reaches of the Northport district including

Peter Janni's lime quarry at Northport, and some of his early day miners. (Northport book).

to the **Electric Point Mine.** It has been said by Dave Magruder of the Spokesman-Review, that Peter Janni purchased 63,000 shares of **Electric Point** at 32 cents a share and sold for one dollar share. He was said also to have lost $20,000 on another mining deal. His major work at Northport was in his own limestone quarry, which he purchased in 1923.

That huge operation employed 70 men turning out 600 tons of limestone for the Northport Smelter – all hand work. In 1968 the quarry turned out 20,000 tons a year with 10 employees and the operation was still much hand labor. The Peter Janni and Sons smelter is estimated to have 20 million tons of limestone remaining and at that rate of production would take 1,000 years to exhaust the supply. Janni's son-in-law, Alex Tyllia, has managed the quarry since 1941.

Mr. Janni could tell of the days when there were 21 saloons for a population of some 2,000. He said "most of the men worked at the mines and the smelter and that most being single, there were some 300 women of professionally easy virtues who worked their trade from a string of brothels that lined the street.

Janni saw his product used at the Seattle World's Fair in 1962 when over 1,000 tons of the white rock and some colored marble chips were used at the base of the Fair fountain to provide reflection; used as surfacing and other decorative landscaping, and on the walls of various buildings including the federal science group.

Having given a great deal to his community, Peter Janni died Oct. 24, 1969. He left two sons and two daughters to remind us of the Janni achievements.

On Nov. 20, 1970, Northport said good-bye to another great community "spirit" Ben Hofer. He moved to Northport in 1900. After 54 years in the community he has been credited with connections at the Kendrick Mercantile since 1916. He eventually became owner of that long-time establishment that used to employ 15 in the boom period of Northport. Hofer was connected with numerous mining operations, including Bechtol and others; sometimes as owner and sometimes as grubstaker. Some of them have been mentioned in this section of our history. Another old-timer that was sorely missed was Magnus Wulff who died March 27, 1975. He had come to Northport via railroad construction and became the "hotel" man of the time. His New Zealand Hotel, built in 1914 was recognized as the gathering place for the monied and genteel of the time. Besides other mining interests, he at one time owned the **Iroquois.**

Mining man and merchant, Ben Hofer was ably assisted by his wife. (Northport book).

Later Developments in the Northport District

On Oct. 12, 1972, Wafford Conrad wrote in the Spokane Chronicle that the zinc-lead industry in Stevens County was being revived by newcomer Scott O. Simenstad, and his Coronado Development Corporation. As president of the firm, Simenstad had entered into purchase agreements, or taken options on the old **Sierra Zinc** mill and seven mining properties. Coronado crews renovated electric motors and did other repairs up to $3,000 at Sierra. Simenstad hoped at the time to mill test runs from the **Schumaker** some 10 miles from the mill, which Coronado was leasing with an option to buy from Tri-Nite Mining Company, of Colville and Spokane. He said the company expected to employ 30-40 men in mining and milling operations.

Richard J. Butters was to be mill superintendent. He had earlier been in that capacity for Asarco at Wallace, then as superintendent at **Van Stone.** Simenstad was also a former employee of Asarco having worked in the **Morning** and the **Jack Waite** mines in the Coeur d'Alene's; and in other places around the world.

Going to work repairing equipment etc. at **Schumaker Mine,** N.E. of Colville, Simenstad said the mine was then operational and he estimated 169,000 tons of ore was blocked out for mining. The old **Sierra Zinc Mine** was held under purchase option. Having yielded over half a million from 1940 to 1955, Simenstad said the audit there was being extended into the **Sierra Zinc Extension** which Coronado also had under purchase option. He estimated that 20,000 tons of ore had been exposed by surface stripping.

He had announced the company was buying the old **Red Top Mine** from Minneapolis interests. Coronado had rehabilitated the mine in May and June of 1971 and he expected to produce 70-90 tons of lead-zinc-silver ore daily. Coronado was also buying **Galena King,** formerly known as **Copper King.** Like too many others, this huge expansion under Coronado came to a close with very little to show for the effort.

New claims filed in Stevens County

In January, and again in September 1976 there were a number of "Big Name" companies stirring up mining interest by filing claims in several districts. The largest group was by United States Steel having filed 422 claims in four different areas, some south of Northport and some near the **Old Dominion.** According to Patrick Graham, writing for the Statesman-Examiner, these claims are called **"PR."** Some of the Fruitland district were called **"Flag",** and the Company was also filing a large number of claims in Pend Oreille County. Slowing down in '77, there were still many filings, some by Kerr-McGee, some by Dawn Mining Company, and Callahan added to it's holdings.

Joseph McNamee, Northport, filed 26 claims called the **"Copper Find"** west of Northport; R.A. LaRocque, Norman Felsman and S. Namisnak, Seattle, filed eight claims called the **"Tririte"** east of Hunters, Inspiration Development Company under Patrick D. Williams of Spokane filed 12 claims called the **"URI"** east of Park Rapids, and Art Bolt, of Colville filed 12 claims called the **"Zale'** north of Colville in Clugston Creek area. These, and many more indicate new mineral findings, and supported the news of work going on at the **First Thought, Calhoun, Van Stone,** around Eagle Mountain and the **United Copper.** Exxon was exploring for uranium; all activity told by Ted Livingston, state geologist from Olympia.

New life for MELROSE

There was a real flurry of hope when in March, 1977 the old **Melrose,** a property of the late Ben Hofer, was suddenly brought to life. Old miners of the area talked about it being the "second **Sunshine."** Located four miles south of Boundary, three unpatented claims belonged to the Paragon Mining and Development Company from 1918-1938, and became Hofer's in 1941. Production was in silver-lead-zinc and copper with 75 tons shipped prior to 1921 yielding gross of $8,000. The mine produced in 1913, 1937 and 1938. A three-ton sample shipment gave gross value of $123.49 with a large mixture of ores tested.

Charleston Resources Ltd., of Vancouver, B.C. had taken over the 2,000 acre property by March of 1977 and were recommending a $190,000 work program. They were working on an agreement with Norex American Ltd. of Spokane, who was doing the development work with Joe McNamee of Northport heading the work. According to the Statesman-Examiner story, McNamee said "You can pick up rocks in the open pit which we've had assayed to contain 69 percent silver. If there is enough of it, you can mine it profitably."

Ralph Ernewin, president of Norex stated, "at full production such a mine would employ 20 to 25 miners and a mill crew of 16." He anticipated a mill putting out some 200 tons per day. Some ore had already been stockpiled by the May, 1977 report indicating, at the price then of $4.80 silver, the company could realize $4.20 per ounce of silver as a net smelter return. Continental Carlisle Douglas Ltd., was said to provide the funds necessary to carry out the engineers recommendations. Little has been heard of this location recently.

We must by-pass dozens of claims, some that became mines, but this great mining district is just that – too large to cover more, and many are still in the low production bracket. A sleeping giant, it is ready to go back to work in every valley and hillside as demand and price opens up mining again.

Cedar Canyon (Deer Trail); Springdale, Summit–

Anyone looking for geographical locations of any of the mines know they can get that from the court-house where the claim has been filed, but for story reason we leave that off. Because many mining activities were intertwined, we will combine the history of those mines in the Cedar Canyon (Deer Trail), Springdale, and Summit districts. Activity from these districts reached as far north as Rice, Wash. on the east side of the Columbia River; south to Hunters and Fruitland, east to Springdale and Loon Lake and north to Valley. Centered between these towns are the mineral laden Huckleberry Mountains where many a location has been made.

Classed now as a ghost town in Stevens County is Cedarville, some six miles east of Fruitland. From the mid 1890's the mines at Cedar Canyon developed rapidly and soon 300 people lived in that canyon, bringing about living necessities, and working populace. Fruit raised in the region, and local gardens supplied those needs. A narrow road from the camp led over the hill to Springdale on which some of the ore was hauled. There was said to be a school open for some months of the year. Most of those living at the camp were working at the **Deer Trail** mines, and the **Cleveland.** "Pioneers of the Columbia," a history by the Greenwood Park Grange better tells of this little village.

On June 18, 1897 the Colville paper mentions Cedarville as henceforth being known as Chloride – with no explanation. A Mr. Keeler was named as a prominent man at the time and **Mountain Queen**

Cedarville, Cedar Canyon, or Chloride, 6 miles east of Fruitland, this very busy little village was populated by over 300 when Cleveland and Deer Trail Mines were in full swing in th late 1800's. (courtesy Pioneers of the Columbia).

Not knowing what steamboat John Rickey operated, the ENTERPRISE is a 1910 example of what boats supplied this area north of Spokane via the Columbia River. (Pioneers of the Columbia.)

was mentioned as one of the claims, or mines. Bunker Hill was in on the early exploration with one reported claim showing 124 ounces of silver and 12 percent copper.

In 1886, a report told of John Rickey running a steamer for eight years from Kettle Falls to Spokane Falls. As mining developed, this served many purposes. Rickey point, just south of Kettle Falls was named to honor this man. In 1889 the settlement of Gifford came about, named for James O. and Sarah Gifford who homesteaded the area before Washington became state. In 1900 a ferry was started by Frank Rail. The steam-powered boat was kept on course by a cable. The first barbershop in the area was opened in Gifford in 1910 with haircuts 20 cents.

CEDAR CANYON (Deer Trail) Springdale, Summit–Hunters etc. –

Many little temporary settlements became part of the history of this area to include Harvey; Bissell, Waterloo, Arzina, Gray and others. Harvey was said to be on a sandy flat and served as a boat landing and ferry site. George Harvey came to the area in 1858 and a creek has been named for him. Mrs. Harvey was said to have opened a store and a post office; with John H., McGee credited with opening a later store in 1883. Arzina was three and a half miles east of Harvey with a post office named for Mrs. Arzina Chamberlain. All of these town sites have disappeared or been swallowed up to larger developed towns. The towns of Gifford, Rice, Hunters and Fruitland, Cedonia, Daisy and Enterprise, have all survived principally as stock raising farmland, and orchards.

Springdale on the inland side of the mountain range became a shipping point after the Spokane Falls and Northern railroad came through in 1889. The townsite earlier known as Walker's Prairie became C.O. Squire's homestead. He had platted the town Nov. 27, 1890, called it Squire City. After the railroad came through they changed the name to Springdale, which was incorporated Jan. 26, 1903. Squire built the first sawmill in the town.

Mark P. Shaffer and Charles Trimble built the first store building in Springdale, in July 1889 and opened it for a general mercantile. The second store was opened by John S. Gray and a third by J.H. Keller. That store of March 6, 1890 was said to be the sole remaining one in the early 1900's. Population at the time was around 400 and a goodly amount of ore was being hauled to this point to be shipped out. A stage did run from Springdale to the Deer Trail area.

A brick yard was established in 1903, and there was much logging going on with population up to 500 in 1905. At nearby Clayton, some one mile southeast of Springdale was the Washington Brick and Lime Company employing 50 in the summer and 35 in the winter of 1893. The **Butte-Anaconda** Mine some 17 miles west, was shipping large amounts of ore. Development and production was good in all the mines until suddenly most mines were closed by 1914 causing tremendous hardship for many in the area who had come just for the mining.

Miners still looking for the "high life" were said to be involved at the Silver Crown (thought to be in Springdale) on Jan. 3, 1924. From the Colville paper, the story goes that an altercation occurred last week between McDonald and Hemmenway in the bar room of that saloon. During the argument both men were in possession of .45 colts. At the end of the argument both guns were empty of bullets. While gathering up the fragments, eight bullet holes were still in evidence in the walls. Holes also appeared in McDonald's back and Hemmenway's leg. Whether accidental, or not, was not known . . . however both men journeyed to Spokane the next day for repairs. "The hole in McDonald had a leak that was too severe and he died," according to the report.

Hunters became an important center of activity. The first white farmer from Rickey Rapids to the Spokane River, and from the Columbia River to high into the Huckleberry Range was James Hunter, who came there in 1880 from Lake Chelan. Involved in it's growth, a post office was opened in 1885; a ferry was installed in 1885 and was first operated by William Franklin and Sarah Anderson – later by Albert and Jenny State.

First stores were in 1890 and 1894. Even the Exchange Bank was opened before 1908 but due to the times, it was closed in 1931. Hunter's Land Company Incorporated had extensive apple orchards and two packing sheds by 1908. The fruit was in great demand by the miners and was also being shipped across the country. By 1910 there was a hotel; telephone service, and the only southwestern Stevens County newspaper the "Hunters Leader." The paper was printed from 1908 until 1943. There were several churches and the town had electric lights and a power plant owned by G.J. Bowen.

Moses C. Peltier was credited with settling Fruitland by 1886. He judged it to have good fruit soil – thus the name. Peltier went on to build a store, and later to secure mail service. Cedonia was also in the vicinity. It's name said to be a short version of Macedonia was chosen from the bible by Marten Scotten and a friend. Scotten was said to be the first postmaster there.

Sam McGee homesteaded Daisy in 1882. He built a store in 1896 and chose the name Daisy after the **Daisy Mine** just to the east. That mine was active for awhile but arsenic in it led to closure. A ferry was owned by Ben Morrison. The town of Enterprise was started around 1920 when a service station and general store was built by Roger Thompson.

Loon Lake was to the east and to the south of Springdale. It's main lay to claim was it's recreational park opened up by D.C. Corbin after he built his railroad through the town. Many of the Spokane wealthy spent summertime at the park – and gradually built summer homes there to get away from Spokane during the hot weather. The population in 1889 was said to be 100. There were three sawmills in the vicinity and one of the greatest industries was ice cutting. The lake had crystal clear water and many an early resident made good winter money cutting ice. Most of that was used in the Davenport Hotel and in the businesses and homes of much of Spokane made richer by the mines.

A get-rich-scheme of the day was reported Nov. 19, 1935 when a mining "boom" on the Addy-Gifford summit came to an abrupt end as 15 workmen filed liens against the property. A Spokane man started activity, hiring every able bodied man in the Summit Valley who wanted work. The mined ore was said trucked to the recently completed Marcus mill, which started operating day and night; however when payday arrived there was no pay. No name was given in the article as to who the enterprising mining-manager might be. (Refer Colville news of 11/19/'35.)

Many of the small town sites mentioned will be near some of the mines of the region to be discussed. Spokane was already the center of mining activity – both from so many in ownership, and it's being the financial center. If one needed backing for development of a good sounding claim, or claims, it was not difficult to get capital from Spokane. You have noticed to this point in our history just how many Spokane men, and companies, were involved in the mines in areas of our coverage. This will continue to be true. Many of the beautiful homes and buildings in Spokane today were built by money from the mines; the city still houses a great many mining company offices and from which activity will spring when the mining industry improves.

Cleveland Mine

Dating from 1892 the old **Cleveland**, or now called the **Santa Rita** mine, 18 miles west of Springdale, and 10 mile east of Hunters, was a good producer. The 12 patented claims showed values in lead-zinc-silver and antimony. Discovered by W.L. Falter (sometimes written Faller), and T.F. "Art" Tinsley (sometimes written Finsley, and Tinsby) and their partner Lingfalter, the mine on Huckleberry Mountain gained acclaim due to a 12-

15 foot wide vein indicating a rich silver-lead deposit. It was good enough that the owners sold the mine in two months for $150,000.

The mine came under the ownership of George B. McAuley and James Monaghan (both of Spokane, and of Coeur d'Alene mining fame), and a C.B. King. Tested early in 1895 to show 40 percent lead and a main ledge showing 25 ounces silver and 59 percent lead, the mine was said "not for sale at any price!" The mine was reported to have produced 1,500 tons before 1897.

In 1896 J.C. Margo, a pioneer prospector and miner of the **Cleveland**, told great things of the Huckleberry Mountain region. On the other side of the coin, Williard G. Jones doing work in development, shot himself – no reason given.

Other old-timers mentioned included Milo and Mary Jane Runyan who came to the **Cleveland Mine** to work. A local news item June 25, 1897 told of James Ladow, John Burden and L.E. Beach making a strike about four miles from the **Cleveland** camp. Burden and Leach (or Beach) discovered a ledge carrying gold, silver and copper that assayed $30.65 from surface material. Beach also made a discovery of a rich gold deposit near the **Copper Giant.**

With so many prospecting, one miner said there was almost a total absence of deserted holes in the Huckleberry range. He went on to say that where muscle was used, and with the honest expenditure of money, returns were gratifying and at the time no great depths had been reached.

Larger shipments were being made by 1899 showing 25-50 ounces of silver; 20 to 65 percent lead, and the ore became richer as the mining went deeper. 30 miles of roadway had been built, and large sums of money spent on development but most of that had been produced from the mine itself.

Al Stiles had reached this portion of his mining travels and by 1894 was working at the **Cleveland** when "Art" Tinsley was still part owner. Stiles also had a claim nearby and with work at both places saw production at **Cleveland** go between $200,000 and $300,000. Under lease to Tinsley, Stower and Higginson, a large work force was being hired in July 1897. Stiles said 300 men worked mines in the district at the time.

By 1903, the mines of all districts in the state were being assessed, the **Cleveland** valuation put at $5,200. A 1920 geological report said shipment of ore was a major cost factor. It was only seven miles to the nearest point, to the Phoenix Lumber Company's logging railroad but costing four dollars a ton to ship concentrates. The only other choice was over a very bad logging road but end cost was 50 cents per ton.

A second 1920 report did say the **Cleveland** had paid $250,000 in dividends but with no records this could not be verified. Fifty tons of concentrate had just been shipped indicating more zinc-lead and iron. Three million dollars in ore was said produced, mostly in the 1920's.

In 1924, the mine was owned by the Santa Rita Mining Company and was said shipping three tons of concentrate daily besides some high-grade ore. In 1942 the company was reported to have produced 4,000 tons. What happened has not been explained but by 1952, the **Cleveland** was said owned by Stewart Compton of Bayview, Ida. and leasing to Frank Marr of Spokane. In between these ownerships Forney Brothers of Springdale had the mine on lease to Clearwater and Associates and it was said that in 1948 it was producing about 300 tons a month.

The 1967-1968 directory of mines lists James LeBret, president of the **Cleveland Mine** and his brother, John C. LeBret was named secretary-treasurer. The LeBret twins are famous for their location of the **Midnite Uranium Mine** in 1954 (of which we will learn more). On Dec. 1, 1966 they were acclaimed as "having done it again."

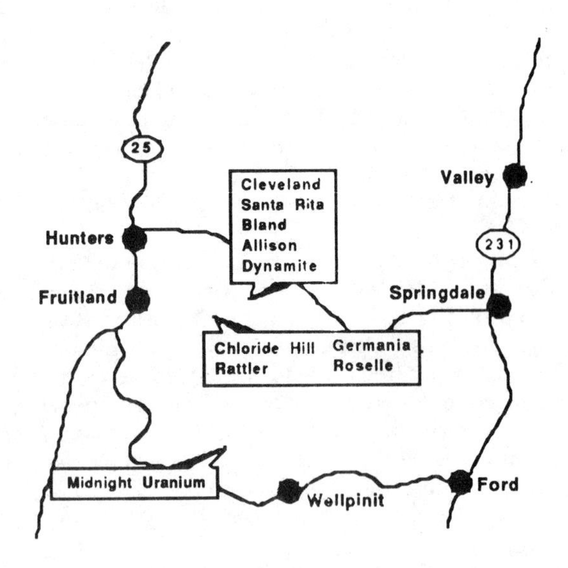

Wafford Conrad in a story for the Spokane Chronicle on the December date said the men had found a new mine but this time values were in lead, silver, copper, zinc, antimony and cadmium. They did not know the extent at that time but high-grade ore was being stockpiled ready for shipment to a smelter. Conrad said 80 tons were ready for shipment when he visited the mine in the Huckleberry's seven mile east of Hunters. The new location was said about 400 feet south of the old **Cleveland.**

As bulldozing exposed ore bodies, James LeBret said grab samples assayed 20% to 40% lead and nine ounces of silver per ton. Another cut showed the bright yellow and green vein which assayed 54% lead; 13% antimony, 2.75% copper, 0.25% cadmium and nine ounces silver to the ton. Burton E. Burnside of Spokane had driven a tunnel.

The LeBrets had been prospecting for three years under a lease and purchase option from Stewart

Compton, of Bayview, Ida., and Walter Johnson, Spokane owners of the 7-claim **Cleveland** property. The LeBrets and associates had spent $135,000 on the venture, to that date. Silver Mines, Incorporated had other stockholders from Spokane. The company had also leased five claims from the Carr Brothers of Valley and hopes were for continued operations all winter.

By March 14, 1968, the Silverton Mines, Incorporated of Spokane had entered into a development contract with the Cleveland Silver Mines who were to spend at least $100,000. The lease was to be in effect until Oct. 1, 1969 – or for an additional two years. John H. Hansen Jr. was incumbent president of the board. By July 3, 1968, Cleveland Silver Mines, Inc. was said trucking high-grade silver-lead ore to the East Helena, Mont. smelter. The company then had a stockpile of 800 tons.

On Nov. 3, 1970, Wafford Conrad wrote for the Chronicle some of the results of earlier surface development. He said some of the boulders glistened with galena (lead-silver). Others were mostly of a dull antimony ore. The new open-pit mining operation at the old **Cleveland** property had about 250 tons stockpiled at the time according to Al Deviny, Spokane, company general manager. He said the high-grade ore was being hand-sorted.

1,200 tons of milling grade antimony had been trucked to the old Northwest Magnesite Company mill which the Pyrometric Corporation of Seattle was operating. As the ore zone was being stripped he comments on the fact the zone was overlooked by former operators of the **Cleveland Mine.** Similar to other mines we have learned about, this one had a mystery boulder – it being of antimony ore found adjacent to the old shaft but not appearing to belong to the area's scene. Plans at the time were to mine all the ore possible from the surface, then to eventually mine the new ore structure from older underground workings, according to James LeBret.

An extension of the **Cleveland,** called the **Bland** was discovered in th late 1800's or early 1900's by Dr. J.P. Turney, A.W. Turner, C.G. Snyder, H.H. McMillan and C.E .Richard, all of Davenport. The claim showed a yield of 52 ounces of silver; 5% lead and a trace of gold. A Jan. 31, 1895, release states that M.B. Runyan sold to G.C. Snyder and J.P. Turney, an interest in the **Bland** and **Allison** Mines, the **Stuart Extension** to the **Cleveland;** the **Dynamite** claim on Hunter Creek, the **Chloride Hill** and **Rattler,** near Fruitland, for $22,000 the sale having been made seven days earlier.

In 1967, Paul L. Jones, Spokane mining engineer credited geo-chemical prospecting for new discoveries at the **Cleveland**. He explained, "This **Cleveland** location occurred in virgin ground about 1,000 feet from old stopes and immediately beneath an existing road heavily used each year by deer hunters. Time and again these hunters have driven their cars across the vein, or have walked the same route without suspecting that rich ore lay concealed barely inches beneath their feet."

Germania Mine

The **Germania** Mine located in either 1894, by J.S. McClean and John Horn; or in 1897, by Hearn and McCoy was at one time owned by the German family that nearly destroyed the world; and at another time by General Electric, one of the most prominent American electrical companies.

Some 25 miles by wagon southwest of Springdale, the mine was reported to have been owned by John McClean, John Horn and Henry Becker from 1906 to 1916. William Sheck is credited with operating the mine from 1906 until he was called into the service in 1914. He was said to have a lease on the **Germania** claims and on **Roselle** from 1910 to the early 1920's, under a company name reorganized into American Tungsten Consolidated Corporation of New York, by 1917.

Foreign ownership of the tungsten mine came about when the local company displayed it's tungsten, or wolframite, at the St. Louis World's Fair. The metal attracted the attention of Germans who immediately came to Stevens County and located, or leased some 1,000 acres of tungsten-bearing ground. In 1906, Sheck opened the **Germania** and worked it for the Germans until he was called to war; and when the German management and help were all called home. Al Stiles told how he was suddenly hired to run the mine until it was closed down in 1916, after having worked for Sheck for several years. Being an assayer, he called it chemical work, and also being capable of management – and on the spot at the time, he did run the mine for a brief time.

Possibly we are not proud of having let the Germans own this mine because there is very little in the mining journals about that ownership. It was the Krupp Gun Works of Germany. During the outbreak of World War I they were said to have shipped this ore by submarine to Germany where it was used to harden steel in the making of cannons – that were eventually turned against us. The entrance of the United States into the war soon brought about seizure of the mine.

Stiles memory of these "Prussians" was not too happy, especially when he knew of the hardships our own people were facing. He said that Alfreid Krupp came to the mine and was not a pleasant person. He told how Krupp preferred Scotch Whiskey to Schnapps, and that the man lighted one American cigarette after another. (Both facts have been repeated in other history of Krupp.) Stiles said all those people "lived like Princes." He told of the furniture and the grand piano. Alfreid Krupp von Bohlen un Halback was the son of the head of the Krupp Company, Gustav Krupp. That man was a Nazi and as a result, Alfreid joined the party.

Scale room, and chemical laboratory such as Al Stiles worked in at GERMANIA, and several other mines in that area. (S-R 1899).

The mine was later opened after a shut-down of five years and in February of 1921 was said having shipped 7,221 pounds of ore with a yield of $3,375. Shipments were made to New York and under a later operation was said having shipped a carload worth $40,000 in 1934. From 1932-1934, there was approximately $200,000 a year produced with 40 men working. Even the tailings were said to show $6.80 in gold, and five percent tungsten to the ton.

General Electric purchased the mine in 1936, from J.A. Scollard for $300,000. That company operated it until 1941, for the tungsten to use in electric appliances and lights; for automobile ignitions, radio tubes, electrodes and other of 5,000 items to include high speed tools for armament factories. Taking the major part of their supply, this last use was not told until sometime later. There were some 100 men working. One was Benton Bailey, a former Colville man who was said to have been assayer for the company for some time. By 1941, 90,000 tons were said shipped with a present rate of production then of 8,500 tons a month. By 1943, Stiles said the mine was declared "worked out." Not so!

The Germania Consolidated Mines Incorporated came into being and held the **Keeth** and **Norton** tungsten properties adjoining but did not control the old **Germania.** The company had 160 acres, plus 320 acres leased ground. A 1945, report to shareholders showed capital of $50,000 and five million shares. The last shipment was said made in 1946, and development work was suspended in 1949. With new ownership H.W. Traver of Springdale was to be in charge of operations – he was a former miner at the property.

Prominent Lind, Wash. wheat farmers made up much of the new corporation: Julius A. Franz was president; Henry J. Franz, vice-president; E.I. Fisher, secretary-treasurer; B.O. Myron and H.G. Loop were directors. Offices were in Spokane. In a report to stockholders the officers told that **Keeth** and **Norton** were said to have yielded 4,600 tons with production said to exceed $100,000. Net returns to the company were reported expected to be $224,544 for the year 1945. This was according to Arthur Lake, mining engineer and geologist.

This 1945 geological report said that at $20 per unit for wolframite at New York prices and paying the wage scale of the day; and with a mill capacity of 660 tons of ore per month, the return could be pretty well assured. Until the 18th century, tungsten was known as a tin ore and was given the name in 1758, because of the weight, then called heavy stone. The company report also told that earlier tungsten was shipped from China and Korea until World War II when shipping became difficult, and when China came under Communist rule. The search began in Canada, and the United States as the need became evident for making high speed tools.

The Germania Consolidated Mines, Inc., developed more than 3,000 feet of open cut, of stoping, shaft and tunnel work. They built a boarding and bunkhouse, a compressor building and machine shop; two garages, and with plenty of equipment had sunk the **Keeth** and **Norton** shaft. Ore price for tungsten had gone from $20.25 for a short ton in 1917, down to $3.15 in 1921, and back up to $64 and $66 by March, 1951. Open and operating, it was then that Traver expected to make his first shipment in 60 days. (Tungsten price Jan. 20, 1989, is $56 to $64 per metric ton unit.)

The German company was not fairing so well. The victorious Allies had confiscated Krupp's properties after World War II but they were returned to him in 1951. Alfreid Krupp had been sentenced and served most of 12 years as war prisoner in the place of his aged and ill father, Gustav. He had been arrested by the American troops just before the end of the war and had been tried at Nuernberg.

After the return of his holdings, Alfreid Krupp immediately set to work rebuilding the steel company but publicly vowed it never again would produce weapons – a mighty step for a company that had grown rich and famous as the cannon maker and armorer for Europe a century ago. That firm had mightily contributed to the arming of Germany in two World Wars.

One of Europe's wealthiest men Alfreid Krupp, the last of the historic Krupp Industrial Empire, died July 31, 1967 of a weakened heart. He was in the process of converting his huge one-family holdings into a public corporation as demanded by the bankers who came to his rescue with "a measly $75,000,000 to maintain his billion-dollar plants" (according to Seymour Freiden, This Week's foreign correspondent). That article written July 9, 1967, before Krupp's death said 110,000 people were employed by Krupp. His second wife had indicated she thought he could still survive on his income and some $300,000,000 he was supposed to have stashed away.

Alfreid Krupp was ordered to sell off his coal and steel holdings by 1959, but Frieden said that was never done "supposedly for lack of buyers." Alfreid did diversify into the making of over 3,500 products from locomotives to mineral water, to orchids - after armaments were expressly forbidden.

The company is said to have a glistening branch office building in the heart of the West German capital on the damp Rhine River Bank at Bonn. Those who were employed by the Germany Company in early **Germania** days might well understand why the only son Arndt renounced his inheritance and any Krupp holdings. He was said to not want any thing to do with this industrial management. He was a playboy and preferred to race cars. (On the other hand, the General Electric Company was well accepted here and were a major asset to the local as long as it was feasible for them to operate.)

On Jan. 25, 1952, the Tungsten Mining and Milling Company received word of conditional approval of a $50,000 loan from the Reconstruction Finance Corporation (RFC) under the government's new Defense Materials Procurement Agency program (DMPA). It was the first made to an Inland Empire mining venture according to the Spokane paper. The money was to be used to purchase machinery for the mill at the **Germania Mine,** according to Paul H. Casey, president and general manager. The company had previously received a $25,987 exploration loan through the DMPA grant but did have to provide working capital to operate the plant until production returns come in.

The mine was still active in 1952-1953, then was shut down until resuming operation around 1965 until 1968. In 1966 Lucky Scarlett Uranium Mining and Development Company took over operation and on Oct. 20, 1967 a 30-ton carload of tungsten concentrate valued at $100,000 was reported shipped to the General Electric plant in the east.

In May of 1981, a news release said that a sale of the American Tungsten mining property was announced. The buyers were said to be W.F. West and J.A. Scollard of Seattle, for a consideration of $250,000. The owners then were named as J.A. and L.L. McLean, of the Inland Adjustment Company. It went on to mention the early development of the property by German interest; and that since, it had attracted the interest of many companies.

DEER TRAIL and DEER TRAIL #2 (PROVIDENCE or VENUS)

One of the better known mining operations in southwestern Stevens County has been the **Deer Trail** Mines. Some of the property claims will also be discussed and include **Idaho, Hoodoo, Runyan Fraction, Jolly Boy, Moonshine, Elephant, Legal Tender, Victor Fraction, Providence, Royal Extension, Happy Home, Splawn** and **Deen** claims. Many of these were totally absorbed into the operation and no longer have any identity of their own.

A number of years ago, Mrs. Herbert Heinz, then of the Aladdin district, told about her father W. Oscar Vanhorn and 11 brothers who lived at the Dalles, Ore. She said some of them, along with her father, came to Spokane and got into a poker game which they won. With this little "poke" some of the brothers came north to Cedar Canyon to prospect.

As in the case of many ore findings in this part of the state, the prospector was also after food. In this case in 1894, W.O. and Isaac L. Vanhorn were pursuing two deer in the canyon. W. Oscar Vanhorn stumbled over a huge quartz boulder carrying galena and immediately gave up the chase and went to prospecting. Pieces of this boulder were later assayed and found to contain 70-80 ounces of silver to the ton. The claim name was obvious, Deer Trail.

As to the correct date of locating **Deer Trail #2,** some reports indicate it was the same day and that the brothers named the second one #2 because of the chase after two deer. An 1897 report says that W. Oscar Vanhorn, and George Gibson, B.O. Gibson, Charles Golden, and Isaac Vanhorn were all involved in the locations of **Deer Trail #2.** History tells that on Aug. 13, 1897; and again on Dec. 25, 1899, the claim was found by two bankrupt farmers. (At least a portion of this story could be correct because the farmers west and south of Spokane at the time were having serious cost and price problems.)

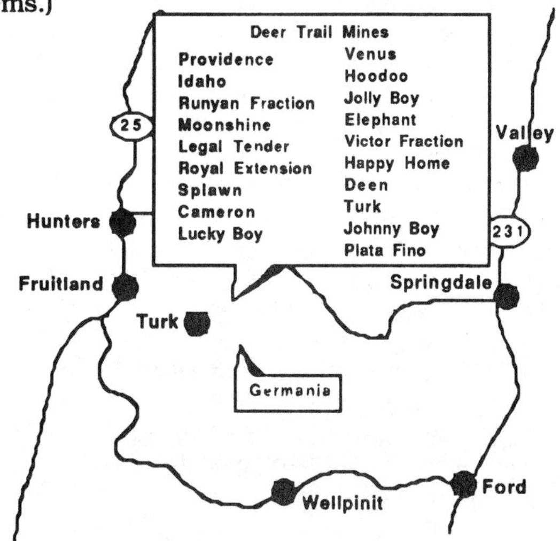

Those who would break the mining laws; jump claims, or bilk the poor miner out of his hard earned profits were warned by a meeting of miners July 9, 1897. These were owners of claims on the Huckleberry Mountain range. They pledged themselves to protect each other's rights and titles where legal locations were made. These old timers were well versed on mining laws. They issued fair warning to rumored "would be claim jumpers" and said they would be driven out of camp. This included "sharks, jumpers, frauds and confidence games." They made prospectors and miners welcome "who are square." This notice in the paper was meant a serious warning to some who followed new locations.

By 1899, one report states that Vanhorns and Golden had opened up a ledge of ore running 10-200 ounces of silver and 15-20 percent lead. They encountered a black sand which they tossed on the dump, not knowing it's value. (Richard Steele termed this a red sand.) He said that W.O. Vanhorn panned some of this thinking he would find gold. What he did find was strings and flakes of native silver– some of the flakes as large as silver dollars and as thin as tinfoil. He sacked up two and a half tons which he hauled to Davenport. After much ridicule, he received enough money to pay the freight and $150 down payment on a later shipment for which he received $1,360.

When the claims were sold to a Spokane syndicate some time later, they worked the sand and netted $200 a ton from 25-500 ounce silver to the ton. The Spokesman-Review Quarterly of 1899 called the sand black and tells how the Spokane purchasers had assays of 250-500 ounce silver to the ton, while the ore body in the vein ran only half that, and less.

By June 15, 1899, a mining company was said formed with capital of one million dollars and at that date was also credited with having paid $35,000 in dividends. The two who located the mines were said to have gone on into the hills prospecting other portions of that rich Huckleberry Mountain area.

Stiles was doing "chemical work" for the **Deer Trail** operation from 1901 to 1903. He worked various of the claims as they were opened and added to the whole. He tells of **Providence** under A.W. Turner's operation, taking $150,000; and $50,000 taken from **Deer Trail #2** under lease to "Jerry" Cameron from 1901 to 1902. He further explained the Cameron operation (which was verified by a Statesman-Index story, that Cameron and 20 men were handpicking ore and shipped $50,000 worth with $30,000 clear money).

Stiles said Ed Crawford was the cook at the camp; Charlie May was the banker they dealt with at Davenport, and that by 1900 John and Mary Parry had also come to the **Deer Trail.** They were more family to the Vanhorns.

Other information tells that the **Deer Trail,** also

Camp of DEER TRAIL #2 – (S-R 1899).

known as the **Venus** or **Providence,** was under the business management of Warren Tollman, a Spokane lawyer, but does not give us the years of operation when it was said owned by Victor and Nancy Allen of Coeur d'Alene, who also were said to have sold to Chapman. Under this operation $200,000 was reported produced from 15 claims. By 1903 that figure was said to be over one million dollars in silver ore removed. Between 1902 and 1909 the mine was owned by Deer Trail Consolidated Mining Company.

The **Deer Trail** claims produced very liberally for four and more years with a quality identical to that of **Old Dominion.** A 1910 report states that the **Hoodoo** on the east side of the mine was added as a producer with a Mr. Newell as foreman for **Deer Trail.** That report said the **Legal Tender** had been located after 1894 by W.B. Runyan, and William Yarwood. (Stiles said Milo Runyan and Wilber Yarwood). Those also became part of **Deer Trail** holdings. Many early claims were said located by farmers from the Big Bend Country, and men from the Old Fort Spokane.

The 1910 report also said that Turk, a settlement two miles northwest was the post office, that the one at Deer Trail had been closed down. It mentions the charred remains of buildings from a forest fire with the desolation of an abandoned camp. At that period of time, the **Germania** was about the only active property in the Deer Trail district. Springdale was the nearest railroad station, some 25 miles distant.

Venus Silver Mines Company bought the Deer Trail group and operated it from 1925-1930. It was around the time that **Deer Trail** reported striking an important vein, the date Feb. 28, 1930. There was a sudden flurry of mining activity with companies willing to put up large amounts of money for development.

The total operation was reported to have produced from 1906-1908; 1911-1912, 1917-1918 and intermittently through 1947. More than three million dollars total production was reported from **Deer Trail.** Alpine Uranium Corporation of Salt Lake City was owner of the property in 1954. By 1925 there had been 21 claims under control of the corporation then operating the mine, the Venus

Silver Mines Company. In 1934, George Vervaeke already involved in area mines, was operating the **Deer Trail** properties.

The **Legal Tender** joined the **Elephant**, and the **Hoodoo** was the lowest working tunnel on the **Deer Trail**, the property at an elevation of 3550 feet. The **Cameron Tunnel** was some 80 feet above **Hoodoo. Victor Tunnel** was 250 feet north of **Legal Tender** and **Providence Tunnel #2** was close by. By a 1943 report, the mine had a 75-ton flotation mill with a tram line running from the mine to the mill. At that date the mine was idle but the mill was being operated on lease by **Lucky Boy Mine** for concentration of copper ores. (We will know this as the **Turk** Mine.)

Hoodoo, one claim of the **Deer Trail Mine** was said owned by Paso Robles Mining Company in 1949. **Moonshine-Johnny Boy** with it's two claims was owned by Barton Brothers of Chico, Calif. in 1941. The brothers also owned an operated the **Plata Fino** a mid-1900's silver-lead find. **Providence** owned by Perdiver Mining Company in 1934 is another **Deer Trail** claim. The mine was said to have produced over $50,000 before 1914 according to a 1956 Washington Minerals Inventory. It went on to produce from 1922 through 1940.

All was more-or-less quiet at this great producer until December 10, 1981 when the Statesman-Examiner told of Madre Mining Ltd., a Canadian company announcing processing of silver ore at the **Deer Trail Mine.** With a new 150-ton mill, this company was running some 90,000 tons of ore left on the surface by earlier efforts dating back to the original 1894. Early miners were said to only ship ore containing 80 ounces or more silver to the ton. Madre Mining was assuming 25 ounces to the ton. Economical operations of today make it feasible to ship at an average 13 ounces silver to the ton. A 1939 report placing reserves at 476,000 tons encouraged underground and tunnel work under Project Manager Don Majer.

Washington State Geologist Wayne S. Moen cited the **Deer Trail Mine** as one of seven in the state that "could become important silver producers." He recalled the mine production of over three million dollars over the years when silver ranged from $1.12 as to as low as 28 cents. He retold the story of the old operation when an entire town stood on the site; when work was by pick and shovel and the ore was moved by horse and wagon. At the 1981 date there were paved roads within three miles; power, telephone, engineering and assay offices, living quarters, machine shops and much other preparation going on for comfort of the workers.

Most of the surface ore being processed lay in huge mounds outside each old claim entrance. Consolidated over the years, the property had more than 1,000 acres in 1981, a portion on deeded land and the rest under the U.S. Bureau of Land Management. Madre acquired the mineral rights in November, 1980. As of Dec. 1, 1980 more than 45,000 tons of the surface ore had been moved close to the mill. As with other older mines, this company has cashed in on the processing of surface ore while they develop new showings.

By Dec. 8, 1983, Madre told of the production phase of their mining having commenced Oct. 7. In an initial 30-day period they had shipped approximately every four days, 30 tons each with eight shipments totaling 240 tons. Smelter returns had not been made but assay samples indicated value of these shipments of 44,000 ounces of silver, 129,600 pounds zinc, and 100,800 pounds lead.

A new ball mill was to be completely installed by December 1983. The vein had been proven to 2,500 feet and much other work had been done. They were driving the tunnel towards the old **Hoodoo** ore body with great encouragement at the showing of 15 to 50 ounce silver to the short ton. The mill at the time was being fed from the old dump material, and the ore from the new vein exposure along the Madre Tunnel.

By Sept. 2, 1984, with the falling prices of silver, Madre was to face the end of their great hopes. Many – too many creditors were in the hole as a result. Norman Thorpe of the Spokesman-Review tells the story in the Sept. 2 issue of that newspaper.

All the work and good news of finding the long-sought part of the **Deer Trail** Vein came to a head July 26, when the Madre Company suspended operations. They had said in mid-June that they had located enough ore "to place the mill on a 24-hour per day production basis." John Sawyer, one of the directors said the cost of operations could not be covered with silver at nine dollars a troy ounce in June, and dropping to seven dollars and $7.45 by the time the mill was shut down. This was Stevens County's first suspension of operation for price with what was known to be a successful property. (It was not long after that, **Hecla** and **Sunshine** in the Coeur d'Alenes follows suit and closed their mines.).

After three years of activity, a string of debt claims had arisen. Eleven businesses had filed lawsuits against Madre in four different courts charging debts ranging from $600 to $12,000. Other suits were to follow. Ponderosa Drilling and Development placed a $1,300 lien on Madre's property; Washington Department of Labor and Industries filed court claims charging Madre didn't pay $129,000 in industrial insurance taxes for 1982 and 1983. The Employment Security Department filed court claims of $39,000, and the company was delinquent in paying 1984 Stevens County property taxes.

Other charges and penalties were involved but Sawyer said "The company has not been walking away from it's obligations – that is why we rarely contest the suits." The company was said to have made local business purchase without ever making even a token payment on the bill. One was to Thunderbird Lubrications, Inc. for $2,000. Sawyer did say that Madre had paid for most of the assets at the mine, which amounted to $12 million (Canadian investment). He said accounts payable and accrued liabilities were less than one million dollars.

Two businesses won lawsuits by garnisheeing Madre's accounts at the Colville branch of Seafirst Bank and the Wheatland Bank in Davenport. Others who tried found no funds.

An old 1939 report said the camp "was once a roaring, shooting, sinful Western mining camp." The 1984 article said "today it is pretty quiet." In the previous fall the company had 40 people working. According to Dean Rielly, vice-president of properties and project planning, that figure had dropped to 14, then down to only a handfull. At the time of the news release Reilly said the company was working on financing for a drilling operation to counteract the two faults they had hit, and to again find the vein. "With the price of precious metals right now, it's hard to dig up the money."

On Jan. 29, 1987, Cortez International Ltd., of Vancouver, B.C., said it recently rediscovered a proven high-grade magnesite ore deposit extending into the **Deer Trail Mine** property. It had been core-drilled and described in 1943. Kevin O'Flaherty, Cortez geological engineer said the reserves represent about 800,000 tons of magnesia, or about 500,000 tons of metallic magnesium. At 1987 price of $1.53 per pound, the ore body has a gross value exceeding $1.5 billion and an estimated minimum net value of $450 million.

Most of the magnesite in the United States is imported, he explained. The deposit would allow Cortez to enter that market. Once the premier silver producer in Washington, the mine shut down in 1985, would be put back into operation when prices rebound, according to Cortez.

In April, 1987, L. George Reynolds, chairman of Cortez International Ltd., said there were plans for reopening the **Deer Trail Mine.** At the time, silver was nine dollars an ounce. The mine was under the control of Oregon Leo Mineral and Timber Company, a wholly owned subsidiary of Cortez International. He explained that a previous operator estimated more than 50 million ounces of recoverable silver reserves at the mine.

A news story from the Statesman-Examiner Aug. 9, 1988, tells that Madre had later changed it's name to Cortez International Ltd., and acquired oil and gas properties in Dallas. The story goes on to say that Madre and Cortez owed money to their auditors at Thorne Ernst & Whinney in Calgary so gave the accounting firm a lien on the **Deer Trail Mine**; according to George Reynolds president and chairman of Cortez, and Ram Industries Inc., the Dallas Company. It was those auditors who ordered sale of certain assets at the mine and which Ram's former shareholders were disputing the right to liquidate assets and had filed suit against the auditors.

The liquidation involved only a small amount of the assets at the **Deer Trail,** according to Reynolds. He said again that Cortez hoped to reopen if silver prices reached nine dollars or they find richer silver deposits. At the time they were trying to find a joint-venture partner for drilling to find other deposits, and to open the magnesite deposit adjacent to the silver property.

Al Stiles and his last 60 years

When interviewed in the mid-1960's, Al Stiles memory proved almost, or even more accurate than mining journals and other bits of history researched. His love of mining gave him a reason to feel proud and tell the production of past years; and what was amazing, was for him to remember the operators and owners of certain properties.

Having come and gone several times into the Canadian fields around Trail; and into the fields, camps, and at the smelter at Northport, Stiles finally settled for the Cedar Canyon district. He was full-fledged assayer by 1900 and his work at the **Cleveland,** and later at the **Deer Trail** included mostly "chemical work" as he called it. He was always prospecting a bit on the side.

Al Stiles was also a good farmer and earlier he had worked on the farm of his future father-in-law Charles Buck. He had met the daughter and on Jan. 7, 1904 he and Ruth Buck were married. The year was too dry and Al said they lost all their money on the farm so he and Ruth moved to the smelter at **Turk.** He had to laugh when he told that the smelter there was not a success because A.W. Turner insisted on using wood for heat.

Al had done a great deal of assay work, and also surveying at adjoining claims when Scheck hired him at the **Germania,** and where he operated that mine the last two years until 1916. In the interview he recalled the **Elephant,** once owned by the Chapman Brothers of Toronto. Al recalled when the mine was run by Dave Layson (or Lasson) of Medical Lake, and said 115 men worked there during 1901. He did assay work for the **Elephant** over a period of time. He said that three million tons of silver ore was produced and he added that the **Legal Tender** was run by Wilber Yarwood and put out 500,000 tons.

One of his friends was "Jerry" Cameron who operated the **Deer Trail** at the time. The 1920 bulletin tells that important tonnage was taken from the **Providence,** and Stiles recalls some 150,000 tons, and 200,000 tons from **Silver Seal** and the **Queen** (better known as the **Queen and Seal).** Something over 250,000 tons came from the **Deer Trail.** Most of these claims later came under Deer Trail ownership bringing production there as one of the best in the state.

Stiles' eyes twinkled when he recalled, and told of Cameron's early mining. Very much without funds, Cameron had approached Charles May, the banker at Davenport. Cameron explained he needed money now but May turned him down until he had "produced." Scrounging what food and supplies he could, Stiles said that 11 men hand-picked the ore that was piled. When it was ready to ship Cameron strung up an old telephone line and called the banker – told him the ore was ready for shipment. May was so impressed that he told Cameron to "write the checks and I will honor them." That shipment brought $30,000 Al said.

Stiles' job at the **Turk** was testing copper ore. He was also doing work for other mines in the area. In 1916 he said he sold a claim for $10,000 and later he and Ruth bought a farm at Fruitland. Several times over the years Al was called back to the **Deer Trail** and on one of those times after he bought the farm in 1935. Daughter Vergie Oens recalled "Dad only lived seven miles from the mine and for three or four years he continued to do assay work. It helped pay for the farm."

His last assays were done in 1957 when he was 89 years old. Vergie commented on her father's care in testing for silver, and how proud she was of his ability to earn almost $1,500 at that age. Stiles prospected and worked his claims with the help of his son Frank until he was nearly 100. His eyesight was good; his hearing not so good and he said because of bad circulation in his feet he never worked much underground. At the time of this interview, and later, he was credited with owning **Saturday Night, Sunday Morning,** and **Frank Long; The Columbia Group** and **The Brash Group.** He said his partner was Roy Armstrong of Republic.

Al H. Stiles was interviewed for the Chronicle when he was celebrating his 100th birthday. A retired assayer and prospector, he still took his daily walks and kept up with the news. (S-C).

This very learned mining man was interviewed by Spokane Chronicle Mining Editor Wafford Conrad on Aug. 18, 1967. Stiles is also recognized for several articles that he wrote for mining publications to include one in 1912; "Deer Trail Mining District" – an article on high-grade silver. It is printed in the Northwest Mining News Volume 8, No. 2, Page 15-16. He visited this writer in her home not long before his death Nov. 2, 1969. At that time this unusual gentleman was still looking for the "big one." He said, "I still think of that one last strike."

After a lifetime of good health in which he took no medicine, never drank or smoked and had one great love – that of mining, Al Stiles had at last found that great bonanza in the sky at age 101. His wife Ruth had pre-deceased her husband by about four years. Besides Vergie Oens at Addy, and Frank Stiles at Fruitland, they were survived by Ann Nilsson, of Spokane, and Lorraine Overmyer of Vancouver, Wash.

Other Cedar Canyon Information

The **Queen Group** was located in 1895 by L.E. and Frank Vanhorn. It was worked by some of the brothers who hired 50 men. By a 1904 report, they were said to have taken out about $200,000 in good ore. It was hand sorted and hauled in wagons 40 miles to Springdale, or Davenport. Most was shipped to Tacoma. Stiles had done assay work for this mine over a period of three years. He had told us some of the silver ore assayed 150-200 and as high as 400 ounces of silver to the ton.

The **Silver Seal** was a low-grade ore, much of which was piled and with hopes of one day becoming profitable. The combined claims came under Silver Basin Mining Company and in 1941 was said to be owned by J.G. Glasgow of Hunters. Total yield until 1940 was said $250,000.

William O. Vanhorn was named president of the Silver Basin. Among other notes concerning him was that his grandfather William H. Vanhorn was one of the Boston Tea Party. This kind of initiative and leadership was evident in the Vanhorn family members as they located and mined many areas of this district.

One location considered in the Kettle Falls district was the **Santiago Mine** said to have been under some ownership of Ralph Emershon Overmyer. That man came to Spokane in 1887 and then to Hunters. The mine near Hunters was said to have had $33,000 spent in improvements – a goodly sum for a gold and silver claim. At one time there was 1,000 tons on the dump but said to be worth only $10 to the ton. Mingo Mining Company was owner in 1915.

The old **Turk Mine,** or **Lucky Boy** was said located by Cooper, Kirkwood and Dr. Hudnut who called it **Blue Grouse.** A.W. Turner and Hubert

Names are unknown but the photo shows some of the mining crew who worked the DEER TRAIL and TURK mines. (Pioneers Of The Columbia).

Davis relocated and named the claim **Copper Butte** and were operating the mine and smelter by 1904-1905. The smelter was said to have burned 800 cords trying to smelt the copper and silver ore. The wood burner type furnace was never successful and they brought a smelter expert Gus Heberlein from Butte to correct the problem. He shipped in coke and "blew" the furnace March 17, 1905. It worked that summer but they did not have sulphide ore to make the matte.

The **Turk, Yellow Jacket** and **Copper Butte** all became part of the scene and was said worked by around 300 men by 1905. The local settlement of Turk had some 80 people living there who took care of needed services. Al Stiles had said he moved to the Turk to be associated with E.B. von Osdel, the superintendent of the smelter, von Osdel was a chemist and Al was an assayer and ore chemist.

Much later the **High-Grade** claim headed by **Henry and Oscar Carstens of Reardan is said to** have taken out $100,000 worth of copper. Many of these claims became a combined operation under one company name. The **Turk** was producing from some of the claims through 1954 under ownership of Alpine Uranium Corporation of Salt Lake. The **High-Grade** became part of the **Lucky Boy (Turk).**

In the early 1900's there was said to be 23 claims owned by C.E. Allen of Springdale. By Sept. 8, 1966 a member of his family, Luke S. Allen, also of Springdale, was completing assessment work on the group of claims including the old **Turk.** The news story tells about the one-time producer they said located in 1885, and mentions the property has been in the family for many years.

Daisy Mine

The history of the **Daisy Silver Mine** goes back many years and in the vicinity of the other mines – this one being 25 miles west of Addy near the summit on the Huckleberry Mountain. It includes nine patented claims. The **Daisy** claim was located in 1887 but with almost impassable roads and lack of transportation the claims were not developed until 1905 when J.J. Browne and W.E. Seelye of Spokane took over that property and other claims.

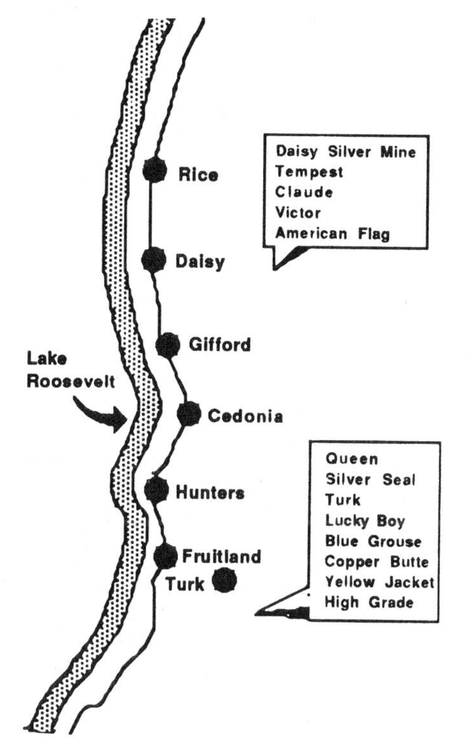

Development went on until 1923 with the building of a concentrating plant. The complex ores of silver, gold and iron could not be concentrated at a profit. Subsequently the Company installed a roaster and erected a bag house to collect arsenic which at that time was profitable. With the decline in price for that product, the operation was abandoned temporarily.

Considerable ore had been hauled by wagon to Addy and shipped to Salt Lake city with values in silver and lead. After much more development and several tunnels, ore from Tunnel No. 4 was said carrying values of 40 ounces silver, five percent to seven percent lead, and the same in zinc. Gold is said in paying quantity, varying from a trace to $1.40 a ton. Several carloads of hand-sorted ore was shipped to American Smelting and Refining Company, in Tacoma in October 1934 with good results in gold, silver, lead and Zinc.

In an April 29, 1967 report from the Daisy Silver-Lead Company, Incorporated, Engineer Charles Goodsell said the mine estimate would be 75,000 tons of known ore with favorable surface showings. Cline E. Tedrow, engineer from Spokane, makes the statement that this gives the property a very favorable start which certainly warrants a major effort to get it into production.

The late Carl M. Lawson had provided this writer Oct. 30, 1967 with the report and a number of small notes concerning the mine. Prominent miner of the Daisy area for many years; he was president of the Silver Lead Company. Among his many interests was the history of mines of the area and in a letter asked if I knew of the **Deadman Mine.** At that time he relates that one partner was supposed to have murdered the other partner and the body was eventually found but the partner who was accused of the murder was never found. This remains an unsolved murder on the Stevens County records today. Lawson was a miner in the Daisy area for over 40 years, and had much history to tell but unfortunately we never got together before his death. His last many years were engrossed in the success of the **Daisy Mine** in which he had such great faith.

The **Tempest** was located in 1888 some 500 feet from the **Daisy.** Combining 13 claims, the property eventually came under ownership as the **Daisy-Tempest** belonging to Mr. Lawson and associates. The property of Tempest Mining and Milling Company earlier, most development had occurred from 1902-1904. These properties are sometimes listed under Summit, and again under Kettle Falls district. Under development by the Lawson ownership, the 1943 report indicated shipments averaging eight dollars a ton.

By 1956, the property listed 16 claims and was said owned by Daisy Silver-Lead Mines Incorporated, of Spokane. It had produced over 2,000 tons prior to 1890 and had intermittent production through 1942-1943. In December of 1968, Silver Ventures, Incorporated of Spokane had been organized to return to production the old **Daisy Mine.** D.W. Watters was secretary-treasurer of the company which then was said to include nine patented claims idle since the 1930's. Work was planned to start immediately to ship "blocked out" ore to Trail Smelter. M. Lorne Craig of Spokane was said president of the new firm which had filed articles of incorporation listing capital of $75,000. Roy J. Wellhoff of Spokane was listed as geologist.

Back in the very early days of this claim P.D. Kearney (of the **Old Dominion Mine**) had filed a claim against L.M. Flournoy and Alice Lou Flournoy with judgement assigned to F.E .Goodall. To satisfy the judgement an amount of $2,905.91 was asked concerning an undivided quarter interest in the **Daisy Lode** mining claim. In March 25, 1893, the Colville Republican said the quarter interest in the mine was sold by the sheriff last Wednesday for $669.45 to satisfy the judgement against the Flournoys in favor of Kearney. No mention is made of the reason for the suit.

The application for patent on the **Daisy** claim was made Oct. 8, 1896. The news article listed the location having been recorded July 21, 1885 and that the nearest claims were the **Claude** and **Victory** or **American Flag.** A 1920 article states ore had been shipped from **Daisy** 25 years earlier. On June 3, 1926, some $12,000 had been raised by the stockholders to get the mine in operation again.

Like so many mines the belief of one or two persons - along with some luck, and plenty of finance, has kept the mine operational. In this case, Carl Lawson has to be credited.

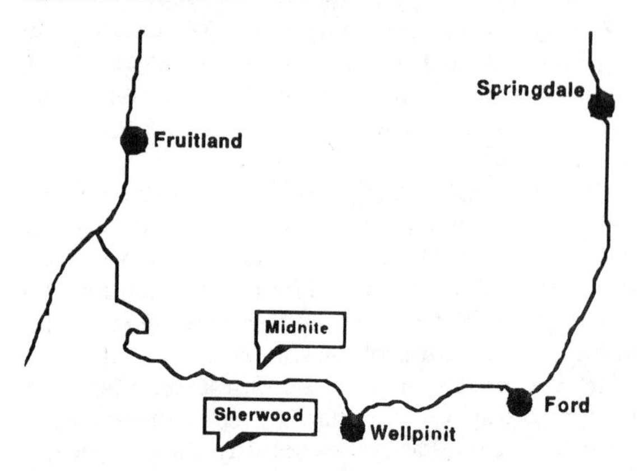

MIDNITE–Dawn Mining Company - Newmont Mining Corp.

There are numerous uranium claims and mines in the State of Washington that we will not discuss because the product is not of the "precious" metals but the mines on the Spokane Indian Reservation have been such producers that some information must be told.

The **Midnite** is located some 10 miles northwest of Wellpinit. The 570 acre Spokane Indian agency land was leased to the Dawn Mining Company of Portland in 1955. John C. and James V. LeBret of Spokane, members of the Spokane Indian tribe, located the claims in the spring of 1954. By 1955, the company was incorporated under Midnite Mines, Inc. with the late Clair Wynecoop as president. Dawn Mining Company was organized by Newmont Mining Corporation of New York, to put the mine into production. As such Newmont owned 51 percent of Dawn, and Midnite owned the other 49 percent. Midnite Mines was capitalized at $65,000.

During 1954-1955, Midnite Mines Incorporated issued 3,785,000 shares with a value of $378,000 of it's common stock for a mining lease within the Spokane Indian reservation. On April 20, 1955, Midnite transferred the lease to Dawn, a Delaware corporation, for 147,000 shares of Dawn Mining Company. The other 51 percent went to Newmont. Dawn's operation cost about $20 million.

Dawn Mining Company was the principal source of income of the Midnite Mines Inc. and substantially all of Dawn's ore reserves were committed under terms of the sales agreement with a utility. Dawn had agreed to sell an additional 3,350,000 pounds of concentrates over the next five years at prices to be determined in accordance with the terms of the contract. Midnite was a personal holding company under the 1954 internal revenue code and was invested in eight other mines, and subleases on Prince of Wales Island, near Kendrick Bay, Alaska.

By June 5, 1955, a flurry of excitement occurred when a bill was introduced in congress to restore some 880,000 acres of land to the Colville Indians. This bill was to clarify the status of certain lands not open to entry but the discovery of uranium on Indian lands (the adjoining tribal lands) brought about the controversy. (The **Midnite Mine** is on "Spokane" tribal land.)

An estimated 700,000 tons of ore reserves was reported at the **Midnite** in 1956. Production was 100 tons in 1954; 609 tons through February 1955, and 94 carloads of ore prior to the end of January 1956. A later report said that from 1954 through 1965, $30 million was taken from the **Midnite.**

Another portion of this huge mine was the 458 acres of allotted land of Ed Boyd's and other members of his family. Ed died and the land on the Indian reservation was leased to Dawn Mining Company in 1956 with payment of $313,444 made to his daughter Lucille Boyd Gallegos, and his son Richard Boyd. She was 30 years old that June and he was just nearing 17 when he became the owner of $159,722, his share. His money was to have been put under control of some trust fund, but that money was said to be the basis of his trouble the remainder of his life.

On March 21, 1961, Richard was found guilty of violation of probation from an earlier charge in January of operating a vehicle while under the influence, and of burglary. He was sentenced to six months in the county jail and a fine of $500 to be prorated at three dollars a day to reimburse Stevens County for expenses. Asked to be heard, Judge Thomas I. Oakshott of the Superior Court of Stevens County, let him. Richard Boyd's excuse was; "If people in this courtroom, or in the state, would have faith in me I could have faith in myself."

Judge Oakshott took time to ask Richard why he had not spent some of his money in helping his people on the reservation, or at least do something constructive for them to show some faith. At the time Richard Boyd's estate was in excess of $200,000 in trust through the Old National Bank in Spokane, with Colville attorney Daniel L. Collins in charge of affairs. Richard's sister died in an automobile accident in Montana in 1961 and left her money to her two children.

Dawn Mining built a $3,500,000 plant at Ford 22 miles southeast of the mine. From 1957 to 1965 they were reported to process 1.5 million tons of uranium ore, according to Earl M. Craig, resident manager for Dawn. An earlier report said the 500-ton mill was expected to process 160,000 tons a year. In 1958 **Midnite** was reported to be the larges uranium producer in the Pacific Northwest. Clair Wynecoop announced in June of 1960 that Dawn Mining company had purchased the **Silver Buckle**

Mine lease for one million dollars cash. Jim Pike, manager for Dawn explained this was originally the Clarence Peters lease and had been assigned to Northwest Mining Company, later to Silver Buckle. At the time **Silver Buckle** was the largest custom shipper of uranium and had an allotment granted them by the atomic energy commission. Dawn acquired the allotment with the purchase.

Silver Buckle Company had purchased the lease from Northwest in May, 1958 and by June 30 had mined and sold 62,379 tons of ore and nearly recovered it's expenditure of $900,000, shareholders were told. Silver Buckle Company was to receive about $247,000 of the million dollar sale after expenditures were taken.

Mining was said halted at **Midnite** in 1964 and the mill closed down in 1965. It resumed operation in 1969. The news release said dividends of two million dollars were expected to be paid by Dawn according to the Jan. 25, 1966 news. It also said Dawn Mining Company had paid $5,917,726 in dividends since July 1958. An extension of contract at reduced prices for concentrate was announced effective from March 31, 1962 until Dec. 13, 1966. It noted that the 1965 shutdown was a result of the fulfilled contract to sell it's uranium concentrates to the U.S. Atomic Energy Commission.

By Jan. 24, 1968, Dawn was looking for sales of their uranium. They found a government market for an estimated $18.5 million worth of uranium oxide. R.B. Fulton told the Spokane Chronicle the firm had agreed to sell two and a half million pounds to two New York power companies starting with Jan. 1, 1970. That represented Dawn's reserves. They had the option to sell an additional one and a half million pounds the succeeding two years, and they also sold 500,000 pounds to AEC. Fulton had been general manger for Dawn from it's start and had supervised the construction of the three million dollar plant at Ford prior to his transfer to Newmont Mining Company in 1958.

Midnite suffered two great losses of their company men when Clair Wynecoop died in October of 1969, and John C. Wynecoop died Aug. 18, 1977. They were two of the six owners of the uranium mine and Clair Wynecoop had been Midnite Mines Company president. George A. Wynecoop, son of Clair, was named to the board and became general manager of the operation. Son Marvin C. succeeded his father John Wynecoop on the board. In 1971 they announced the **Cleveland** silver mine was brought into the company as another investment.

The errant, willful, rich boy Richard Boyd had succeeded in bringing about several more charges against himself when on the night of Jan. 21, 1969 he choked to death on a large piece of meat at the Kon Tiki night club in State Line, Ida. He was said a resident then of Coeur d'Alene. A full-blooded Spokane, he had been reported to have donated some money to the Pacific Northwest Indian Center in Spokane.

At the time of his death he was reported to be worth up to $1.3 million. The news stories included reference to his jail sentence in Colville when he was charged with breaking probation on a burglary charge. Known to have money back then, he was asked why? He confessed that "It was just something to do."

Richard Boyd was reported to have been released to the Morning Star Boy's Ranch at the time in 1957. His money was in trust and investments and royalties had piled up. When he died he was survived by his 83-year-old grandmother Sadie Boyd, of Wellpinit, but his will included only $250 a month for life for her. $500,000 was to go to Morning Star; and very large amounts to other relatives. In March, 1969 Sadie Boyd contested the will alleging that Richard Boyd was mentally incompetent when he executed his will in June, 1968. His Indian trust property was willed to the Spokane Indian Tribe. His grandmother filed because she said she had in large measure raised and nursed and cared for the deceased during his entire life and that his father had died when he was six months old, and his mother died when he was five years old. She claimed to be the guardian from that time on. She had also cared for the sister Lucille – and that in fact she had cared for and reared various members of four generations of the Boyd family, all without recompense. It went on that she worked at whatever menial employment she could find to provide food an clothing for them, with no financial help.

Other heirs joined her in seeking to have the will set aside. The claim was that Richard's mental, emotional and legal history and "bizarre and unusual" conduct contributed to the allegation of mental incompetency. It also claimed the will "was accomplished through undue influence." Superior Court Judge Ralph P. Edgerton of Spokane, cited the Old National Bank, executor of Richard Boyd's estate, to appear in court that March 28 to show cause why Mrs. Boyd's petition should not be granted.

Tibor Klobusicky became a director of the board and later president of the company. He had come from Czechoslovakia and also Hungary where he had been an assistant superintendent of large mines from 1940-1949. He fled from Communist rule and with his family eventually arrived in the United States. His three degrees helped him and in 1951 he served as administrative mine superintendent for **Highland Surprise** in Kellogg, Ida. For eight years he was exploration engineer for **Bunker Hill.** Earlier the **Midnite** was said to have a known and measured un-mined uranium ore reserved value at about $500 million. In 1971 it had produced about 47,000 tons containing 300,000 pounds of uranium oxide valued at $3.25 million. James LeBret was president of **Midnite** and in

1970 said $7,007,000 had been paid in dividends. In December 1976, Klobusicky said, "The only possibility of depressed prices would come if the U.S. Energy Research and Development Administration forced an allocation of uranium."

The mine was shut down in 1981 and in March 14, 1986 was still facing the problem of filling up their dumping pond. They applied for a state license to ship as much as 11,000 barrels of mildly radioactive soil from New Jersey. Dawn Mining Company still stored uranium tailings in the now-defunct mine. The soil was to come by train from New Jersey to Spokane or Reardan and be trucked to the dump. The dirt contains very low levels of radium 226 said Dawn, and state officials.

The company did promise to put at least $193,500 from the project into the state "perpetual care and maintenance fund," depleted by closures of several uranium mines. Attempting to decide about the license, the state was also worried that Dawn would declare bankruptcy, leaving them with a huge cost of upkeep for the pond. If it isn't filled, the state fears radon gas will escape. State Representative Dennis Dellwo, Spokane, said regarding the controversy of not allowing waste to be brought into Hanford that "it appears as if we're opening our back door." (if allowing it at Wellpinit).

Referring to New Jersey, Dellwo asked, "Why the hell don't they want it?" (themselves)

The soil came from underneath homes in Essex County, New Jersey where the U.S. Environmental Protection Agency found unhealthy levels of radon gas in 1979. The gas was traced to radium manufactured at a nearby U.S. Radium Corporation plant in the 1920's. Waste from the plant was used to fill in an area where homes were built later.

EPA and New Jersey have been digging up the soil and storing it in 55-gallon drums and metal boxes according to the Spokesman-Review March story. In December, the state Nuclear Waste Board decided not to allow the soil to be dumped at Hanford commercial site because of it's low radioactivity. On the other hand, the Dawn pond meets the requirements of being at least 10 feet above the water table on land that will be unoccupied and have a perpetual source of funding for maintenance. The pond has been licensed as a uranium tailings dump for years since the mine was shut down and the mill closed in 1982.

The application did not say exactly how much soil would be dumped said Nancy Kirner, supervisor of waste management of DSHS's Radiation Control Division. They had been notified that 2,700 barrels were ready for shipment. Kirner said Dawn had given the state wide ranging estimates of needs from 22,500 cubic feet in 4,000 barrels to 12 million feet in 11,000. "You're starting to talk about a lot of trucks!" she said.

On Feb. 14, 1988 the problem still existed and David Bond of the Spokesman-Review wrote about it. Bob Nelson who lives in Chewelah was the operations superintendent who told Bond of hopeful solutions. Dawn wants to return the land to it's natural condition, not by hauling away the radioactive tailings, but by importing new material to the site. There are four uranium mill-tailings dumps. They want to have unfenced area with grasses and trees and a place where the hunting ought to be good, "and the game won't glow in the dark." Dawn has proposed to accept 10,000 barrels of the New Jersey dirt – material they say is less radioactive than the mill tailings already on the site.

Being paid to bring in the New Jersey dirt for fill seemed like a natural solution for Dawn, because Nelson said the company lacks adequate funds to reclaim the pond. But the idea of importing even low-level radioactive material from New Jersey was too much for the neighbors and environmentalists, who turned out in force to oppose it. Dawns application was initially denied in May 1986. Dawn then offered to spend $1.50 per cubic foot of material accepted; and would pay the state an additional 50 cents per cubic foot, and then would close the facility permanently. Nelson said "they can buy beer with it if they want to."

The dirt was to come up for bidding but due to delays, Dawn was not allowed to bid. Another dump site in Tooele, Utah, similar to Dawn's, was bidding. The dirt was contaminated with radium from a now-defunct manufacturer of luminous watch dials, Jim Staples of the New Jersey DEP said. The question seems to be whether what is already at the dump site is not worse than the solution ,but difficult for the "neighbors" to visualize, or give permission to change. The situation is still in "limbo."

SHERWOOD – Western Nuclear Company

The **Sherwood Mine** was actually discovered in 1955, by two prospectors who after much exploration and mining 305,000 pounds of the ore, let the lease terminate in 1964. On Sept. 21, 1967, Western Nuclear, who had taken up the property, said new ore was found beneath the former open pit workings. It was said under the old dumps. Drilling reports gave an estimated 10 million pounds worth about $80 million. At the time it was said that Silver Buckle had been the former operators.

On Nov. 2, 1975, the Spokane Indian reservation was "teetering on the brink of a new mining boom." Edward W. Coker Jr. of the Spokesman-Review wrote that Western Nuclear, Incorporated, a Denver company was developing the **Sherwood** which could be much larger than the **Midnite.** $10 million was to be spent on development for the open pit works and the mill.

The key to the new operation was when President Ford announced in June that he would seek Congressional approval for private sponsorship of all new uranium enrichment plants. Under the Ford plan private industry would be encouraged to sell billions of dollars worth of enriched uranium to nuclear plants around the world. Up until his announcement, Yellow-cake (uranium oxide) hovered between six dollars and eight dollars a pound. After the announcement and in 1975 that price went to $22 per pound and was up to $32 by 1980.

On July 15, 1976, Executive Director for the Spokane Tribe Glenn Galbraith said actual production could be as much as 18 months away and the financial impact on the tribe as much as four years away. Western Nuclear a subsidiary of Phelps Dodge Corporation indicated they had staff requirements of 135 for the 2,000-ton-per-day plant and mine. The hope was to hire from the tribe but Galbraith said he wasn't sure that the tribe could provide 50 percent of the workers. He pointed out that the opportunity was there for those who would work.

He explained the tribe had a lease with Western Nuclear and the returns were to be used for a 1,700 acre irrigation project to cultivate more acreage. There was the need for a larger sewage treatment system, and a need for a water system .He also discussed the additional traffic onto the reservation and said there would be need for more law enforcement; schools and housing. "We will be able to upgrade our school out here," he said.

James H .Stevens, superintendent of the Bureau of Indian Affairs of the Spokane agency said "generally speaking - mining is a boom or a bust situation." He pointed out that examples could be seen all over the west. He did remind that ore is not a renewable resource and according to predictions the **Sherwood** reserves would be used up in 10 years. "One thing the tribe will be able to do with the money received from the mines is to buy back lands on the borders of the reservation," he said.

In 1977, the **Sherwood Mine** opened. It was termed a $20 to $40 million project and another report said they expected to produce $25 million a year. On May 10, 1981, Larry Young wrote for the Spokesman-Review that it looked like a giant's sandbox. At **Sherwood**, big scoop shovels dug up to 15,000 tons of waste rock daily to recover each day's output of 2,200 tons of "hot" ore. Some the size of coffee cans, some as big as a room but Paul Blair, resident manager said about the operations, "what we're doing is like picking the raisins out of a fruitcake."

At the time of this story the **Midnite** was on production from the 1977, discovery of reserves that would yield 100,000 pounds of uranium oxide. Randy Anderson, a geologist pointed toward another larger pit at Midnite (some 20 miles north of **Sherwood**). He said "radioactive ore here is like pearls on a necklace - we're always looking for the next pearl." He added that "we have to look harder here than over at Western Nuclear."

The **Sherwood** being the largest of the two mines employed nearly 30 and represented a $50 million investment from mill to tailings pond dam. The mill processed 2,000 tons of ore daily. Spokane Tribe's general manager Clifford Sijohn said, "we welcomed the mines on the reservation because of our high employment." The Tribal Chair, Alex McCoy did say the council was disturbed about the impact on the reservation lifestyle.

The Tribe sought federal aid for assistance for more schools and housing but were told there was no funds. They were also very concerned about mine closures which Dawn did from 1956 to 1964, due to price. The $32 price had fallen to $25 by this time in 1981, and it had been as high as $42. (1989 price is around $12.) On the plus side was employment quotas in their lease agreements which the tribe wanted increased. Indian workers were to have first consideration. Royalty payments were also received. Larry Young said "the tribe is secretive about these payments but it is known that in the year before, that Dawn paid $1.3 million." He added, "if **Sherwood Mine's** royalty agreement is comparable (being four times as big), the payment would be in the $4 million range." - but no one will say.

Blair did point out to Young the 50-gallon drums each having 900 pounds of yellow cake in it and although worth $20,000 apiece at the time, no one had stolen one. (That is how strong the taboo is concerning uranium.) Entering the mill was a surprise as it looked more like a winery with it's vats and the walls "squeaky clean because of the radioactive dust and possible radon gas."

At that time Dawn was building a new four million dollar tailing pond, Thompson said it would be the first one like it in the United States. He explained "there never was a radioactive leak but the problem was high sulfates and the government would not allow a discharge if not natural, even if it is not harmful."

The 1981 article went on to explain that Dawn was collecting $5.30 for each pound of U308 sold to be used for the land reclamation fund which will restore the area to forest and meadows. One slope had been covered with dirt, fertilized and seeded with grass to hold the soil and Dawn had planted 13,000 seedling trees on the slope. (Now retired as executive director, Glenn Galbraith told this writer on Feb. 7, 1989, that the New Jersey dirt deal had been given up.)

At the 1983 time, like all nuclear workers, Thompson wore his blue badge which was sent to Hanford each week and read by computer. He said no one has ever had a high reading and asking the guy who makes the check how they compare with badges worn in the Deaconess Hospital X-ray department, he was told "your badges are 10 times lower."

The mines are both shut down now "not for lack of ore", explained Galbraith. He did say that the tribe accepted a proposal for the Western to operate the mine again and provide management, with labor coming from the tribe. This was done in December, 1988, but no further information is available.

Kettle Falls and Meyers Falls Mining Districts History

Without the actual geographic description, these mines are often mentally miss-located because the old Meyers Falls is the town of Kettle Falls since 1941. Most of the Meyers Falls mines were located on Gold Hill just to the north of the city. The early town of Kettle Falls was flooded out by backwater of the Columbia River when Coulee Dam was built and the people moved up higher and joined the Meyers Falls people with the name eventually becoming Kettle Falls.

In Circa 1814 or 1816, Northwest Co. built a grist mill which was operated for 14 years. Located own at the falls on the Colville River, it was torn down and a new one built in 1830. L.W. Meyers came from Canada; secured the mill and site on Nov. 19, 1866, and operated the grist mill until 1872. A later one was operated until 1889. This mill served a large territory to include those at the actual falls of the Kettle Falls area; and up and down the Columbia River. In 1890, the town of Meyers Falls was platted by Jacob A. Meyers, and G.B. Ide. The village had existed since 1862, and the name was given in honor of L.W. Meyers.

Not far east of the town was the St. Francis Regis Mission built around 1870. In 1902 there were nine teachers, and 150 scholars. The large ranch was expanded in later years and a school and settlement for the "Sisters" was built across the road. In most recent years the farm was operated and owned by Earl and Evelyn Enright. They sold a portion to Leonard Fuhrman; and State Representative Steve Fuhrman and his family now own and live in the home last associated with the Mission property.

Miners and prospectors going and coming from Republic to Colville branched off onto Gold Hill. **Right Side,** or sometimes called **Gray Eagle** was located and owned by Stambaugh and Nelson of Kettle Falls. They alleged that Bob Veatch "jumped" the property and leased it to J.A. McJunkin. A suit was brought by the owners against McJunkin which suit also contended there was several tons of ore at the depot belonging to them.

This property said to be one of the more prominent claims on Gold Hill, had over $2,000 spent on development, according to a July 23, 1897 report. By Sept. 3, 1897, **Right Side** litigation had expanded into five lawsuits. An editorial comment: "This bids fair to be a duplicate case of the litigated cow — one claimant gets the horns, another the tail, while the lawyers get all the milk."

Patsy Clark, more noted for his mining properties in Rossland, B.C., and his **Republic** Mine in Republic, had also done some prospecting and grub-staking in Stevens County. He had been working the **Right Side** and on Jan. 8, 1897, he was said to shut down the work temporarily. The shaft had been sunk to a depth of 60 feet. John Argus and P. Hollahan had been working at the mine and on May 17, 1897, they leased the **Right Side.** Sorting rock on the dump for shipment they were said to have struck fine ore just 20 feet from the old shaft.

Gray Eagle and **Koyote** were reported in 1920 to be a part of this property and George Schenk and Jim O'Neal were other names mentioned. The mine produced a small amount of copper in 1918, and under lease to Earl Gibbs (of the **Bonanza** and **Old Dominion**), he took out some 50 tons in 1954. His brief lease was from Chamberlain and Miller. assays were in copper, silver, gold, and lead.

Niagara Mining Company was said by the Statesman-Index on July 2, 1897, to be negotiating with **Gold Hill** in an incorporation. By July 9, 1897, that paper said that Galbraith and Staler were preparing the papers. A news release in the same paper June 25, 1897, had said that "an enforced lull is on in mining deals – there are legions of locaters (prospectors) but miners can be counted on one hand. Owners of properties have failed to show faith and buyers are fighting shy. Capital is cautious – when mines are developed there will be no trouble getting capital to do the rest."

Niagara had just shipped one carload of ore to Nelson, B.C., and the company assured a carload a week July 16, 1897. On August 13, A. McNeilly had contracted with **Niagara** to deliver that ore to the depot. He would transport by pack animal to the first bench and from there to the station by wagon. A carload was shipped to Trail smelter that week. Under the name **Gold Hill,** the gold, silver and lead property was owned by G.L. Duckworth in 1942.

Gold Reef, or **Benevue** was another old timer on Gold Hill. On June 14, 1915, a strike was made, said to be a vein three feet wide and eight inches – with high-grade gold on a foot wall. No production was reported and in a 1920 report the mine was owned by Mr. Nunn. There were four claims and values were said later to run in gold from $3.50 to $28.50 a ton. A later report said by 1935 that $100,000 in ore had been taken from the mine – we have no verification.

In September of 1969, a news report said the mine will open soon according to Harm H. Schlomer, president of Gold Mines Incorporated. That corporation purchased the deeded ground from Anton Backus, a former TV and screen star, of Santa Barbara, Calif. Schlomer said they hoped to hire two or three men for this small operation and added "this is a small operation but could show a good profit if worked on a small scale." **Gold Reef** had last been worked in 1946-1948. Dr. Fred M. Handly, geologist on the faculty at Washington

State University said, "the mine has exceptionally favorable geology to produce rich and continuous ore bodies - given intelligent management, and suitable milling equipment."

Just a few of the many other claims on Gold Hill and that vicinity include: A.O. & F. owned by W.H. Oakes, W.B. Aris and E. Blackmore. Assaying $122 in gold and silver to the ton, they expected to make shipment Feb. 1, 1897. **Mother Lode** was reported putting on a force of men May 14, 1897. They were also negotiating with L.W. Meyers for water power from the falls where they contemplated a smelter that summer. On May 28, the paper said C.L. Betts of the J.I. Daniel Company in Spokane, was the secretary of the Mother Lode Mining Company. Oatman and Judson, large owners, had gone to work in earnest and had completed a first contract by July 16, of that year. They planned then to begin another tunnel 50 feet deeper. No returns were available for this claim.

Raymond Brown, S.A. Milliken, and R.C. Richardson from Spokane, were working **Birds Nest** on contract; and William McArthur and Charles Fish leased the **Sparks & McMillan** Mine in later years. **Black Jack** Mine was owned by C.W. Gardner and James Black of Meyers Falls. William Carmichael had some association. That mine was capitalized for two million shares at 10 cents each and according to a Dec. 25, 1899 news report they were considering a pending offer of $75,000.

Other claims that made news included the **Chloride Hill** and **Rattler.** A certificate was filed March 27, 1897, in Colville and in Lincoln County March 31, 1897, - to Lillie M. Richards, formerly Lillie M. Coleman - a forfeiture notice for $200 for labor and improvements. It was signed C.G. Snyder, A.W. Turner, E.E. Plough, and H.H. McMillan.

Those claims attributed to Kettle Falls lay more to the south of that first settlement. The former townsite, and the present one are at, or near the crossroads of two major highways and have been the center of a great deal of history. The earliest to visit this area were the transient Indians who came from all directions to "the" Kettle Falls on the Columbia River, to fish for salmon. Standing at that great and powerful falls, they used wicker baskets, or spears to catch the salmon which were dried on the banks and made ready for winter food. Chief Joseph of the Nez Perce tribe spent his last days on the adjoining Colville reservation.

Immediately following the Indians were the Hudson's Bay people who established "the" Fort Colvile (note spelling) near the falls, in 1811. David Thompson was one of the earlier ones to come north through the Colville Valley; saw the mighty falls, and canoed down the Columbia River. St. Paul's Mission was established nearby in 1826 on the grounds that Father DeSmet had visited the Indians much earlier. He,and Father Ravalli are credited with the Mission being where it was active until 1871.

Kettle Falls on the Columbia River where the Indians fished for salmon until the Coulee Dam backwater destroyed that portion and formed Lake Roosevelt. (Courtesy Pat Graham).

A photo of portions of the HUDSON'S BAY COMPANY Fort Colvile of the 1820-2 (Photo courtesy Pat Graham.)

The old St. Paul's Mission at Kettle Falls prior to restoration in 1940. (Photo courtesy Pat Graham.)

Restored in 1940, and again in 1951, St. Paul's is an important part of an interpretive center being built on grounds overlooking where the Hudson's Bay Fort Colvile (one L, Lord Colvile) was built. The history and exposure of that most important fort was done by David H. Chance from the University of Idaho, who headed the team who did the archaeological work; and wrote about the "digs" and their finds.

It was Angus McDonald's aid at the mission who first found "colors" in the Columbia River and started the first search for wealth northward up the banks and mountain-sides of the Columbia River valley. (Angus McDonald was chief trader at the mission from 1854-1871). And then the race was on!–

In the wake of the earliest prospectors and miners came the Chinese. By 1865, there was said 300-400 doing intensive placering along the banks of the Columbia from Hunters, north to the Canadian border. Many of the earliest men to arrive in the area were looking for gold but from that group began the settlements of families along the river banks. In 1888, Marcy H. Randall was said to have built the first cabin at what would become the first Kettle Falls village. Major interest at the time was a way to convert that tremendous force of water from the falls on the Columbia, into electric energy.

There were those who visualized a "killing" in this new country with all it's potential. One such enterprising corporation was the Rochester and Kettle Lake Land Company with an engineer in the hardware business in Rochester, New York, John Goss and his partner in the plans W.B. Aris, a promoter from Rochester. The two are credited with platting the townsite on Aug. 14, 1889. Forming a corporation with a capital of $500,000 they proceeded to promote the area and to go about the building of a beautiful hotel said the largest west of Butte. Cost of the hotel was at $18,000 and furnishings were said to cost $9,200.

D.C. Corbin was building his railroad northward and the corporation anticipated that he would "have to" build through Kettle Falls. W.B. Aris was general manager and had elaborate plans for the town, counting 1,000 population by 1891. Orville Dutro tells that lots were selling there for $200 and jumped to $1,000 and $1,500.

For reasons of his own, Corbin by-passed the town and went up the river a different route. Along with that great loss; the awful, jolting ride from the east – much of it by carriage, cooled the interest of those who were ready to finance this proposed development. By 1903, the "grand plan" was a failure. Lots dropped in price as low as 50 cents. One historian tells that Mrs. Martin who operated

The handsomest hotel west of Butte was built at Kettle Falls by ruthless land developers. Open by late 1890 the entire investment was lost by 1903. (Kettle Falls Souvenir edition– S-E).

the hotel, was offered $90,000 but she was holding out for $100,000 when the bottom dropped out of the development plans. There was also said to be an offer of $300,000 for the hotel, for 147 lots and two houses – all became a total loss. Some 40 houses were moved to new settlements. By mid-September the death-palor hung heavy over the town and John Benson was reported to have committed suicide.

Following this debacle the town was slowly built again with mining people, and those interested in farming and raising orchards. By 1941 their roots were torn loose again when they had to move up to higher ground with the flooding of the area from the back water of Coulee Dam—and it's forming of Lake Roosevelt. Combining Meyers Falls and Kettle Falls has been a good economic move with the town now becoming a very active center of business and recreation.

Some of the Kettle Falls district locations became active mines, many others remain claims. **Sugar Loaf**, or **Vanasse** had some 12 claims on a State lease by Luther "Luke" Vanasse, of Colville. On May 28, 1897, the property was owned by Paul LaPlante and Sigmund Dilsheimer, a Hall Mines investor. It was near the Mission. Showing high assays, LaPlante bonded Dilsheimer's interest for 90 days with three men working. No report of later activity except that at one time John R. Cook was interested in making a deal.

Acme is one of the better known properties. It was located around 1887 and has two patented claims **Acme** and **Dora**, and one claim **Columbia View Lode** by assessment. The mine is located near the Heidigger ranch close to Rice, Wash. Some ore had been taken out by 1891. On May 7, 1897, the local news reported that J.D. Gookey of Harvey, Wash., reported a strike at **Acme,** said six feet of solid mineral with assays $150 in gold, silver and lead. The claims were bonded to the operators who predicted finding a great value.

A week later the news stated two shipments had been made to the smelter and the owners were listed to be W.H. Hulbert, I. Kaufman, Sig Dilsheimer, and E.B. Burdick. The following week's news said a road was being built from the mine to the river road. They planned to have a steam hoist soon and then would commence to ship ore.

A report on Dec. 25, 1899, said the "strike" vein was 64 feet wide with no walls and assayed $60 to $88 to the ton. A 215 foot tunnel, then another one was built. At that time it was announced that the property originally owned by Dilsheimer and some Portland and New York capitalists, was bonded to a San Francisco party who was very pleased with the purchase.

Some mining continued and on June 23, 1901, the first misfortune occurred when William L. Aldredge was killed in a blast at **Acme.** He had come to Stevens County in 1891, to be near a brother, James F. Aldredge. Better luck was reported Aug. 6, 1915, when W.F. Stevens of San Francisco was said to have taken an option to claims owned by the Dora Mining Company, for a consideration of $30,000.

Very little work is told in the early 1920's and the property eventually was owned by Louis Strauss, Colville merchant. (Sig Dilsheimer was Strauss's cousin and both names come up in different mining investments.) By the 1943 report some 1,600 feet of work had been done but at the time of investigation the winze was partly full of water and not accessible. Assays were said then to be $44 in gold, silver and lead. Since Mr. Strauss's death in the late 1970's, there has been one further report of activity; a brief lease showed little results.

Blind Discovery had been a great hope and was added to the **Silver Queen** property around 1904. J.F. Sherwood and P. Larson were listed as owners, with others. The **Silver Queen** was the better known property south of **Blind Discovery,** and **Vulcan** and **Fannie**, was also listed as a Larson property. The **Silver Queen** was located in 1892, by G.I. Budd and his brother J.J. Budd, and in December of 1899, was said to be the property of Silver Queen Mining and Milling Company with 33 claims. It was two and a half miles south of old Kettle Falls and assayed very high-grade ore – some selected samples showing 3,000 ounces silver and no assay was less than 101 ounces silver to the ton, with some lead. J.F. Sherwood was president and Peter Larson was secretary of the corporation. In October, 1896, the company had a meeting to increase the stock from 640 shares to 500,000. At the time Larson was said to be president with J.P. Fogh, H. Nelson, and J.J. Budd some of the officers. (Larson may be Lewis P. Larsen of Metaline Falls).

On Oct. 25, 1907, the Ark Mining and Milling Company of Ymir, B.C., took a working bond on **Silver Queen.** In 1915, 14 tons were produced. The mine was down for awhile and in 1923 was said started up again. In 1936 it was leased to Continental Silver and Smelting of Seattle and by 1937 reported 1,700 tons mined. In 1939 another 250 tons were reported. In a Sept. 9, 1937 report the company said the new mill was "grinding out silver, lead and copper," according to J.J. Budd. In 1941 the mine operated a short time with a 50-ton mill and full air mining equipment. It was down again in 1943 but total production was said to be $18,000. It was started again and in 1949 reported 841 tons and on Jan. 6, 1967, was said working after several weeks being idle. D.C. Wakefield, was said owner in 1950.

Mining Tragedy

Of the many Kettle Falls district mining properties a very real tragedy occurred at the **Eagle And Newport.** Owned by Mr. and Mrs. John Marty, and H.E. Ross by the early 1940's the two unpatented claims and 160 acres of ground had been worked for years. Listed under Silver Eagle Mines, Incorporated by the Rice couple and a Spokane mining firm, the company was capitalized at $500,000. That figure was set by Big Nine Mine and Mineral Incorporated and the Marty's – each of whom owned half interest.

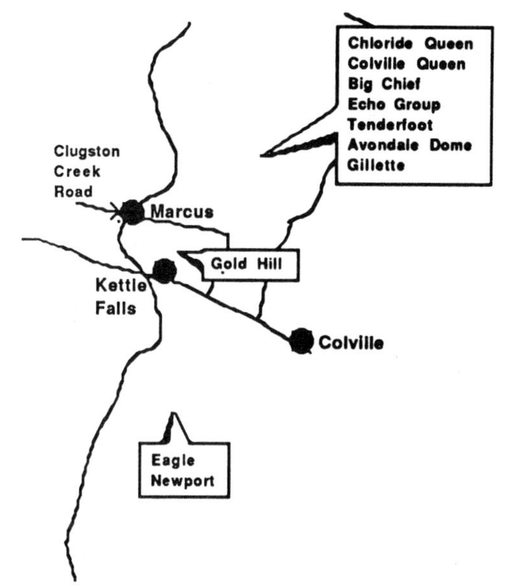

Extensively developed they wanted to put the mine on the stock exchange May 16, 1968. The original claim was located around the turn of the century and about $300,000 was spent by former owners sinking a two-compartment shaft 228 feet and tunneling about 200 feet on a vein showing values in silver, copper and gold. On June 17, 1968, Leonard H. Porathe, of Opportunity was president; Victory V. Wakefield, Spokane was treasurer, John Marty was vice-president; Mildred I. Rockne, Spokane was executive secretary, Harold H. Rusch of Spokane and Jo Baxter were also officers and directors.

On Dec. 13, 1968, John Marty told the Colville Chamber of Commerce that he estimated between $500,000 and one million dollars would be spent on the property, with a mill and even a smelter in the plan. He announced the stock was for sale. Marty and Big Nine were holding one million of the five million share at 10 cents each with the money to go for development. He told the chamber that the mine had a shaft and rail, ore car and hoist, plus 1,000 feet of drifts and cross-cuts.

Later he told that the investment at the mine by May 16, 1969 was over three quarter of a million dollars. On Jan. 9, 1970 it was said they were hoping to expend funds to re-ladder an old shaft 180 feet, and pump the water out of the workings to that level. They would also re-timber a tunnel and drive a 60 foot drift at the 180 foot level.

It was that work that led to the worst mine disaster ever known in Stevens County. On May 22, 1970, John Marty (76); Gerald D. Newton (39), both of Kettle Falls, and Louis Koerner (63) Rice were all found dead at the mine.

The official report made by Robert H. Price, mining and explosive inspector for the State Departament of Mines, said that all three died of asphyxiation by carbon monoxide gas. The gas was caused in the mine by a small gas driven water pump trying to pump out that water at the 180 foot level.

Sheriff A.E. "Dutch" Holter's office was informed about 8 p.m., after Mrs. Marty had walked three miles to the mine when her husband failed to return home. Members of the Colville fire department were called to the mine. Volunteers Chief Bob LaPlant and fireman Jack Citkovitch using oxygen equipment, went into the mine. They found Marty's body, and the other two a few feet further down. The small gas pump used to pump out the water had left monoxide gasses, killing the three well-known miners.

Clugston Creek Mines– CHLORIDE QUEEN or COLVILLE QUEEN

As prospectors became more numerous some were spreading north from Kettle Falls, and some south from the Ferry-Stevens County dividing line. In 1887, the **Chloride Queen** was located. There were five claims listed with ore in lead, zinc, and silver. Shipments of some 400 tons were made to the old Colville Smelter and later four carloads were sent to Tacoma and Everett smelters with silver assaying the highest at that time. On Clugston Creek, the mine was included in the Colville district.

In 1914, two carloads went to Northport smelter, and on Nov. 15, 1923, a silver-lead strike was reported in the local paper at **Colville Queen.** Another carload was shipped to Kellogg in 1924. At that time it was discovered that there were two apparently separate ore bodies; the lower one principally silver, and the upper one lead. Colville Queen Company took over the property and for five years there was little or no work done.

In 1943 there was 5,000 feet of work reported by inspection; five or six buildings and an old mill. Nothing was in use at the time but in 1944 a concentrating plant was said operating on a three shift basis. From 1949 to 1951, Clugston Creek Mining Company was reported leasing from the Colville Queen Mining Company and by 1951, Clugston Creek mining Co. was said owner of the **Big Chief** group of claims. They had also purchased the old **Chloride Queen.** The announcement was made by Thomas LePage, of Spokane, president of Clugston Creek Mining. He said his firm acquired 420 acres deeded ground and now have 700 acres under lease, bond, or have paid a substantial cash payment.

June 19, 1955 James Keeley Colville, was said to have purchased **Colville Queen** for $25,000 cash.

Principal stockholders were H.D. Baker, Redmond, Ore., Harold Gleason, Spokane, and M.A. Rodman, Colville. In 1965 and 1966, the old property was being developed by Engineered Mines, Inc., with Robert Hundhausen from Hayden Lake, Ida., president; Daniel Collins secretary, and James Keeley part of the firm. 1,100 acres of deeded property and the **Echo Group** in addition to **Tenderfoot, Big Chief** and **Chloride Queen** were being worked.

A new open pit mine was expected to go into operation in the spring according to an announcement by Hundhausen Dec. 18, 1965. The property would include the old **Chloride Queen.** Bulldozers were stripping at **Tenderfoot** and opened an area containing commercial values in lead, silver and zinc, according to Hundhausen. Those values were also found in a fissure vein in another part of the property. Associated with him was Keeley, and also the late Clair Wynecoop of Wellpinit.

For a short time Bunker Hill had leased **Colville Queen** and **Big Chief** and discovered a small ore body. Hundhausen had agreed the open pit would work and Engineered Mines, Inc. was organized to engage in general mining operations. Articles were filed in Olympia by Daniel Collins, Colville attorney, and capitalization was set at $50,000. Incorporators were Hundhausen, Keeley, Wynecoop, and William Schauls of Kettle Falls. Collins was the secretary, Hundhausen the president and Keeley the treasurer.

On March 6, 1967, Clugston Creek Mining Company of Spokane levied an assessment of two mills per share on outstanding capital stock for expenses, according to Mrs. LePage who became president of that firm when her husband died in September, 1965. The company had entered into a contract with the Engineered Company to explore the Clugston Creek's Stevens County property; with an option to lease it for 30 years and give Clugston a royalty of seven and a half percent of any net smelter returns. Earle M. Peterson was elected trustee.

The Gibbs Mill at Palmer Siding had been operated for a short time by Calix American Corporation, and Arco Industries Incorporated was now operating it. There was to be 50 jobs there according to an Aug. 22, 1969 report. Ottis Pruitt was general manager; Carl Wilde, mill superintendent and Carl Shields, mine superintendent. Ore was being trucked in from **Blue Bird Mine** near Kettle Falls, and **Daisy Mine** near Daisy. Negotiations were going on for ore from the Clugston Creek properties of Keeley and Hundhausen. Arco also said they were negotiating with Clugston Creek owners, to purchase their property. Fortunate for Keeley and Hundhausen that deal fell through as it was not long before Arco was in court on Income Tax problems.

In October, 1970, certain trouble was evident with the Clugston Company and a stockholders meeting was being called. Orville L. Moe was asking voting proxies and said he and his father Clarence had some time ago purchased and optioned controlling interest in the Clugston Company from the late Mrs. LePage, and others. By November 1970, the Clugston Company had increased it's number of directors; and put Orville Moe in as president. Thomas J. Dillon was made vice-president; E.W. (Enid) Conrad, secretary-treasurer and Robert E. Kovacevich named legal counsel.

On Aug. 8, 1976, William D. Roberts was the new president of the Clugston Creek Mining Company. The two mill assessment levied in 1967 was dropped. At the time there were 2,426,042 shares. Since that time, the property has been sold again for the timber and mineral rights. Actual output from the 102-year-old mine was never available but it persisted, and operation was carried on for most of those years.

There were numerous other locations on Clugston Creek. **Avondale Dome** formerly known as **Tenderfoot,** and Also **Gillette,'** was said by 1941 to belong to Lloyd Jacobson and Eric Ecklund. Several carloads had been shipped prior to 1890. **Uncle Sam** was once known as **Eureka,** or **Surprise.** Jesse R. Hall was interested in this property when he lived east of Bossburg. Around 1896 he bought the Colville Standard newspaper where he was said to have been the editor. It could not have been much challenge because he sold in a year and went mining. He had come to Kettle Falls in 1894.

Hall's property was on the west slopes of Jumbo Mountain said near the head of Clugston Creek. Optioned by New Leadville Company in 1924, a chimney of lead was mined and shipped. Renamed **Hi Cliff,** the two claim were said owned by J.D. McDonald of Evans. There were three tunnels, some ore in the bunkers but no figures on production. **R.J. Mine** was relocated to form part of **Hi Cliff** and some ore was shipped. That property was said located in 1900.

Silver Chief in the late 1920's, was said to be owned by Hazen & Jager Spokane morticians; Biesen, of Spokane, and Herb Minzel , former Colville mayor and Ford dealer. Showing silver and lead, there was a 1,700 foot tunnel. Davis, and Pomquist and son were promoters. Their remark was pretty universal amongst miners, "Had to get to the hanging wall." (to find ore). On July 31, 1930, a crosscut at the end of the 1,500 foot tunnel hit a three foot vein of silver sulphite ore – very high-grade. No reports of production were given and the claim was closed 1957-1958.

Silver Maid Mining Company (the mine of the same name) was purchased around 1896 in October, by H.J. Earnest and J.E. Ettinge, from F.B. Goetter of Colville. Price was said to be $1,000 and the reports were that **Silver Maid** had a three foot

One of the freight lines to mines of the vicinity, and to haul passengers, was generally driven by George Wiley of Kelly Hill. (Stevens County Historical Society).

solid lead and silver ore vein, and was shipping. By March of 1897 Henry Earnest was considering an offer for **Silver Maid.** No production reports are available for this location. Goetter was a prominent Colville druggist who became interested in numerous mining properties.

Bossburg

Two of Stevens County's major mines, the **Bonanza** and the **Young America** have been discussed, and were in the Bossburg area. The town had a population of from 400 to 600 between 1901 and 1903. It is 20 miles north of Colville. First platted, as Millington, it was later named after C.S. Boss a leading citizen. From 1897, through 1901 it was most prosperous due to mines in the area. It was also a shipping center and in the later 1890's was said to have all kinds of freighting outfits going both north, and south. It had telephones, a newspaper, saloons, a general store and two hotels. Sadly, after the boom, the town very nearly closed down.

Some of the better known local citizen miners were J.M. Bewley who on July 6, 1897 was said prospecting and developing properties near Bossburg. In the 1907 Northport News of Nov. 20 we learn about B.W. Chapin a resident of Bossburg who crossed the plains of Montana with Captain Fisk in 1864. Chapin had found the **Gold Coin** claim near Helena, near the present **Spring Hill Mine** on his way to Washington. Locally he was interested in a number of locations. William J. Gilpin of Spokane was said to have discovered the **Columbia River Marble** in 1900, some three miles from Bossburg, and William Stearns located **Little Giant** near that town. In 1934, placering was going strong on the west side of the Columbia River. A property owned by the Indian Department was said to have produced $17,500 from less than two acres of gravel three foot deep, by placering.

Flat Creek and 15 Mile Creek

LaFayette Mining Company Incorporated had headquarters in Kettle Falls but their property was listed in the Northport district. By May 21, 1897 R.B. Lane and the Ledgerwood Brothers of Colville had established their holdings at the head of Flat Creek along with the **X-Ray** Group. On June 11, 1897 F.A. Edgerton was listed as manager of **X-Ray** property and a heavy stockholder in **LaFayette.** Development work was well on it's way and the work force increasing when the **X-Ray** Group was sold to an English syndicate for $35,000. That was sold to an English syndicate for $35,000. That report Dec. 25, 1899, said the property assayed well in silver and some gold.

Forest fire had driven the Ledgerwoods, Frank Goodwin and E.D. Miner from camp in 1896. The loss of buildings and supplies had required a new start in 1897.

The **Talisman** was another local property along with the **Napoleon.** The **Napoleon** was first mentioned from the Republic side of the Orient district. It was owned in December, 1915, by British Columbia Copper Company. There are two claims just outside the town of Boyds. Values run in gold with a massive body of iron. Three were 4,000 feet of tunnels, and a large tonnage had been shipped to a smelter as flux. Portland Cement took over the property and was mining the surface zone oxides for use in the cement industry. Tonnage was approximately 2,000 tons a year.

The property had been mined to a considerable extent,ore shipped to a Greenwood, British concern. A mill had been erected in 1910, to extract the gold. By 1934, a report did say the mine had shipped a large amount of low-grade ore.

Fifteen Mile district was in line for prospectors coming south after the Colville Reservation was thrown open for mineral search in 1896. The district is in a triangle between Kettle River, the Columbia and the border. The first claims made were the **Ibex** and **Antelope** by John Hall. The **Orpha** Group followed with values in gold and copper, and the **Minorca Group** located by King and Wells. 50 tons of ore was said mined from the **Antelope** assaying as high as $65 a ton but the claim was abandoned by 1941. About 100 tons of ore were reported shipped before 1920 with values in gold, silver and copper from **Minorca.** The claim was said located in 1897. **Homestake No. 1** was another location. 100 tons were shipped by 1920, with assays running $17 gold, $8.60 copper and $3.15 in silver. With locations happening rather often, 23 prospectors got together at the Flat Creek Store in 1898, and organized the Fifteen Mile mining district.

Marcus

To arrive from the south, into any of these districts one had to go through Marcus. The town came about first as "Old Fort", then Whites Landing, and eventually named Marcus after Marcus Oppenheimer, a prominent citizen and merchant at the time. There had been a general store in the village and by 1885, a post office. James Monaghan and Oppenheimer are credited in the Statesman-Index with being the founders of the town April 23, 1897. Marcus and his brothers, Joseph and Samuel were the merchants who dispensed goods to many of the miners going and coming through the area. A boat had traveled the Columbia for 12 to 15 years and brought much of the merchandise to the store. Other shipments were made the slow route across land from Walla Walla. Besides the many mines bringing in thousands of dollars, there was a mill at Marcus. The 60-ton plant was being built in 1935, at a cost of $20,000. William Brown who was said to waste a fortune from the **Silver King** in Nelson, B.C. started a ferry on the Columbia, and placered on Rogers Bar.

Marcus with some 2,000 population around 1900 was said to be the largest town in the area. Like Kettle Falls, it was covered with water when backwater came from Coulee Dam in 1941, and the people re-built their town up higher. It is still the crossroads to many of the mining districts mentioned here.

FIRST THOUGHT MINE – Orient Mining District

What would become Stevens County's largest gold mine was located shortly after the Colville Reservation was thrown open in 1896. Those who discovered the **First Thought** sold the location to a Bossburg prospector for a days wages. That man, P. Burns did a small amount of work and sold the property for $25,000. Development continued for nearly two years followed by some three years being tied up in litigation.

The mine is between two and three miles by road northeast of Orient on First Thought Mountain. It has 13 patented claims and 50 other acres. It was an almost continuous shipper from 1904, through 1910, with values mostly in gold ore. It was estimated the mine produced over 40,000 tons of ore having a value of over $650,000 averaging $15.50 to the ton in gold and silver. (Remember this was based on $20 gold.) From 1900 to 1911, Alex Sharp was general manager. (He died in Vancouver B.C. in 1921.) A 1904 report gave assay values running from three dollars to as high as $47 to the ton in gold, and for tax purposes the mine was assessed at $50,000 in 1903.

The property was eventually developed with three tunnels, a great deal of underground works; a 15 horsepower hoist and among other things, a 15 horsepower gasoline engine that ran the wire rope tramway 12,880 feet. The ore was carried in buckets on the tramway from the mine to the bunkers on the Great Northern tracks a short distance from Orient. It cost the company 33 1/2 cents per ton to transport the ore and load it onto cars. (Extreme caution should be taken by anyone wandering around the area – there are two tunnels from the surface connecting with the shaft and drifts – both to the right and left of the vein. Numerous open cuts, tunnels and shafts have been made on other claims of the property making them extremely dangerous to any trespasser!).

By 1904, Orient had five saloons, three livery stables, a bank, three hotels, two mercantile stores and a post office was opened there by 1907. A Kettle River Journal weekly newspaper of the

The town of Marcus and the old railroad bridge, 1911, before the entire area was flooded by Coulee Dam backwater in 1941. The Sheppard homestead is seen in the distance. (Stevens County Historical Society).

"Pierre Lake and Kettle River districts" dated Dec. 30, 1905 was turned in to the Statesman-Examiner in Colville. That Kettle paper was short of existence but long on "homey" type advertising such as: "The OK Saloon. Our Fountain head; Blue Grass Bourbon. A trial will convince you this whiskey is OK . . . I desire to please and that in view will always handle the best grades of wet goods." G.A. Dahl, proprietor. (We will assume this is Gustav Dahl who came to Orient in 1904-1905.) Other ads of interest was one for "Piso's Cure for Consumption. Cures where all else fails. Best Cough syrup. Tastes good, use in time. Sold by druggists."

A recent news story on Aug. 15, 1985, about Marge Dahl, Gustav's daughter-in-law who married Fred Dahl, tells that her grandmother Sarah Margaret Noble was a nurse for Dr. R.S. Wells the doctor we read about in earlier Northport history. Sarah Noble also had a boarding house in Northport. Marge and Fred Dahl lived their married life at Orient and he worked at the **First Thought Mine** until it closed down in 1942. Marge lost her husband Fred in 1974. Being one of Orient's busiest, she was a postmaster and is also involved in the First Thought Days held each year in August since 1976. It draws many of the old timers who worked at the mine and knew friends, and still have relatives in the area.

A 1923 news report said that work was progressing on re-timbering the mine but problems arose and the county had to take over the mine. On Aug. 25, 1933, they voted to accept an offer from J.A. Maginnis, of Seattle to buy **First Thought** for $15,000. By Oct. 20, 1933, the mine had to be put back up for sale by G.E. Gilson, the county treasurer, for lack of payment by Mr. Maginnis.

By Jan. 17, 1935, Russell Parker was operating the mine on a lease from Stevens County. He was the father of Mrs. Bill Parrott, and in-law to the late

FIRST THOUGHT MINE prior to 1912 (U.S. Geological Survey).

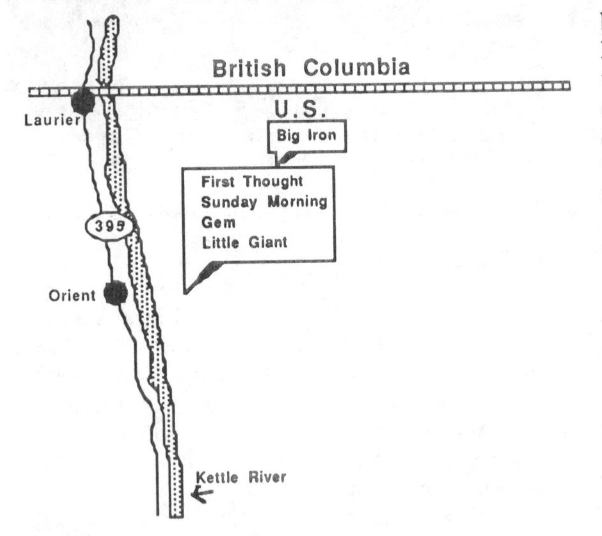

Bill Parrott, prominent Colville merchants. Bill also worked at the mine for Mr. Parker before opening his store in Colville. On April 24, 1936, Parker had ordered a 50-ton flotation mill for the mine. On May 27, 1936 a news release tells of an agreement for selling the mine to him for $27,000. That was concluded that week by Stevens County commissioners.

An earlier report from the Trail Smelter said the **First Thought** shipped 5,045 tons – making it the biggest independent shipper to the smelter. Knob Hill from Republic had only shipped 4,021 tons at the same time. By that fall, mining was really coming alive in all the district. In 1939, a diamond drill hole had encountered 175 feet of mineralization below the working level at **First Thought** – another encouragement. By 1941, the mine was said to have produced $1,350,000 in gold ore but by 1943, it was closed down for the war effort.

In the fall of 1968, Fred E. Costich of Springdale sent a letter to this writer stating that he owned four claims in the group and had for over 10 years. Unfortunately, he was another that time limited ever talking to him about the company.

In 1970 Silver Champion of Spokane was listed as owner of **First Thought** with Cline Tedrow, mining engineer and president of Silver Champion saying they were studying open-pit mining. The Company owned the mine from April 19, 1971, to 1976. That June there was a report of shipping ore having been found with values in gold, silver, copper, and lead, at $60 a ton.

Orient district had a number of claims to include **Sunday Morning Mine.** This one was three claims one mile from Rockcut. Values were in copper, lead, silver and gold showing $82.68 to the ton. Officers of the Company included Sig Dilsheimer, president, Dave Barman, vice-president, H.G.

Kilpatrick, secretary-treasurer. He was prosecuting attorney of Colville in 1896 and Barman and Dilsheimer had the Barman's Store. J.F. Demault was an added trustee. Capital stock was for one million shares of which 400,000 was treasury stock. The company was incorporated Oct. 2, 1899 for $50,000 at five cents a share.

Gem, on the north slope of First Thought Mountain and belonging to Paul Anderson, of Colville, assayed as high as $1,000 in gold and averaged $40 to the ton. **Little Giant** had one claim owned by George Bryant and Charles Taylor, of Orient. Some shipments were made with values in gold, silver and lead but no returns reported when first mined in 1898. A 1920 report gives the mine five claims with assays running $10 to $23 mostly copper, and in veins that run in bunches. A later assay showed gold and copper running $30 gold and 20 percent copper. First located in July, 1897 the mine made small shipments by 1914, but by then the shaft was said full of water.

On May 21, 1897, Jewell Mining Company of Kettle Falls incorporated with capital stock $750,000. George F. Batorff, Alonzo Moe, both of Kettle Falls; J.A. Willburn, F.M. Brown, and C.E. Chamberlain of Harvey, were all trustees for their mine on Toulou Mountain near the **Scotia.** That mine was discovered May 28, 1897, six miles from Bossburg. The miners were reported to have struck pay dirt at 225 feet. They sacked the ore and sent it to the Marcus mill with average values in gold, silver and copper running $200 and over to the ton. By 1943 **Scotia** was owned by the county. The several tunnels were flooded by 1914 and 1920. Some shipments had been made and Clyde M. Hawk was credited with once operating the mine. The Jewell Mining Company property is no longer listed.

By May 21, 1897, there was much news of the day – all "up-beat" about the mining activity. There was much going on in the district and more and larger pack trains than in the past. **Double Standard** was working three shifts, and **Yellow Jacket** was working every day. One report was that citizens on Onion and Deep Creek woke every morning to the sound of Jack Hammer and blasts of powder. There were good reports from all districts at the time.

In 1965 the L. & S. Mining Company from Wenatchee was developing a gold-silver property in the Orient mining district. The open-pit mine with reported ore reserves of 100,000 tons was announced by Silver Crown Mining Company of Spokane. The firm stockpiled several hundred tons of ore and had plans of operating a mill at the site. Little has been heard from this endeavor since that date.

BIG IRON – Pierre Lake District

Perhaps the most interesting part of early-day mining history are the people associated. Ownership of the **Big Iron** was one of the more interesting. The mine was said located in the late 1890's or early 1900's, by Boyle Brothers. It became the property of Colville attorneys H. Wade Bailey, Irving Jesseph, Herman Wentz and Lon Johnson. It had been leased several times to different parties and considerable ore had been shipped. Some of that work, and shipments were made by Colville mining man Archie Wilson, (also associated with **Buckhorn Mines** in Okanogan County).

There were several claims, and locations were still being listed from 1929 through 1950. There was a road to the mine and two cabins. In June, 1926, the **Big Iron** had 150 acres and was listed as nine miles from Rockcut near Pierre Lake. The Big Iron Company was incorporated. The mine was shipping 2,000 tons of ore a year to the Northwest Magnesite in Chewelah, which contained 90 to 95 percent oxides. Six dollars to nine dollars a ton in gold was also contained in the ore but was not saved at the time.

On Dec. 24, 1935, the mine closed for the winter after shipping 3,000 tons of magnesite ore to the magnesite. It had worked consistently for the past eight months and between 1924, and 1937, there were 35,000 tons mined and shipped. Some production was reported through 1941. Around 1939 there was said to be about 500,000 tons of ore in site with values in gold, silver and copper running as high as $50 in gold.

Since that time, and into the mid 1950's through 1970's, the mining camp became more of a place for R&R. Many an official found the quiet, the fishing and the hunting more interesting than the mining. It was the solitude that appealed to Edward R. Murrow. He became interested in the place when he came here with his brother Lacy Murrow, a former director of the State Highways – and he bought the place.

Most who read this history today will remember Edward Murrow the broadcaster. Born in a log-slab house in Pole Cat Creek in North Carolina in 1908, he rose to fame throughout the world. He will be remembered as one of the truly great men of the 20th century, his life devoted to freedom. He was also well-known to many in Stevens County as he visited the area in partnership at the mine with his brother Dewey Murrow, of Spokane. They planned to develop the property as more demand for the quality of ore came about. Murrow, his brother Dewey and son Casey visited in August, 1959.

He was also known to Former Colville florist Mrs. Vic Barnes, who attended school with him at Edison North in Seattle, when he lived in the next

Dewey J. Murrow, of Spokane (left) visits with his brother Edward R. Murrow, then of Washington D.C., in Spokane June 4, 1962. (S-R).

block from her. She knew him as Egbert, which name he changed later to Edward. She said he had to work his way through high school and during the last year was a school bus driver for Edison. He went from there to graduate at the head of his class in 1930, from Washington State University, at Pullman, working in the woods in the meantime to finance his education.

Fame came to Murrow in World War II in his CBS broadcast each evening "This is London" where he reported from roof-tops under the German blitz, and also from B-29's in bombing raids over occupied Europe. Following this his programs "Person to Person" and "See it Now" were seen throughout the Free World.

Although he was making $300,000 a year in TV, in 1961 he accepted President Kennedy's appointment to head the United States information service at $21,000 a year. A dedicated man, while working he was stricken with cancer of the lung. It was removed October 1963 and he returned to USIA but resigned in 1964. Edward Murrow died April 30, 1965.

In 1956, the **Big Iron** was said to be owned by the late Judge Lon W. Johnson. On Dec. 22, 1967, Copper Chief Mining Company uncovered some gold and traces of silver at the leased **Big Iron.** L.E. Neilson, was company president. The organization included L.X. Magney and two Neilsens. It was capitalized at $200,000 and at that report the Copper Chief Company planned to acquire property in the county.

People of Colville and the Colville Mining District

The majority of earlier Colville mining history has been told around the location and development of the **Old Dominion, Bonanza, Young America, Dead Medicine** and even the effect **Silver King** in Nelson, B.C., had on the area. It was regarding that mine that J.H. Young came to Colville in 1885, with James Duncan. Young was said to have bought an interest in the **Silver King** which later sold for $1,500,000 much of that returning to Stevens County, and Colville.

Interested in real-estate and in other mining property Young purchased and modernized by 1903, the Rickey building. In 1910, he had D.H. Kimple build a beautiful home off north Main Street overlooking the city. Young was said to have built a livery stable and to run a stage service bringing in miners, and the goods needed by them and other citizens. With his wife Ann, they lived and entertained in the home now shown by the Stevens County Historical Society. After Young's death, his widow Ann married Louis Keller in 1915. "Lou" was said to be a wealthy Cincinnati hardware man who came to Colville in 1907, eventually bought the Rickey building, and ran the Keller Hardware until his death. The Keller estate turned the home and property to the Historical group who feature special "open houses" and regular tours through the Keller House and Museum.

Another person, vastly interested in the area and especially it's "first citizens" was Mrs. Mike "Christine" Kitt a member of the Calispel tribe. The daughter of Peter and Susan Mullan, she was the granddaughter of Captain John Mullan who blazed the trail and built the first road from Fort Walla Walla, Wash., to Fort Benton, Mont. She was born at Valley, Wash., lived in Colville for all of her married life, and died Sept. 12, 1973. Mr. and Mrs. Kitt were highly respected members of the community as were their mothers. (Also spelled Kalispel)

Another who made very little "stir" in Colville but came from a noted name was Mrs. James "Margaret" Williamson Miles who later moved from Colville and died in Spokane Jan. 22, 1970. Born in Calgary, she was the ninth child in a family of 14. Her father was a native of Ireland and an early gold miner inthe northwest, coming across the Rocky Mountains by wagon train in 1849. He was Sam Livingston, for whom the town of Livingston, Mont., and the range in the Rockies called Livingston, are named; also the Livingston River. He was one of Calgary's "first citizens." Heritage Park there is on some of the land he once owned and his home now is part of the park in that city. A building dedicated in 1966 also bears his name.

Mrs. Miles came to Colville in 1920 and helped build her home that still stood near the city park at Colville until 1986 or so. Her husband was a rancher and was killed in a car accident in Colville. She is survived by two sons, James H. and the late

Prominent lumberman Fred Draper advertised his community each year with a photo on his calendars. This one December 1952 was probably taken around 1912 – looking "down" the street from North Main to the south end of Colville. (Stevens County Historical Society).

She is survived by two sons, James H. and the late Walter W. (better known as "Chief" and with whom this writer went to high school; and she also knew Mrs. Miles); and a daughter Mrs. Cecil L. Kohlieber, and grandchildren.

The world renowned Dr. Alfred Strauss came from Germany in 1892, and at age 11 years, made his first American home in Colville, Wash. The oldest of nine children, his Uncle David Barman, (whose wife was a sister of Alfred's mother), wanted Alfred to come to Colville to help in his store – then called the Boss Store. Later taking the name Barman's Store it was the largest mercantile north of Spokane and continued to be so until it closed in August of 1987.

David Barman, and a cousin, Sig Dilsheimer were interested in numerous mining properties around the country along with having had some interest in **Silver King.** One of their last mining possessions was the **Sunday Morning.** Another, the **Acme** went to Louis Strauss who came from Germany after Alfred, in 1902; and owned the Barman's Department Store until his death. Son Robert Strauss took over what was then two stores, including the former Keller property,

Mrs. Peter (Susan) Mullan, daughter of Capt. John Mullan, and Mrs. Kitt, Mike's mother. Photos in Colville. All are thought to be members of the Calispel (or Kalispel) tribe. (Stevens County Hist. society and Helen Dubois).

The Keller house (former Young home) was built in 1910 for $12,000. A terrible fire Oct. 7, 1911 burned the roof, and much of the interior. Totally rebuilt, and after Young's death, his widow married Louis Keller. The Keller estate turned the beautiful home over to the Stevens County Historical Society. (Photo Statesman-Examiner).

Alfred Strauss attended the Cushing Eels Academy in Colville because there was no high school here in those early days. That school ran from 1897 until 1901, when a regular high school was built by county funds. He then went to the University of Washington in Seattle to later become a "favorite son." Really left handed, they didn't approve of using that in Germany so he became ambidextrous, a great help as a noted surgeon. His earlier interest had come from association with Dr. L.B. Harvey and Dr. Peck in Colville as a boy.

Graduating from the University, Dr. Strauss did his internship in Chicago where for 63 years he practiced until his death. He was honored in 1951 by the university as it's "Alumnus Summa Laude Dignatus," for his medical work, the highest award the university offers. He is known the world over for significant books and articles he wrote concerning cancer research.

Colville Mining District

There are so many interesting people in the area who in some way were associated with early mining – but like the many prospects, and claims, we cannot tell about all of them. Of the numerous claims in Stevens County, there are fewer in the Colville district. J.W. Burrus, who owned the Little

Left to right, about 1912, Mike Kitt and his wife Christine (granddaughter of Capt. John Mullan. L-R, son Jim, daughter Mary, and son John (man standing behind Christine unknown). (St. Cty. Hist. Society and Helen Dubois).

An early photo of the TULARE Dolomite Mine east of Colville. The administrative building and office are gone along with any outward appearance of the mine. The maple tree in center front by the small office is still standing according to Carlson's grandson Ken Howell, who loaned the picture.

Davenport restaurant in Colville, had property out near Aladdin. He had installed a 50-ton ball mill in June, 1934, after years of development work. The first shipment was sent to Bunker Hill that year, with ore averaging $20 to the ton of lead, silver, zinc, copper and gold. Around $1,000 was produced from this property.

By Jan. 18, 1968, that property of nine claims was listed as **Alschlecht** Silver Mines, Inc., of Spokane. Capital was said to be $50,000 and the company listed Cline E. Tedrow, Edwin W. Alban, both Spokane; Raymond Schlecht and Halvor Knudtzon Jr. of Longview. The lease and option to buy was from Joe R. McNamee and Associates of Northport. The **Morning** east of Aladdin drew limited acclaim and made less history but seems of interest in 1989.

Tulare

The property that benefitted a few of the local citizens looking for work was the **Tulare** Mining Co., with their dolomite property just east of Colville. W.P. Bartlett of Portersville, Calif., came here in the early 1900's as superintendent and developed the mine. In 1915 Eric Carlson came to be foreman. The mine was owned by Crown Zellerback of Portland and the raw material was being shipped to Crown Willamette Paper Company. Later the company converted to burnt lime producing 650 tons of the calcined product monthly – and averaged 12 cords of wood used daily to burn in the kiln. The product was also used for mortar for building purposes.

There was a bunkhouse, an office and cookhouse and the company employed from 25 to 30 men to include old timers like Wes Hawthorne, Forrest Buchanan, James Slentz, Albert Keough, Burley Keough, Amour and Claud Slentz. It was from this property that Joe Dupuis had the first contract to haul the product to the railroad. He and Roland moved thousands of tons between 1916 and 1919.

Though not considered a precious metal, the jobs were precious. An ad of the day told that Willett Brother's new Ford cars were priced $93.37 higher – the touring car then selling for $524.03. On March 8, 1918, 30 men went on strike resisting being asked to work nine hours a day for eight hours pay.

The day came in the mid-1920's when the town was very aware of the **Tulare Mine**, some four miles away. A real blast was set off with what was reported to be 7,250 pounds of dynamite. It loosened approximately 30,000 tons of dolomite rock. The market kept going down and the mine was closed down around 1945.

Stevens County Newspapers

Newspapers played a great part in spreading the mining news and the Stevens County Miner served the people well by publishing mining notices. J.B. Slater was publisher of the once-a-week paper established Oct. 17, 1885. The Colville Republican followed on Aug. 18, 1892. The Colville Index on Nov. 9, 1893, and the Stevens County Standard by Sept. 12, 1894. Said to have supplied the news for over 81 consistent years, the Statesman-Examiner came about in 1948 with the late Charles T.

Some of the crew at the TULARE Dolomite Mine east of Colville, around 1928-29, including Manager Eric Carlson, Sig Haggmark, Tony Lindgren, Joe Jacobsen and one unknown. (courtesy Eric Carlson's daughter Agnes Howell).

Graham as publisher. 104 years since the first newspaper, the Statesman-Examiner is still telling the news of what little mining activity is going on in Stevens County. Pat Graham is the publisher.

The Stevens County Mining Association elected David E. McMillan as their president on June 17, 1936. By Jan. 19, 1942, the mining men in the Colville area were meeting to discuss the proposal of the Prospectors and Mine-owners Association to establish an ore-buying station in this section of the country. In most recent years the mining people of the region have become members of an Association whose meetings are held in Spokane each December, The Northwest Mining Association.

SCHUMAKER MINE – Triton and Tri-Nite Company

Putting these operations into their right perspective involves several Colville mining people, and a few "outsiders". The first important knowledge of the **Schumaker Mine** was in 1943. At that time it was owned by Northwest Zinc Company, had six claims and four men were hand-sorting and shipping high-grade zinc.

The late Colville restaurant man Darrell Newland had been in the mining game for some time as president of his Triton Company. Cline E. Tedrow was vice-president. By Jan. 25, 1952, along with Karil J. Rausch, Jerry Waters, Lester J. Somerlott, Thomas Parks, and Newland, they were doing business under lease as Pioneer Mining Company. The **Longshot** claim was said located by George S. Ferguson and George S. Watson on May 20, 1940 according to a Statesman-Examiner release of Jan. 25, 1952. The claim was some eight miles from Colville and north of the **Old Dominion.**

At the time, the mining property was under a suit filed in Spokane Superior Court with Martin Mikkelson against George S. Watson. The suit was against Watson but involved Pioneer Mining Company because of the lease of the **Longshot.** The complaint was that Mikkelson furnished tools and machinery for work on the claim and that Watson, said to have a half interest, promised to convey a quarter interest to Mikkelson. Mikkelson said the lease with an option to Pioneer came about as a result of his backing. He stated that Watson had been receiving rents, royalties, and profits and had refused to give him the promised one-fourth interest in the claim.

Mikkelson asked that Watson be required to abide by their agreement, and that the court find Pioneer's lease not binding on Mikkelson, and that Pioneer be restrained from making further payment to Watson pending trial. The suit was apparently settled satisfactorily because in 1962 Triton was doing exploration work at the **Pioneer.** On Feb. 11, 1962, Pioneer Mining Company was installing new machinery, according to president Newland. The company had produced and shipped several tons of lead-silver ore in the past year.

A new corporation to be considered by 1960-1961, was the Delles-Sullivan Mining Company with Joe Delles and the late Pat Sullivan co-owners. They were mining the **Joe Creek** claims, some 14 miles northeast of Colville. Diamond drilling on the seven claims, they had a strike and hit an eight foot zinc-lead vein along which four diamond drill holes had been made. "It looks like an extensive ore body," said Delles. The property was just north of the **Triton,** owned by Darrell Newland.

On Dec. 15, 1961, Delles and Sullivan Mining and Milling Company, put 1,250,000 shares of their common stock on the market – the first public offering. 1,000,000 shares had been sold or accounted for privately, of the original 4,000,000 shares at five cents per share. The new offering was at 20 cents a share and the money would be used for mining equipment. 6,000 tons of lead-zinc milling ore was stockpiled, according to Delles. A purchase agreement was on the Gibbs **Bonanza Mill** at Palmer Siding.

On April 13, 1962, an announcement said the **Bonanza Mill,** quiet for the past four years, would operate with an agreement between Triton Mining, and Delles-Sullivan, the owners of the mill at the time. It would take from two to three months to mill the ore from Triton's **Schumaker Mine.** The crew at the mill were some of the old-timers Warren Liebman, mill superintendent; Art Vining, Howard Schueck, Norman Bryant, Art Bolt, Dean H. Lynn, Leslie Johnson and Art Loiselle (many of these have passed away at this writing). Delles-Sullivan added two additional directors to their board June 13, to include Everett W. Stark from San Francisco, and Grant Cowie, Spokane. At the time Sullivan was president; Dr. Edmund Gray vice-president, and Delles was secretary-treasurer.

By 1962 Triton had purchased a percentage of **Schumaker Mine.** Newland said there was $48,000 worth of ore stockpiled at the mine and at the time they were dealing with Goldfield for the **Sierra Zinc Mill.** Triton purchased the 400-ton mill from Goldfield for a figure said to be $50,000 and moved some of the machinery to the new site at **Schumaker.**

By mid-year it was decided to organize Tri-Nite to take over the **Schumaker Mine,** and sell buildings and equipment to Tri-Nite for $17,500. Triton was also to convey the **Joe Creek** property which they had taken over earlier; 400 acres in the **Schumaker Group,** the **Baker-Ramminger** millsite, and leased properties from the Erickson lease. Triton would keep it's **Galena Knob** (a claim located by Gus Schweitzer in the late 1800's and later owned by W. Lon Johnson); **Pioneer Mines,** and the Uribe Barite property.

Tri-Nite was to be formed by R.J. Hundhausen and L.L. McLean, and to be capitalized at $400,000 divided into four million shares at ten cents per share. Hundhausen and McLean were to assign four leases adjoining the Triton **Schumaker** – leases covering approximately 800 acres for which they would receive 400,000 shares of Tri-Nite stock. Each of the four leases carried a five percent net smelter return royalty to the owners, Victor K. and Mary L. Lawson, and James E. and Florence Keeley; and the owners likewise had a minimum advance royalty totaling $5,000 for all four leases, according to the news release of June 15, 1962. Triton was to reserve five percent royalty on net smelter returns of the **Schumaker claims,** and receive 1,500,000 shares of Tri-Nite stock.

It was explained at the stockholders meeting in July that the **Schumaker** had been under development for over two years and Triton had blocked out large bodies of lead-zinc ore. They had 4,100 tons milled at the **Bonanza Mill** but returns were not yet available at the time. In August it was made known that Tri-Nite had started to move the **Sierra Mill** but ran into legal problems. They eventually received full registration and the right to sell stock for financing such an operation. They controlled 1,400 acres around **Schumaker** at the time.

In 1964, the Sidney Mining Company announced the sale of their plant to Tri-Nite and it was to be removed from it's Idaho property. In the meantime Newland announced that Triton had purchased 50 percent interest in the old **Casteel** two and a half miles south of Northport, from Leo Sanders of Spokane, and Claude Sanders of Colville. The property had some production around 1900, Newland renamed it **Silver Crown.** On Sept. 20, 1966 he announced an ore strike assaying 42.8 percent zinc, 218 ounces silver and grossing $132.32 a ton.

By March of 1966, Calix American had entered the Stevens County mining picture. Along with other properties already discussed, they announced they had bought 51 percent interest from Tri-Nite. The late Clair Wynecoop, of Wellpinit was president at the time, of Tri-Nite (along with also being president of **Midnite**). He said Calix had agreed to spend up to $150,000 to further develop **Schumaker.** They were currently drilling at the property at the time. Calix also purchased the **Bonanza Mill** from the Delles-Sullivan Company where they planned to process ore before continuing the move of the **Sierra Mill.** Tri-Nite had benefitted from 100,000 shares of Midnite stock and along with fixed assets including the **Sidney** mill, they reported cash assets on December 1966 to be $1,091 and total liabilities $49,914.

William Cullen a Tri-Nite counsel said Calix was to retain 65 percent profits until it recovered it's capital investment, then gave Tri-Nite 65 percent of the profits until it had been paid $350,000; thereafter net operating profits would be split 51 percent to Calix and 49 percent to Tri-Nite. That fall Calix had 10 men employed at **Schumaker** and was hauling 3,000 tons monthly to the **Bonanza Mill.**

On Oct. 17, 1967, a Spokane news article said Calix had $10,000 in zinc-lead concentrates stockpiled at **Bonanza Mill.** They were to have been shipped to Anaconda, Mont., for refining but Anaconda's plant was closed by a copper strike. The concentrates were from ore from the **Schumaker Mine** and were called of "good grade." In

August a management contract had been signed with Raemart Management Company Ltd., under which the latter agreed to buy 100,000 Calix shares at 20 cents each and 100,000 shares at 30 cents each.

Over extended, and lowered prices, Calix was reaching the end. As in dealings with Calix already discussed in previous chapters, Tri-Nite cancelled their lease by September 1968. They had cancelled the mill lease much earlier "for non-performance" according to William Cullen, Tri-Nite's counsel. At that time Calix said they would let Tri-Nite use "free of charge" the **Bonanza Mill** which Calix had contracted to buy, to concentrate the 20,000 tons of ore from **Schumaker.** Tri-Nite planned to sell the concentrates to cover the cost of the **Sierra Mill** they had purchased in July, 1962.

Another "failed" attempt began under Coronado Development Corporation of Tacoma. Scott Simenstad, president of that company said in December 1972 they had a contract with the **Schumaker** property. As mentioned before, they were also redoing the **Sierra Mill.** On March 20, 1974, New Wellington Resources Ltd., of Seattle, said they had entered into a joint venture with Coronado on the **Schumaker Mine.** They said they would have had a 25 percent interest in the joint venture, with Simenstad as project manager. The New Wellington firm said the mine could be put into production at about 150 tons a day after six weeks of preliminary work, and probably at 200 tons a day after more work. the ore was to be concentrated at the old **Sierra Zinc Mill.** Coronado had leased the property with an option to purchase from Tri-Nite in 1972. At the time they said there was a reserve of 30,000 tons of ore that would assay at 8.40 percent zinc and 1.08 ounces of silver.

Also involved in too many properties; and the price of zinc falling drastically, and Coronado was "out!" This was the last major operation in Stevens County. There is much searching going on – and some development, all awaiting the day that minerals will become more of a necessity to the buying concerns. **As if** that day had come, Cordilleran Development Inc. announced first shipment of ore from **Schumaker** to Trail, B.C. July 11, 1989

Pend Oreille County – Pend Oreille Mines and Metals– Grandview – American Zinc, Lead, and Smelting Company

Pend Oreille County mining history will be the last piece of the jig-saw puzzle this has become. In June, 1911, Pend Oreille County became that part of Washington State, the last chunk bitten out of the huge area that was originally Stevens County; and had some of the last developed mines – but some of the most productive.

As early as the 1850's and through a portion of the 1900's, placering was done on the Pend Oreille River and it's tributaries. The name has been controversial being of French background but the Indians called it by the name, meaning ear pendent. The Kalispel tribe has dwindled down to some 200 members and have been located on their reservation on the east side of the river from Cusick, since 1914. In this more-or-less isolated part of the country there must have been an unspoken peace between those tribal members and the slow, but steady approach of the prospectors and miners. (Richard Steele claims that by 1894 the Pend Oreille Gold Mining and Dredging Company had taken half a million dollars from their placers.) (Often spelled Calispel; the tribe is seldom heard from and made no objections to gold removal.)

Newport, Idaho – Washington

Courtesy of Charles I. Barker, secretary to the Newport Chamber of Commerce, we learn that Newport was officially launched and had a post office by 1890 – but that it was still considered an Idaho town. It was just inside the Washington line and west of the Pend Oreille River where that river entered Washington from the Idaho side. the town came about as a "new port" for the steam boats and tugs to dock. Mike Kelly erected a log store building on the banks of the river, and mail was delivered to the post office by buckboard from Rathdrum, Ida.; or through Blanchard Valley, on to Albeni Falls, then ferried across the river to Newport, Ida. The first postmaster was R.I. Towle.

Soon there were two general stores, a hotel, eating house, blacksmith shop, barber shop and two saloons. The Great Northern Railroad was in operation by 1893 and because of the great amount of timber in this new area, cutting wood became one of the money earners. Delivered and loaded on the cars to ship, cord wood sold for $1.50 to $1.75 a cord and hand hewn ties brought 30 to 40 cents, also loaded.

By 1909, the Great Northern had built it's depot on higher ground to the west. Charles Talmadge had bought 40 acres of land from the Northern Pacific Railroad for a town, and had it platted – and Newport, Ida., was left high and dry and was called "Old Town." It took several years for the post office to understand the move but eventually allowed the change and Mr. Towle became the first postmaster in the new location. "Old Town" was later incorporated under that name but had no post office. It did have two hotels, several dance halls – and houses of ill fame. Because of no Sunday closing on that side, and no sales tax, that small village prospered.

Jessie Cass Scott had started an eating house in Newport, Washington to cater to the mining men, the local business, and especially to the railroad. She called it "The Cottage House" in 1892, and by 1904, had expanded by erecting a three-story, 23 room building which she operated until her death in 1940. She served meals to the railroad crew and passengers. The engineer would blow one long blast, then several short ones to indicate to her how many would stop for meals. Leaving the train stand, they would walk up through the woods and Ben Brown, the colored chef would have meals well under way on their arrival – costing 25 cents for all one could eat.

Around 1900, hundreds of homesteaders came into the area to settle lands southwest, and northwest "down" the river for a 50 mile distance when the land was thrown open for homesteaders. (Yes, down the Pend Oreille River going north – one of the few rivers in the United States that flows north.) Mills soon were being built the length of the heavily timbered region. Lumber was loaded on 60 foot barges and pushed, or towed to Newport by the larger steamboats, and two powerful tugboats. It was then transferred on dollies to the yard in Old Town, seasoned and planed at two sawmills, then shipped. Millions of feet were barged to Newport from 1900-1909, and hundreds of barges of cedar telephone poles arrived in Newport for yarding; the poles later shipped by train. In 1909 the old Corlis steam engine from the Diamond National Sawmill was put at the entrance of this newest of Washington cities – the engine no longer needed when Public Utility came into the county.

13 steamboats, three tugboats and two passenger launches made Newport home for over 20 years. All freight, passengers, and mail going north down the river had to come by way of Newport. There were steamboats from 43 to 200 feet in length, many were stern-wheelers. The **Ione, New Volunteer, Newport, Ruth,** and **Spokane** were names of just a few. Harry Sadler operated five horse drawn drays and two passenger coaches to take freight and passengers from the boat landing to the depot in Newport in 1906-1907. The horde of gold seekers took advantage of this great means of transportation, some of them fanning out around Newport but most going down the river to more towering mountains . . . and already the word of new ore findings.

At one time Newport had four saloons and encouraged that business because it had imposed a $1,000 license fee. Barker said that it was a common sight to see two to six customers sitting on beer kegs in front of the various saloons, sleeping off a jag and dead to the world. As they were not disturbing the peace, town marshals never even woke them – it would have been poor business to molest such good customers who helped pay that huge license fee.

By 1910, Idaho and Washington Northern Railway was being built down river to Ione, Metaline and Metaline Falls. There was said in all of the County to be 2,500 employed in various construction projects. It was also easy to understand why this area wanted to become a county ,with a county seat of business. One had to travel over 50 miles of rough dirt road to get across the mountain to the County Seat at Colville, or go miles south to Spokane, and again north. As mining claims were located, this trip had to be made, as well as for any other official business.

There were some mining properties located in the Newport district and Charles Barker and Associates were said in 1941, to own the **General MacArthur** or also called **Glass.** Some 31.5 tons of hand picked ore had been shipped with values assaying in lead, silver, gold, copper and zinc at that early date. There are other small properties but most of the mining is in the Metaline district in the north end of Pend Oreille County.

In 1952, the Spokesman-Review published a book written by Ralph E. Dyar, "News For an Empire". In it he told about the early 1900's, "while

The major source of transportation. Left: Steamboats on the Columbia River. Right: On the Pend Oreille River going towards Metaline Falls (S-R 1899).

on a trip north to investigate some mining property, "I went through a virgin stand of timber in Pend Oreille County, that was about to be cut." He extolls the beauty and was so unhappy about it's being cut that he went back to the Publisher W.H. Cowles and told him of his distress. Cowles told Dyar to try to purchase the property at the best possible price. Dyar did that, and got the price reduced on the land and timber rights with the end result being that 100 acres of this beautiful drive through Pend Oreille County was turned into a State Park. Cowles was one of the largest financial sponsors.

Since 1896, Metaline Falls has been the center of placering. When that town and Ione became official, 1909-1910, they were also becoming the most active with rail business, and with the arrival of many prospectors and miners. Head of the Idaho and Washington Northern Railway development, capitalist F.A. Blackwell (sometimes said of Pennsylvania, and sometimes of Spirit Lake, Ida.) saw all this magnificent white and yellow pine which led him to putting in sawmills at Ione and Cusick. This was in the 1900-1915 period when Diamond Lumber opened their mill at Dalkenna, the mill that later became the property of W.M. "Bill" Sinn, of Pend Oreille County then; now of Colville. Mr. Sinn tells of the horrible forest fires that in the mid 1900's ravaged much of his timber. At that time timber burned and the cost of fighting was charged to the owners of the property. His cost for this particular fire was near $15,000, which he did not have. Going to the bank, he recalls the kindness of Betty Thompson with the Ione State Bank, who swung sentiment in his direction and got the bank to loan him up to $2,500 toward the bill, far more than was a customary loan.

What he was trying to explain was the kindness and helpfulness of people at the time when most were very hard up. Some of the better known, and in most cases loved "characters" who lived in the county were in the lumber business. Yes, he knew well the lady logger, Mrs. Bob Hershey who had her own logging outfit - "and her language was as strong as any logger of the day". Also remembered was Sadie Ragsdale and her daughter - "Sadie could cuss-out any one in the woods - and turn around and dress up in black silk and lace on Saturday night, dance the night away and few - if any, ever knowing who she was."

One of the favorite "characters" and known hospitality for miles around — for anyone who was ever down on their luck, was "Tiger Mary" who ran her hotel and saloon at the little village of Tiger some four miles south of Ione. Nothing but a single building stands today but there are still living, those who have forgotten her last name, but cannot forget her. Sinn said she could really cuss-out anyone who tried to put anything over on her, but would favor anyone she chose to help. She lived her own way of life, generous to a fault and she died around 1950.

In 1903, a limestone cave was discovered. Called Gardner Caves, it was at one time sold to W.H. Crawford who deeded it to the State of Washington in 1921 for a State Park. It is 1,250 feet in length and 270 feet deep, and it has the largest stalagmite column in the Pacific Northwest, along with numerous other formations. There are many legends about the discovery of the magnificent cave but it was actually found in the search for ore. It is lighted and accessible in season, 11 miles north of Metaline

In 1920, prohibition was on. These towns were pretty well populated with "tougher breed" of logging and mining men, and cement workers. "Can you imagine those hardened fellows lined belly-up at the bar ordering near-beer?" Sinn said there were bootleggers, and the "runners" from Canada had no trouble at all finding a sale for their deliveries. As a result of this, there were a few locations, and dealers who were easy to contact for a bottle, or a case. Common sense does not permit using names and places.

Mining seemed a bit slower in the area because of limited gold but to Lewis P. Larsen, a capitalist, and the founder and builder of Metaline Falls, there was greater wealth in other minerals. He was born in Denmark in 1876 and came first to Utah where he mined. In 1897, he came to Spokane and was in mining in Wallace, Idaho. His name is better remembered in association with Larsen and Greenough mining companies. Still, and always a miner at heart, he went to the Metaline area and prospected the hills in all directions. In 1900 he was connected with the **Last Chance** east of Metaline Falls.

In 1905 he discovered deposits of cement rock at Metaline Falls. Seeing the potential, he interested F.A. Blackwell in the prospects of a cement plant. Colonel Harry Trexler who was in the cement business in Mason City, Iowa became a large stockholder and president of the organization to build such a plant. Blackwell, also became a large stockholder and vice-president. Larsen was to provide property and water rights to the extent of $75,000; $50,000 to be taken in stock, for which he would give his land said to be his **Defiance** and **Sullivan Creek** claims on which he had acquired a townsite. Blackwell was to see to rail transportation.

Anthony Bamonte has just published his book "Metaline Falls" which tells more of the town, and Larsen, but in a nut-shell - the Inland Portland Cement Company was organized and a one million dollar building was put up. This was to be the largest and most complete cement plant in the northwest. Mills and buildings have floor space of several acres. On May 17, 1909 the Company was incorporated with capital stock on one million dollars divided into 10,000 shares of par value stock at $100 a share. In operation by 1910, the plant's early production was 2,000 barrels a day.

Lewis P. Larsen's dream plant at Metaline Falls - the Lehigh, now the Heidelberg Cement Company. The plant was closed the summer of 1990. (Pat Graham).

The plant paid it's first dividend in 1913 of $494,000. In 1914 it had it's first large contract to furnish the United States Government with 10,000 barrels of cement. The Company even built a steam-heated brick hotel at a cost of $15,000. It is said still in operation by Teen McGowan.

In March, 1914 the plant and property was bought by Lehigh Portland Cement of Allentown, Penn., for $1,006,000. It was not automated in any way until 1923 when the first tractor was put into use. Until then, horses were used to haul stone to the crusher. It did build a 5,180 foot cable tramway used to transport the crushed stone and shale. The tramway is still in use.

Bamonte honored Al Schaeffer who was originally from Allentown, Penn., - and who was plant manager for 22 years of the 39 he worked there. He was said to have made the most changes in overall conditions. On Dec. 15, 1977, a German company, Heidelberg Cement Company, bought the Lehigh plant for some $85 million. 93 percent of the stock was tendered at $25 a share. Franz Lanka is current manager.

Safety records over the years are rather outstanding. According to Bamonte, only three men have been killed since 1909 in this huge operation. One was from falling rock in the quarry; one was crushed in an elevator, and the third was suffocated when he fell into a material bin. The cement plant itself has been the nucleus of the town for 79 years. There are still 59 working at the plant and it is producing 200,000 tons a year. There are said to be sufficient reserves to last 100 years. The company did provide cement to Boundary Dam, and some for Grand Coulee, Hungry Horse, Dworshak Dam and other leading projects.

As Metaline Falls grew it became apparent that a bridge was needed to cross to the town-side. A bridge was built in 1920 across the Pend Oreille and it was toll operated for 10 years Up to that time a barge, or boat had to be used in the summer; and cross on ice in the winter.

Larsen's home in Metaline Falls was designed by the famous Kirtland K. Cutter, the house still standing and belonging to Dr. Byron Hanson—who is offering it for sale. The beautiful home is offered at $85,000. Cutter, who designed many of the fabulous homes and buildings in Spokane, also designed the Metaline Falls school, used since 1962 as storage.

Lead had been discovered in the county by 1896. One of the earliest bits of information more in general is on Nov. 12, 1896 when John R. Cook, of Rossland, B.C. traced the Rossland belt to Sullivan Creek on the east side of the Pend Oreille River. He found the same character of ore as the great LeRoi. Paul LaPlante was interested with Captain Cook in the Metaline district and Sullivan Creek properties. (They were both involved in properties in Kettle Falls shortly after this date.)

The earliest published claims seem to be the **Cliff**, the **Conquest**, and the **Comstock**, all of the 1899 period. The **Cliff** shipped a carload a week and that was packed on horses 15 miles to a steamboat landing below Box Canyon Dam north of Ione. The ore was largely lead with some silver. The claim is 11 miles north of Metaline Falls. In 1925 it was owned by San Ramone Mining Company. With one patented claim, it was said owned by Drumheller Investment Company of Spokane from 1935-1945. 848 tons of ore was milled in 1944, with values in zinc, lead, silver and gold. By 1950-1952, **Cliff** is credited with five claims patented and owned by Sullivan Mining Company, of Wallace.

The **Conquest** – or **Kootenai** was located in the last of 1889, near Marshall's Landing. It was said owned by a Spokane Company and regular shipments were being made but results were not told. Values were in silver, lead and copper. In the Bead Lake area, the mine was eight miles from the road at Newport. It had 170 acres of patented land, and two claims, and was said owned by Newport Mining and Leasing Company of Metaline Falls in 1951. 32.2 tons of ore was shipped from 1917 through 1930.

The **Comstock** was developed to the shipping point but was idle around 1919, becoming active sometime later. Litigation delayed plans at first. It was some eight miles northwest from Newport and was also owned by Newport Mining and Leasing in 1951. A carload of lead, silver, copper and zinc ore was shipped in 1930 with returns of $1,650. Ore was shipped intermittently from 1917, through 1937.

Larsen's claim to recognition is not only in the launching of the cement plant and all of it's benefits, but went on to greater heights when with his claims, and those he came about, the Pend Oreille Mine and Metals came into being. Some of the property was known about as early as 1867 but Larsen's was made public by 1910. This became the largest producer of lead and zinc in Washington and it has been published that from 1942 to 1952, 20 percent of our nations lead and 13 percent of it's zinc was produced from the Metaline's. This great mine has been closed since 1977, but there are very active plans afoot at the time.

The **Josephine,** or **Clark** group may have been his first set of claims to form the nucleus of the Pend Oreille Mines and Metals Company. Located

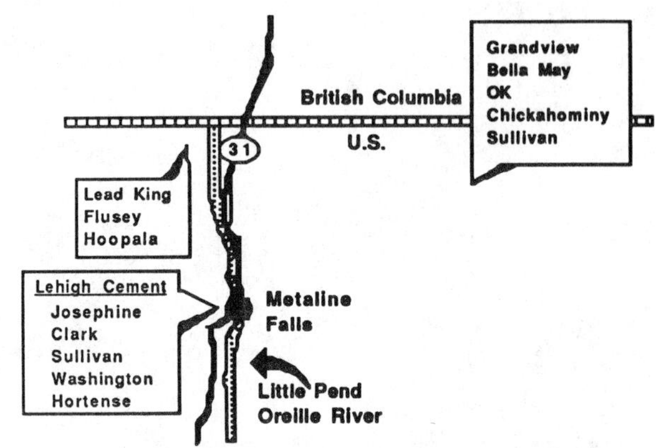

on the west side of the river, early reports of 1915 said this mine had produced 3,111 tons with returns of $8.76. This property of the Lead-Zinc Company from 1912 to 1924, became a Pend Oreille Company mine in 1934. This huge concern became a mass holding company of claims, of which we will name only a very few. They had 100 acres of deeded property aside from many claims. Pend Oreille Mines and Metals Company also owned controlling interest in Reeves MacDonald Mines Limited, in Canada in 1934.

The **Josephine** included 15 claims in 1924, most of those patented. It was said to have increased to 20 claims by the time Pend Oreille took it over. The **Clark Group,** formerly belonged to Charles W. Clark, president of Metaline Lead Company; and there was listed 17 patented claims. It included practically all the claims on the west side of Metaline Falls, with the exception of **Sullivan** and **Washington.** This property was said worked from 1907 to 1919, under lease and bond by the Lead Zinc Company, under direction of Larsen. Total net smelter returns prior to 1919 were said to be $275,000 from 40,000 tons of ore mined. Values were 11 percent zinc, and two percent lead. The ore was milled and 4,000 tons of concentrates were shipped, the concentrates averaging 52 percent lead. At this time zinc prices were said very high with some of the product bringing over $100 a ton, most was over $50 a ton.

Hortense also became part of this operation. The ore was trammed largely from open pits and glory holes to the mill on Flume Creek. Concentrates were hauled by truckload from the mill to ore bins over a mile distant, which were located on the Metaline road overlooking the river. Here the concentrates were sent by aerial tram across the river to the railroad near the cement plant.

By 1956, there was more than one mile of underground workings on these particular claims. The company operated a 750, and 2,400-ton flotation mills and they had their own complete camp. Known ore bodies were 5,000 feet long, as much as 700 feet wide, and up to 100 feet thick. (Reported as early as 1941.) Not including the 1919 report, – by 1935 the **Pend Oreille Mine** had produced

243,000 tons; 187,000 tons by 1949, and 273,520 tons by 1951. From 1952-1955, 12,950,584 pounds of zinc, 6,045,915 pounds of lead, and 16,041 ounces of silver was produced.

Larsen had liquidated his Inland Cement stock by 1915, apparently finding mining more lucrative. He has been considered one of the most successful in that endeavor, in the State of Washington. By 1915 and until 1926, he had the **Lead King** which was brought under the Pend Oreille Mines ownership from 1928 to 1945. On the Boundary road north of Metaline Falls, and one mile west of the Pend Oreille River, this mine was one of the first locations to be made on the west side of the river. It had six claims and was up to 10 when it became a Pend Oreille holding. There had been two or three carloads shipped in 1918, and a small shipment was made in late 1925, assays showing led, zinc and silver. Larsen was president and his friend Jens Jensen was vice-president.

Larsen also had the **Flusey Group**, or **Hoopala**, mentioned earlier. It had two patented claims in 1924 and was owned by the Flusey Lead Company, with Larsen as president. East of **Lead King** on the west bank of the river, the mine assayed in lead, zinc and copper. By 1944, this was under control of Pend Oreille Mines.

The **Washington** claim on the west side of the river was just below the bridge and above the falls. Production was said in lead, zinc, and silver and some was mined by Lehigh Cement for use in their product in 1936. It had been part of Larsen's **Lead King** Company's property from 1924-1926, and came under Pend Oreille holdings in 1943. **Evolution** was another formerly owned by Larsen. **Metaline Metals**, six miles from Metaline Falls, and with three patented claims and 39 unpatented, came under Pend Oreille properties in 1949; along with **Z-Canyon**, or **Silverado**, with 18 claims in 1943, and **Schallenberger** about the same time. **Torrential-Sphinx** was added in 1941 and **Yellowhead** had been in 1930. **Riverside**, though abandoned in 1924, was added to Pend Oreille Holdings by 1943.

With such a mass operation it would not be possible to cover every claim and it's production. A surprising note was made by one geologist "Any attempt to find the ore through diamond drilling, with the irregular pockety occurrences of the deposits, leads to a considerable amount of wasted work."

Mr. Larsen had such plans for his Metaline Falls, and his **Pend Oreille Mines and Metals**, that among other improvements he had built a five-story brick building to have 55 apartments. It was erected in 1928 but construction had to be halted in 1930 due to financial hardships for Larsen, according to a portion of his **"Metaline Falls"** book by Bamonte. That writer goes on to tell that Larsen planned the first floor with high ceilings. It was to be a shopping mall with an elegant restaurant, a grocery store, barber ship, and a number of other shops – all to be leased to tenents. Not to miss a single bit of potential income, he "planned to hire a team of prostitutes to recycle the money he paid the miners". When he finally completed the building around 1948 the laws had changed eliminating prostitutes, and the planned shopping mall. These apartments have been vacated since 1979 following the shutdown of the **Pend Oreille Mines** in September, 1977.

By 1949 prices were going down and this huge operation was considering closing. In May of that year, zinc was 11 cents and lead 13 cents. A news report Aug. 12, 1949 said the American Mining Congress convention was slated for Spokane Sept. 20-30 and would visit the Metaline district which extends both sides of the border. They would visit the **Reeves MacDonald Mill**, rated the newest and most modern mill in North America, then return and visit the new 2,400-ton mill at the **Pend Oreille Mine.**

A feature of the operation was the inclined shaft provided with a concrete roadway for travel of rubber-tired tractors handling men and supplies, with a belt conveyor for bringing ore to the surface. Another report March 6, 1953, gave **Pend Oreille Mines** credit for being fourth in the state in silver production. Other increased producing state-wide was said due to the amounts of lead and zinc ore mined in Pend Oreille County.

A Dec. 22, 1950 news report included new additions to the **Pend Oreille** multi-million dollar operations – to include the 2,400-ton concentrator; a large warehouse and change house, office quarters, powerhouse, machine shop, crusher house, conveyor system and concrete ore bins. Mr. Larsen had seen all of these improvements and more before his death July 14, 1955. The mine came under the resident managership of L.M. "Len"

Loading of a 12-ton truck by a front-end loader under ground at Pend Oreille Mines and Metals. (D.N.&R.)

Kinney, and the Washington directory of mines for 1965-1966 names Jens Jensen as president.

GRANDVIEW MINES – American Zinc, Lead and Smelting Co.

Not only did Pend Oreille County have the largest producing mine in the state, they also had the second largest – **Grandview** – American Zinc combine. As new claims were being discovered in the early 1920's, the **Bella May** was located. High grade chunks of galena were said dug from near the surface and hauled by wagon to Metaline Falls. The **OK** claim was mentioned; followed by the **Chickahominy** claim, then the **Sullivan** group and the **Grandview Group** – a mine to receive great recognition and be second in production in the state.

Some two miles northeast of Metaline Falls, five claims of the **Grandview** were said patented before the 1900 period. By 1956 there were 11 patented claims. The mine was listed under American Zinc, Lead and Smelting Company, of St. Louis, Missouri by 1937, and leased from Grandview Mines, Inc., of Spokane in 1929. The operation grew and into 1952 American Zinc was leasing mineral rights on 9,000 acres in Pend Oreille and Stevens Counties. This included **Grandview, Hidden Treasure, Red Top, Mohawk** and others, the last three mines more important for lead content and the **Grandview** for it's zinc content.

From the time of discovery and through 1951, there was many thousand feet of development of underground workings at the **Grandview.** By 1952 they had a 750-ton flotation mill and six dwellings. Production was listed from 1924 through 1955 and just from 1940 to 1951 – 1,254,000 tons of zinc, lead and traces of silver came from this great mine.

A third large operation to be considered was the Metaline Mining and Leasing Company of Spokane. Along with Grandview Mines, Inc. and Metaline Mining, the company leased the **Hidden Treasure** from Thomas Doughterty and Associates, of Metaline Falls, in 1952. The 10 claims assayed in lead and zinc. This combine also leased the **Red Top,** or **Bailey-Hanson** in 1952. The 19 unpatented claims were the property of Raymond Bailey and Associates. Also a lead, zinc mine, American Zinc eventually leased both the **Red Top** and the **Hidden Treasure.**

Bunker Hill was interested in this newer area and in 1930 had the **Lead Hill,** or **Bunker Hill.** By 1945 it came under American Zinc's holdings. Showing lead, zinc, and silver, a small carload of hand picked ore was shipped in 1937. There were 1,230 tons milled in 1951; and 9,570 tons in 1952 with zinc becoming the prominent ore. The mine consisted of 50 claims and had much development including a compressor; shop buildings, and 100-ton ore bins.

Dumont, or **Bluebird** became one of American Zinc's properties by 1952. Some seven miles northeast of Metaline Falls, the mine had 48 claims. Ore was zinc, and lead and in 1952 the property was being explored by means of a $120,000 diamond drilling program.

By November, 1963, the production from American Zinc was said to be 800 tons daily with 80 men working. It was then said to be the State's second largest. On Sept. 6, 1964, the **Grandview** was reported to be exhausted and the operation was shut down. Production total had dropped to 1,881 tons compared with 10,793 tons in the previous 12 months. It was during these last years that Ralph E. Calhoun had died and his son William Calhoun was sent from the American Zinc home office in St. Louis to become assistant manager at **Grandview.** He was also sent on to the **Anderson-Calhoun** mines in Stevens County before American Zinc closed that operation. Grandview, American Zinc operations come under Washington Resources, of Spokane at the present time.

Metaline Mining and Leasing Company mines were adding to the tremendous production from Pend Oreille County. **Bella May,** one of Metaline Mining and Leasing's better properties was a mile and a half south of Metaline Falls. From 1923-1924, over 250 tons of practically pure galena made up the shipment put together by Oscar DeCamp and his associates. It showed ore in zinc, lead, silver and cadmium and from 1937 through 1943 the mine produced 208,678 tons of ore. One adit was 6,400 feet long and there was reported to be 12,000 feet of underground workings by 1956. **Blue Bucket** – or **Kroll,** was listed under this same Spokane firm's ownership from 1932 on. From 5,000 feet of underground workings by 1956, the mine produced from 1906 to 1950. Totals were 175,000 tons by 1942; 200 tons in that year and 56,131 tons 1937 through 1941, of zinc, lead and silver.

Metaline Contact was another of their properties from 1949, along with five more unpatented claims at **Riverview. Diamond R.,** a 1932 Century Zinc location became Metaline Mining and Leasing property as early as 1936. First shipments were said made in 1918. Sterling Silver-Lead Company made locations in 1915, and in 1932 as **Sterling,** the 10 patented claims came under this same operation.

Sullivan Mining Company of Wallace, Ida., had also entered this area and had the **Giant-Flying Squirrel.** Belonging to James Ehle, of Metaline Falls in 1941, Sullivan had the two unpatented claims by 1952. It is a zinc, lead property. Much work was done but no reports of production were made. This company also had the **Sullivan** by 1924, said part of **Josephine.** The **Robert E. Lee** became another of their operations when they took over the five patented claims in 1950. This had at one time been a property of the Drumheller Investment Company, of Spokane. No report of production was made from any of Sullivan's properties.

The local citizens of the area had many claims and some were good producers. No production report was available for B.T. Beaty's six unpatented claims the **Beaty.** Moraldus Pierce had one patented claim on his **Haliday** in 1941, and not to be out-done, Miss E. Thompson had 10 unpatented claims at **Iron Cap** in 1941. She and her associates were from Ione. Johnson and Bailey had the **Jenney Dee-Snowshoe** of seven unpatented claims. **Alger** and **McCuellough** came about early in the picture. Of that property, Ed Alger of Newport was leasing **Key Fraction** to Newport Mining and Leasing in 1951. This, along with his **Alger** and **McCullough** were some of the copper locations that for some time did give great hopes. There are a number of other smaller claims around Newport, and many more in the north end of Metaline district.

There are a few properties that could be classed as silver claims, the **LaSota,** or **Silver Crest,** or **Bromide** is one. The mine is on the saddle and flanks of Aladdin Mountain and has 16 unpatented claims belonging to the late F.P. LaSota, and C. Clark, of Metaline Falls. Principally silver, lead, zinc and copper, there has been a fair amount of work done. One quartz vein was two to six foot wide and traceable for 450 feet but no report is made of production.

F.P. "Phil" LaSota was one of the interesting men of both Stevens and Pend Oreille Counties. For many years he operated a sawmill in the Deep Creek part of Stevens County. His obituary tells that he homesteaded in 1909, at the tiny village of Tiger and was married in 1917. Coming across the hill to the mill business he worked the Deep Lake area for some time, the mill itself located at the Forks of the Aladdin, Deep Lake, Northport road. Carl Fredrickson remembers when LaSota went broke on that job leaving several of his help short of wages – but the reason was unusual. He had over extended himself to buy two caterpillar tractors to log by pulling loaded sleds over snow. This was the winter of 1923 and unfortunately there was very little snow. Try as he would, he was determined to pull as many as six to seven sleigh loads of logs – first with one "Cat", then with two. Most of the time he was pulling in the mud. Knowing how serious this was, owners of property gave him an easement to go through their property where it was more flatland – and hopefully more snow. Many pitched in and shoveled snow under the sled — the end result just did not produce desired results, he lost his "Cats" and LaSota went broke.

Not giving up, he started a small mill on what is now called the Charlie Maki place. LaSota and some of the men had to pull part of a former mill to that location by horse and sleigh – a terrible job because of the awful weight. The place was heavily timbered and most of the logs were skidded right to the mill. Times were still bad and LaSota could not pay for the stumpage so he gave the lumber to Maki – and then came the fire of 1926, taking everything including the un-insured lumber. Elsie Isaacson White tells of that fire and how she cared for her two-year-old sister, Helen Maki Fredrickson.

LaSota's time in Stevens County ended and he was said to have become a chef at the Davenport Hotel in Spokane, followed shortly by having a store in Metaline Falls around 1934. In 1935 LaSota was named postmaster of Metaline Falls, from which he retired in 1955. Thelma Rohlf remembers LaSota as playing town Santa Claus there for 35 years. He died Feb. 4, 1980 and his wife died in 1981. As to his mine – family members visit Metaline Falls often and we were told they still take care of the assessment work on the **LaSota.**

There were other locals who owned mining property; J.W. Bustua of Metaline Falls had six unpatented claims at the **Last Chance** (there are three Last Chance's still listed as mining properties in this district). A.L. Lakin and Ed Tanghe owned eight claims at the **Meade:** J.D. Nelson had six unpatented claims at **Nelroe,** and former merchants E.O. Dressel and Associates had **Red Crown** – and then there were many, many more.

By 1950, Box Canyon Dam had been built and there was plenty of power for the mines. A report by K.T. Reike, manager of the metallurgical engineering for Singmasters and Breyer, a chemical and metal engineering firm from New York, was published July 31, 1939. He said the increased production rate from 11,500 tons of ore a day in Pend Oreille County, to an estimated 20,000 tons per day can be absorbed by rising consumption. He said the rising production will require an electrolytic zinc smelter and an acid plant and also a fertilizer complex. He stated that 7.38 percent of the world's known zinc reserves and 10.25 percent of the world's known lead reserves are in Pend Oreille County – and he said there would be no difficulty in attracting the estimated 100 million dollars required for establishing the plants required for future increases in production. (This was made known before Grandview-American Zinc closed down).

Production was reported at $146 million between 1902 and 1969, from all of the mines of the county, with peak production at $7.5 million in 1957. The figures included 18 million pounds of lead and 22 million pounds of zinc. By 1964, production was down as a result of zinc at 14 1/2 cents a pound. There seems also to be a shortage of workers limited only to lead production. By then, 63 percent of the total lead and zinc mined was from the huge **Pend Oreille Mine.** A 1965 report was 640,604 tons of ore milled with results showing 0.72 percent lead, and 2.32 percent zinc with receipts of $3,258,486 from the concentrates.

Boundary Dam was completed by 1968, after

years of argument as to what damage the backup of the water might do to the mines. Heaviest employment was in 1964-1965. The county has many wild animals including deer but in the 1950's they started their first herd of elk by implant, and mountain goat were next, followed by caribou in later years. The mountain goat were planted generally north of Metaline Falls and towards the Boundary Dam. One goat did not want to go wild and during most of the dam construction period it wandered across the bridge at Metaline Falls – usually meeting, and following along with any car that crossed.

The **Pend Oreille Mine** closed down September, 1977, but had earlier come under Pintler Corporation, a wholly owned subsidiary of Gulf Resources and Chemical Company, owners of **Bunker Hill**, and the **Pend Oreille Mines.** There were some 80 men suddenly out of work of nearly 140 regularly employed. Bunker Hill made the statement the mine would be closed indefinitely. This mine had actually put Pend Oreille County in business close to 100 years ago and the mill was processing 2,400 tons a day when it was closed down.

On Nov. 17, 1988, some of the best news that end of the county could receive is that Resources Finance Corporation of Toronto, Canada, has an 18 month lease to conduct examination of the **Pend Oreille Mine's** potential. President of that company, Douglas H. Nicholson said the company likes what it sees so far. "You don't step up to the plate and spend a million dollars on an assessment program unless you feel quite positive about the project," he said.

There is great stirring around the old **Pend Oreille** office with Jack Morton in charge. Nicholson said there would be 100 to 150 employed at the mine and he was depending on the local labor pool. **Bunker Hill's** strike led to the closure of this mine which was supplying material to the Idaho mine; and the metal price was down. Under this test period Resource Finance is paying $150,000 for a 12 month option renewable for six more months for an additional $50,000. If the company decided to exercise it's option it can acquire 100 percent ownership of the mine for $1.25 million Nicholson said. Zinc price was up to 72 cents when the story was written by Michael Murphy and Sean Jamieson, for the Spokesman-Review. Nicholson said the operation would be engineered to 40-cent zinc. "If the decision is made to proceed with full-scale development, we could start by the end of 1990," Nicholson said in July, 1989. He announced 400 million gallons of water being pumped out of the mine.

Known over quite a wide area in the mining industry, former Bunker Hill and Pend Oreille Mines and Metals Company president Frank G. Woodruff died March 1, 1989, in Park City, Utah. He became president of these two operations in March 1970 and February 1971. While in this office he moved Bunker Hill's corporate headquarters from Spokane to the Company's main office building in Kellogg in 1971. He became president of the Gulf in May, 1975, and retired from that several years ago.

With the very probability of opening up the **Pend Oreille Mines** again; and the building of the Ponderay Newsprint Mill at a cost of $300 million, just outside of Usk, the County's current assessed valuation will take a great leap – and the county seat at Newport will be extra busy.

This last of the "Colville Country" counties takes it's geographical place in the State of Washington and our area of interest becomes complete. We have recalled many, many names of those people, and those claims that made up the mining history – our "Gold Seekers" have covered many miles. Prices are a bit unstable at this time in 1989 but miners are a "tough lot" and the future indicates a "win" for some; but for many who challenge this gamble, it will only be a "lust".

CONTINUATION OF A STORY WITH NO ENDING . . . March, 1989

Alot has happened in the mining field in the past year, but just to give a brief update I will tell you:

It would take another 100 pages just to skimthat **Bunker Hill** finally got back to work in September, 1988. Closed since October, 1981, the new ownership headed by Jack Kendrick was reported working 137 men as against 2,300 when the mine was closed by Gulf. The present operation is only a very small part of the original **Bunker Hill** complex but was reported profitable again March 1, 1989 with production of 1,000 tons a day, working up to 2,000. Workers are said to be on a profit-sharing, or bonus salary. With the price of zinc up to 95 cents a pound, a Kellogg stockbroker probably called it correctly when he said "zinc was poor man's silver." Development is also proceeding at the 72-year-old Crescent.

One problem that keeps rearing it's dirty head is the request to have the U.S. District court in Boise to unseal the files on the Gulf Resources and Chemical Corporations lead poisoning of Kellogg children. The case came to a head back in 1974 when a fire at the smelter's bag house allowed excessive lead emissions. Rather than go to the expense of repair and replacement, Gulf allegedly operated the smelter for several months according to an article written by David Bond, Oct. 16, 1988.

According to that story, 20 years worth of lead came out of the stack within a single year. The case went to trial and Gulf was charged several million dollars for health damage to children involved. Because of the large money settlement, many in the area chose to turn against those families when **Bunker Hill** was suddenly closed by Gulf. There is a certain public outcry to have files reopened that Gulf ordered kept closed, the outcome of which will settle a certain amount of question as to responsibility of a huge cleanup. Again according to Bond, writing Feb. 8, 1989, the parents of the injured children, a number of local newspapers, and Bond, are all behind the opening of those files so that the Environmental Protection Agency can honestly establish responsibility for the work that is so necessary.

Kellogg has been called "America's dirtiest city," but according to Ray Chapman, a 30-year resident, former **Bunker Hill** personnel worker, and a reporter with the Kellogg based Shoshone News-Press, he says now some 20 years later, that the city is cleaner and looks better. Everyone is getting into the act of making Kellogg an alpine village. The excitement of re-doing the city came with the announcement of a $12 million expenditure going into the building of a gondola to Silverhorn, the ski hill that many have known for years at the **Bunker Hill** site. Said to have some of the best snow in the world, the ski lift will be three miles long with base and top stations.

In a city that had lost it's Ford and Datsun dealers, it's Safeway, and Montgomery Ward, the architect George Olson who is building the ski lift, encourages those living in Kellogg by saying they must turn their shops into quality showplaces. David Bond, one of the area's greatest boosters again writes that the Silverhorn could become capable of rivaling Vail, and Sun Valley; and Kellogg and the ski hill could draw as many as 5,000 people a day. He suggests calling this the "Jackass Bowl," in deference to the "town the Jackass built," title often used of Kellogg (even if it was a burro).

Now the problem is that the ski gondola is to be built through, and over the properties of several Wardner residents. John Mathews, who owns one of the 10 Wardner properties that lie in the gondola's proposed route fails to see how those towers over his property could improve it's value. This problem will have to be solved as progress is made towards the building of the gondola with a $6.4 million federal grant to finance a part of the cost. (called **Silver Mountain**).

Jack Kendrick reminds that this is still a mining town, but like the residents of Kellogg, and those of Wallace who are working so hard to encourage tourist trade, another year will bring a whole new picture of this area. The towns are cleaning up – and "fixing up"; the hills are greener with new plantings and the Silver Mountain ski gondola will be ready for use by Thanksgiving 1990. Hagadone Corp. has entered into some future plans.

Sunshine, and **Lucky Friday** were each closed down due to the price of silver in the spring of 1986. **Lucky Friday** reopened June 1, 1987 and **Sunshine** by Dec. 11, 1987. **Sunshine** claimed it had reached the "break-even" point by Dec. 11, 1988, and paid each of it's 400 workers a $250 bonus. They expected to hire their full 450 soon. This was quite a different report than one made by David Bond, Nov. 8, 1988 when he said life was some changed for Michael Boswell, who took over **Sunshine** from the Hunt Brothers.

Bond went on to explain that price made the difference and that the Company no longer had it's silver, suede-upholstered jet which they on occasion leased to Arab sheiks, and diplomat Henry Kissinger, when Boswell wasn't globe-trotting. The company also leaves idle the mountaintop mansion overlooking the Shoshone Country Club. At the time, Bond quotes Boswell as saying that "Sunshine is in better financial shape today than it's been in probably a decade." By Dec. 28, 1988, **Sunshine** was trying to buy out a plastic firm for some $865 million. Boswell said the reason being was that mining, and oil and gas have been depressed by low prices. Expected to have the deal closed by Jan. 31, there has been no such announcement made.

Lucky Friday was employing 175 in December of 1988. On Sept. 24, 1988, they did announce profit sharing payments to 140 workers that averaged $1,028 each, said to raise their base pay from $9.71 an hour to $15.29. Hecla, who owns **Lucky Friday** announced the opening of **Yellow Pine Mine** at Stibnite, Ida., July 1, 1988. The company expected to produce 100,000 ounces of gold over the next four years and had contracted **Pioneer Metals Corporation** of Vancouver, B.C., to process 400,000 tons of oxide ore the mine will produce annually. By October of 1988; and Jan. 28 1989, **Pegasus** announced a large purchase of shares of Pioneer Metals Corporation.

Suddenly **Hecla** announced on Feb. 17, 1989 that they had a mine but no processor in the wake of a falling out with Pioneer Metals. **Hecla** said this would probably idle the mine for all of 1989. Was there any connection? Mine manager Hans Geertsema said that Pioneer and Hecla ended April 12, 1989. Hecla announced the building of a $3.9 million heap-leach facility at **Yellow Pine. Pegasus** has proposed mergers and some have been cancelled.

On Oct. 1, 1988 it was still involved in a case brought in 1987 against Frank Duval and Milton Zink, along with Hobart Teneff, concerning multiple securities violations against the men. They maneuvered to have **Pegasus** acquire **Florida Canyon,** a property they controlled. **Pegasus** acquired full ownership of **Florida Canyon** and have since developed the property as a working mine. Teneff and Zink resigned their positions at Pegasus and settlement involved several millions of dollars returned to owners and shares returned to source. The case was not entirely settled at the 1988 date.

By Oct. 9, 1988 **Asarco** was said by Michael Murphey, in the Spokesman-Review, to be rolling up their sleeves for a fight on the battleground that covers Lincoln and Sanders counties along Montana's northwest reach. The battle was to be with United Mine Workers of America. Asarco is mining the **Troy Unit** north of Noxon, Mont., and bitter problems have arisen between the company, and their help, who think having the union behind them might solve some of the problems. Some 350 people were working the mine at the time. Threats had become very serious and reports that **Asarco** has taken at least $40 million in profit from the **Troy Mine** has not improved the situation.

That company was threatening not to open **Rock Creek** Mine, another property where some of the men anticipated finding work but one remark was made that, "If the price of **Asarco** developing **Rock Creek** is low wages and unhealthy working conditions, a number of the pro-union miners say, so what if **Asarco** pulls out." With the unemployment record at the time being 10.7 to 12 percent, Troy Mayor Robert Kensler said "we have no complaints, they still employ 350 people up here."

In the better news department; Echo Bay Ltd. plans to open two gold mines and construct a new milling facility in Ferry County by early 1990, according to a Nov. 23, 1988 report. Estimated production over a seven-year life of the mine is 729,000 ounces gold, according to Brian Labadie; and hire as many as 200. Work is going on at this time at both the **Kettle**, and the **Overlook Mine**, where Echo Bay said they had spent $13 million developing the project and would spend another $47 million.

Harry Magnuson and Frank Duval told Aug. 5, 1989 of the $185 million tunnel they are starting at Libby; and owe for in 1991.

In January, 1988, Bema Gold Corporation announced a new gold mine near Arco, Ida., expected to produce 450,000 tons of ore with a yield of 17,500 ounces gold that year. Meridian Gold Company also announced a gold deposit found in central Idaho that could produce more than two million ounces. In March, 1989, Hecla Mining Company announced they, and a Canadian firm were exploring reopening the **Gold Hunter**, an old mine adjacent to Hecla's **Lucky Friday**, and not having operated since the depression. And yes, Appleatchee Riding Club in Wenatchee does not mind at all being paid off by Cannon Mines, partly on their ground.

Van Stone, in northern Stevens County is opening under Equinox Resources, says Ross Beaty from his Vancouver, B.C. office Aug. 22, 1990, price dependeing. Zinc was 65¢ January 3, 1991.

AND THE STORY CONTINUES . . . June, 1992

A portion of works at the **Bunker Hill Mine** was reopened in 1988 by **Bunker Hill Limited**, made up at one time by Jack Kendrick, Harry F. Magnuson, Duane Hagadone and Jack Simplot, who dropped out later. Great effort was made to keep the mine in operation but low price, and heavy debts saw the old mine, and smelter closed down Jan. 25, 1991. It's innards torn apart and sold, neglect, and clean-up problems hanging heavily, the concern was in bankruptcy by the end of January, 1991. The gondola and ski hill above the mine leaves some pride for this great old producer of the 1885's.

Only two Coeur d'Alene mines remain operational; **The Sunshine** and **Hecla's Lucky Friday**. The nation's largest silver mine operating as **Sunshine Precious Metals**, filed chapter 11 March 10, 1992. Low price forced cutting of staff, and leaves little encouragement as to job longevity. Mines need $5.00-$5.50 per ounce to stay in business. Silver as of June 8, 1992 is $4.05. **Hecla's Lucky Friday** is still "hanging on" with hopes from a new find near that mine; at the old **Gold Hunter**. It operated from 1902-30 and produced a total of 8.8 million ounces of silver, and 174 million pounds of lead.

Another of the nation's largest silver mines, the **Galena**, owned by **Callahan** and operated by **Asarco** will be closed by mid-July by reason of low prices and some hint of labor problems.

Coeur d'Alene Mining Co. acquired **Callahan** and the **Calladay** operations in late 1991. They are also affiliated with **Echo Bay**, operators of two good gold mines in Ferry County's Republic, WA **CDA Mines Corp.** and **Echo Bay**, is also said to have paid $20 million for the **Kensington** in Alaska; and **CDA Corp.** was committed to spend $13 million in exploration. We have no later information on this partnership.

In partnership with **Crown Resources; Echo Bay** anticipated production of 110,000 ounces of gold annually from their two Ferry County mines and mill. A news release of April 15, 1992 states **Echo Bay** applied for environmental permits to mine two gold deposits, the **Key East** and **Key West.** A number of jobs will be available from these, and other activities in Ferry County.

The one serious and unfortunate accident was at their **Overlook Mine** near Republic, when a dynamite explosion killed two miners November, 1991 and damaged a tunnel.

Possibly one of the largest gold finds in the United States is the **Crown Jewel**, out of Chesaw and the Buckhorn Mountains of Okanogan County, WA. A February, 1990 geological gold reserves report gave 842,000 ounces of gold with a minimum grade of .102 ounces to the ton of rock. **Battle Mountain** paid $5 million March 15, 1990 for a 51% option, and were said to pay another $5 million by January, 1991 to **Crown Resources. Battle Mountain** anticipates a facility capable of processing 3,000 tons of ore a day by July 1993, and the mine operational by 1994. More jobs should be forthcoming.

Pegasus took control of **Panegeas Pauper's Dream** the end of June, 1989 near Helena, and renamed it **Basin Creek Mine.** That company has done considerable research which also included some in Stevens County, Washington. **Coeur d'Alene Mines** planned to extract 135,000 ounces of gold by heap leach pads over five years at their **Thunder Mountain Mine,** (near **Yellow Pine Mine)** in Canyon City, adjacent to **Lightning Peak,** and east of McCall, Idaho. The mine was operated June 7, 1989, and was "down" by January 15, 1990, virtually depleted. One report was 12,444 ounces of gold; down to 16,595 from an earlier report.

Hecla Mining Co. may extend life for **Yellow Pine** gold mine. **Barrick Gold Exploration, Inc.** will further explore and develop the property mined during the last three years.

An additional two to four million ounces of gold are anticipated in a feasibility report. **Barrick's** 70-percent interest depends on a $7 million option according to June 1, 1992 news.

Cyprus Minerals, a molybdenum mine on Thompson Creek, S.W. of Challis, Ida., employed nearly 300. It had opened in October, 1983 and closed when it saw the market go down to $2.70/pound July, 1987.

This, and so many other Idaho mines are down because of price. **Sunbeam Gold Mining Corporation** on Jordan Creek, also listed as Yankee Fork of the Salmon River in Pinyon Basin; **Black Pine Gold,** at Cassia City, Ida., expected to produce 50,000 ounces of gold annually by mid 1991. The 10,500 acre site has been purchased from **Noramdan Exploration** for $6.5 million. **Grouse Creek Mine, Inc.,** five miles southwest of Challis should produce late 1993; it is estimated to yield 100,000 ounces annually of gold. One report indicated 17,000 ounces of silver and 720,000 ounces of gold.

Meridian Gold Co. was extolled by Gov. Cecil Andrus of Idaho, as "maybe the largest single gold deposit ever found." In central Idaho, near Salmon in Lemhi County, the first finds were November 29, 1988. Geologists estimated the site at more than 36.9 million tons averaging .055 ounces of gold per ton, and containing 2.03 million ounces of gold. Development plans for the $60 million project have been scrapped until prices go up.

One of the great "oldies of the area" is the old **Delamar Mines,** leased from **Sidney Mining Company** to **Nerco Minerals.** 227 claims lay in the Silver City district of Owyhee County. Great production is expected to be reported for 1991 and 1992. Past production of **Delamar** has been in the millions.

With many properties closed down due to price ($338.70 oz. gold June 8, 1992) there is still some placering going on. On Oct. 20, 1989 an unusual partnership included a Chinese corporation, **China National Nuclear Industry Corp.,** and **Cimco Mining Corp.** of Spokane on a $2.7 million surface gold mining project near the Idaho-Montana border. **Cimco** had earlier that year begun a large surface placering on the east fork of Eagle Creek, 20 miles north of Wallace, Idaho. They reaped more than 25 ounces of gold nuggets a day, valued at about $9,200.

With so much bad news, Stevens County will be glad to know the large lead-zinc mine **Van Stone,** south of Northport is scheduled to open the 15th of June . . . This grapevine news indicates around 75 will be working again, with the product shipping to **Cominco** in Trail, B.C. Ross Beaty of the **Equinox Corp.** said earlier — "price dependent". Zinc at this date is 66¢ to 70¢ a pound. **Equinox** participates with **Pend Oreille Mine** in that county where diamond drilling goes on.

Nostalgia is an important part of history and Cathy Free of the Spokesman-Review is reviving many fond memories with her tri-weekly columns. Murrays **Bed-Room Mine;** the Molly b'Damm Motel (minus ladies of the night), where Jack Hull will offer you a room to relive the past of many old characters; and other stories making the area come to life again, — EVEN MINUS THE GAMBLING AND THE BROTHELS.

And — the argument continues as to who will own, and who will display the fantastic 11,200 year old Clovis points found earlier in East Wenatchee.

References Used

Various Panorama Country publications of the Statesman-Examiner; **Inland Empire D.C. Corbin and Spokane** by John Fahey; (permission for use was received from any source that was used).

The Spokesman-Review Quarterly, July, 1899.

Mineral Rights and Land Ownership in Washington, Circular No. 36 – 1962.

Department of Conservation, Division of Mines and Geology Publications – July, 1965.

The Mineral Industry of Washington, Bureau of Mines Minerals Yearbook – reprint from 1965.

Washington Geological Survey, Geology and Ore Deposits of Republic Mining District, Bulletin No. 1, 1910.

Research material from Barbara Dahn, Seattle, regarding material acquired for her own book.

Hecla Mining Company information courtesy Joseph Suveg, Unit Manager Republic Unit.

The Knob Hill Mine, by J. Maurice Slagle, 1964

with Reference Data from Republic News-Miner, **Northwest History and personal observations.** (Deletions, additions, and corrections by A.R. Patterson, vice-president and general manager.).

"Three Cheers for Chewelah," by Bruce Johnson, The Tacoma News Tribune and Sunday Ledger, September 9, 1973.

Many news stories from the local Colville publications to include the present Statesman-Examiner.

Many news releases from the Spokane Chronicle, and the Spokesman-Review (titles and writers listed separately).

Books from the Washington Library – courtesy Colville Library:

Bulletin of the University of Wisconsin – by William J. Trimble, 1914

Northwest Mining Journal – 1909

Mining in the Pacific Northwest – by L.K. Hodges

History of North Washington – by Richard F. Steele, 1904

Washington – sponsored by The Washington State Historical Society, compiled by writers of the WPA.

Pioneers of the Columbia – published by Greenwood Park Grange (for cross-checking) **Tales of the Pioneers** – Mother's Club and Stevens County Historical Society

Interview, and bits of later information from Alfred Stiles of Arden, Washington. Pertained to Montana, Rossland and Trail, and many parts of Stevens County mining history. Information about Hecla's Republic Unit Mine from Josef Suveg, unit manager. Report about Newport from Charles I. Barker. Mining papers and associated business reports loaned me by Judge Lon Johnson. Metaline Falls information from Thelma Rohlf and Jack Morton. Fact verification from "Metaline Falls" by Anthony Bamonte and from **"Lost Mines and Treasures"** by Ruby El Hult.

Numerous individuals to include: Judge W. Lon Johnson, Mr. and Mrs. Albert Beaudry and Sid Wurzburg in behalf of Joe Moris, John Citkovich, Jack Citkovich, R.T. Jacobson, H. Wade Bailey, Richard Boucher, Carl Sauvola, Earl Wheeler, Roy Davidson, Vickey West, the staff of the Stevens County Auditor's office, Wilma Jacobsen, Carl and Helen Fredrickson, Elsie White, Al Nugent, Mrs. H.I. Minzel, Mrs. Vic Barnes, Mrs. Herb Heinze, James Keeley, Milton Flugel, Delbert Scoles, Vergie Oens, William "Bill" Sinn, Clips from his book by Dale F. Underwood of Winter Park, Florida, and personal letter from Mrs. J.H. Knapp. Personal letters from many members of the Hall Family.

"Old Dominion" – Spokesman-Review – April 8, and 15, 1928. Quoting E.E. Alexander (one of the locaters), and Joe Moris.

Letter from Mrs. J.H. Knapp Sr. regarding her grandfather John Keough, and the Dead Medicine Mine.

Article about Henry L. Day's death – in Spokesman-Review March 22, 1985.

Norman-Thorpe – Spokesman-Review – Nov. 24, 1985 – **A page in history – or a real Phoenis About Knob Hill mine.**

David Bond – Spokesman-Review – 1986 – **Griffith didn't think much of property bought from Day Corporation.** – A later article quoting Bierly when Hecla took over Day holdings.

David Bond – **Knob Hill Labor** problems – Spokesman-Review — April 27, 1987

Statesman-Examiner – May 6, 1987 – **Knob Hill back to five day week.**

Echo Bay announcement – July 25, 1987 – Spokesman-Review

Bert Caldwell – Spokesman-Review Aug. 1987 – **Drilling rigs at Overlook. Development on Echo Bay's Granny** – Mayor Chadwick remarks.

Oct. 16, 1987 – Crown Resources – **about Granny, Lucy, and Brutus claims.**

Nov. 4, 1987 – Statesman-Examiner – **Letter to editor about placer gold** – Fred J. Richardson.

Mine under Republic?? – May 17, 1988 – Yes – Oct. 26, 1988 – Spokesman-Review

June, 1988 – **Echo Bay – Waste water problems** – Spokesman-Review.

Tree figures – Spokane Chronicle

Larry Young – **Mount Tolman story** – Spokesman-Review

"Okanogan Highlands and Echoes" – Community Development Committee – Molson, Chesaw.

Cecil M. Ouellett – Feb. 1972 – **Red Mt. Mine – Trinity Town.** Spokesman-Review Sunday.

David Bond – Aug, 1986 – **Cannon Mine** – Spokesman-Review

"Northport Pioneers" – Compiled by Northport Over Forty Club.

Pend Oreille Profiles – By Lee Taylor

The Peoples History of Stevens County – by Fred C. Bohn and Craig E. Holstine. Published by Stevens County Historical Society, 1983.

People of the Falls, by David Chance – 1986 – Kettle Falls Historical Society

Chewelah Independent – Oct. 25, 1907 –**Coal near Valley**

Statesman – **Coal on West Side** – from news report of August 3, 1893.

Statesman-Examiner – 1943 – story about Alcoa findings as written by Hunting.

Statesman Examiner – **Alcoa Secret Is Out.**

Colville Republican – Nov. 19, 1892 – **Enterprise mine owned by Jack N. Squier**

Colville Republican – May 14, 1897 – **Red and Frisco Kid in Region**

July 30, 1897 – Colville paper – **Steps to be taken for a new burial site – Northport Smelter taking present ground.**

1930's – Colville paper — **about Otto Ronka and sons leasing Electric Point.**

"Tales of the Pioneers" – Mrs. Joe Garvey tells of Boundary and 1908 homestead – and about

Electric Point etc. etc. mines.

Richard Steele - **Bio of Daniel L. Zent** — who came here in 1898.

Wafford Conrad - Spokane Chronicle - **About Calhoun Mine**

"History of Onion Creek" - Mrs. Ivy-Gus Anderson, told of George Van Stone.

Wafford Conrad - Spokane Chronicle - **Tells about Simenstad, and Coronado Development and Sierra Zinc Mine.**

Statesman-Examiner - **About McNamee and the Melrose Mine**

"Pioneers of the Columbia" - tells of Cedarville

Colville paper about Cleveland Mine - June 25, 1897

Seymour Frieden — This Week foreign corespondent, July 9, 1967 - **about Alfreid Krupp and his death** (former owner of Germania Mine).

Mrs. Herb Heinze **tells of her father W. Oscar Van Horn finding Deer Trail** - an interview with her in mid 1960's.

July 9, 1897 - Colville paper - **warning against crooks "jumping claims" etc.**

Richard Steele on Van Horn and Golden - opening Deer Trial.

Spokesman-Review Quarterly 1899 - talking of black sand and huge Spokane assay at Deer Trail.

Dec. 10, 1981 - Statesman-Examiner - tells of Madre Mining Ltd., at Deer Trail Mine.

Washington State Geologist Wayne S. Moen tells of Deer Trail and some of seven mines making up important silver producers - and some of the old towns.

April 1987 - Cortez International reopening Deer Trail - Spokesman-Review

Statesman-Examiner - Aug., 1988 - **Madre Changes name to Cortez International.**

Wafford Conrad interviews Al Stiles - Aug. 18, 1967

Report from Daisy Silver-lead Inc. - April 29, 1967 - tells of potential.

Carl M. Lawson - letter, information, and history - Daisy Mine — Oct. 30, 1967.

March 25, 1893 - Colville Republican - lawsuit regarding Daisy claim.

July 21, 1885 - location of Daisy mine recorded, Colville news - also 1920 article tells of ore being shipped. - another June 3, 1926 - money being raised by stockholders.

Dawn Mining and Richard Boyd - Spokesman-Review.

March 28, 1969 - R.B. Fulton told Spokane Chronicle about Dawn Mine.

David Bond - Spokesman-Review - wrote of Bob Nelson telling about reclaiming Dawn Mining land.

Edward W. Coker - Spokesman-Review — Nov. 2, 1975 - Teetering on new mining boom - opening of Sherwood Mine.

Glenn Galbraith - phone interview about Dawn and Sherwood Mines and reclaiming land.

Larry Young - May 10, 1981 - Giant Sandbox - Sherwood Mine _ Spokesman-Review.

Niagara Mining Company - **by Statesman-Index** - July 2, 1897 - **negotiating with Gold Hill** Same paper - forfeiture notice - March 27, 1897 — Lillian Richards — Snyder, Turner.

Orville Dutro - tells about Kettle Falls in a story in Statesman-Examiner.

Marty Mine - May 16, 1968 - Dec. 13, 1968 - and May 22, 1970 - Development and the three miners die. Statesman-Examiner.

Idaho:

The Geology and Ore Deposits of the Coeur d'Alene district - by Frederick Leslie Ransome and Frank Cathcart Calkins - 1908.

Idaho Bureau of Mines and Geology, Moscow - Bulletin No. 18 and No. 22

The Coeur d'Alene Mining District in 1963, Pamphlet 133.

Sketch portraits of Men Who Made Idaho - by Irvin E. Rockwell

Gems of Thought and History of Shoshone County - compiled and edited by George C. Hobson - 1940.

History of the State of Idaho, by Cornelius J. Brosnan

The Bunker Hill Company - A story of Progress from Mines to Metals - and other publications concerning Bunker Hill and the CDA Mining District.

Courtesy of Elmo B. Thomas - geologist for Bunker Hill

Alfred E. Nugent - exploration geologist for Bunker Hill

Ray Horsman - project engineer for Bunker Hill

The Coeur d'Alene Mining War of 1892 - Robert Wayne Smith - 1961

By courtesy Richard G. Magnuson - limited bits of information and reference for verification from **"Coeur d'Alene Diary"**

Information from Wallace Idaho Chamber of Commerce

The pamphlet **Historic WALLACE, Idaho** - Also from Idaho Mining Association (Norm Radford, Pat and Sherrill Grounds). The Wallace Miner, July 9, 1987 - Information from Wallace Chamber of Commerce.

Brochure - **Sunshine, Kellogg operations, from Sunshine Mining Company** - Kellogg

Brochure - Asarco - Idaho Silver Mines, Galena-Coeur from Asarco Incorporated, Wallace.

Coeur d'Alene History (from an old Greenough Book no longer available).

John J. Lemon - Spokane Chronicle - July 23, 1970 - **Tenderfoot supply carrying camels.**

Janice Ruark - Spokesman-Review - **Ghost Town in Florence, Idaho** - July 17, 1981.

Mildretta Adams, Homedale, Idaho - **"Historic Silver City:**

Sarah Pugh (Jarrett) - **Florence and Buffalo Hump** - Spokane Chronicle

John Beasley - **Golden Chest,** Aug. 1987/ and Jan. 1988 - Spokesman Review.

Sherry Devlin – quoting Darby Stapp – **Chinese at Pierce, Idaho** – Spokesman-Review.

Wallace Free Press – **Mullan not afflicted with houses of ill repute.** Rowland Bond – **Molly Berdan from Murry** – Murray newspaper.

Wendell Brainard – **Mounds in Murry – in "Historic Wallace"**

Keith Goodman – 20 people left in Burke – Sept. 11, 1983.

Wallace Miner – History – and fire story.

Doug Clark — Spokesman-Review – **Canyon Silver Mine** – Oct. 16, 1983.

Barton Preecs – **Sunshine and Silver Syndicate** – Oct. 7, 1979 – Spokesman Review

Clare Nichols – **Iron Mask Mine**– Spokesman-Review, 1984.

Janet Jensen, **Bonner County – Compton White – Jeannot Hotel** – 1980, Spokesman-Review.

Tucson, Arizona news report February 11, 1984 relating to story of Wyatt Earp resigning as deputy sheriff of Pima County Nov. 9, 1890.

Canada:

Mother Lode and Sunset Mines – Boundary District, B.C. – **Canada Department of Mines – Memoir 19,** N. 26 Geological Series –1913.

Reprints from 1895 edition Rossland Miner.

Craig Weir, editor of Cominco Magazine – photos and copies of the magazine.

Miner Printing Company Ltd. – Printers of Rossland Miner.

"Rossland – The Golden City – published 1949 – Rossland Miner Limited.

Rossland Chamber of Commerce – **"On the skyline of the Kootenays"** – and other publications.

Nelson Chamber of Commerce – maps and brochures.

Archie and Erna Coombes, Rossland – stories from memory and her writings.

Personal letters from members of the Hall Family relating to Silver King Mine.

Yellowstone Park – from February, 1989 issue American West – story by Erwin A. Bauer.

Montana:

Dorothy Rochon Powers – **stories of Anaconda** published by Spokesman-Review.

Brochure – **Tour Butte, the "Richest Hill on Earth"; Hell Roarin Gulch,** and other information.

"The New Enchantment of America" – Montana, by Allan Carpenter.

Spokesman-Review – Research on Henry Plummer.

Dennis E. Curran – Spokesman-Review, (for Associated Press) – **Uptown Butte.**

Neil J. Lynch – **Butte history**

Spokesman-Review – Photo demolished Great Falls stack.

Spokesman Review – E.P.A. Closure of Anaconda

Mayor Don Peoples – Butte story, Spokesman-Review

Montana Resources – Spokesman-Review

Jardine Gold Mine – Jan/ 1985 – Larry Wills – Spokesman-Review

Magnuson and Duval purchase Borax property – by David Bond, March 1988 – Spokesman-Review.

Dennis E. Curran, Dec., 1986 – **Treasure State – not just one industry.** – Spokesman-Review.

Norman Thorpe – **Montana Tunnels** – Spokesman-Review – and **Pegasus opens Montana Tunnels.**

Bill Sallquist – **Pegasus accuses officers of fraud** – Oct., 1987 – Spokesman-Review

Gary Langley — **Mining in Montana** – Spokesman-Review.

Associated Press – **Mine near Lincoln, Mont.** Feb., 1988 and **Blackfoot Gold Mine.**

David Bond – **Stillwater Mine**– March 1987 – **Platinum** – Spokesman-Review.

CREDITS, Added names of those who helped with research and corrections:

Much information from the lage **Dept. of Natural Resources** "Super" **John Link.**

Updates from **Herb Buffan**, retired "super" of **Van Stone.**

Linda Terry with information about her grandfather, **O.L. Richardson**, and the **Hall Creek Mine** at Inchelium.

Harry A. Sherling of Oroville, Washington for photos and information.

Philip M. Lindstrom, Mining Consultant at Silverton, Idaho for corrections and additions.

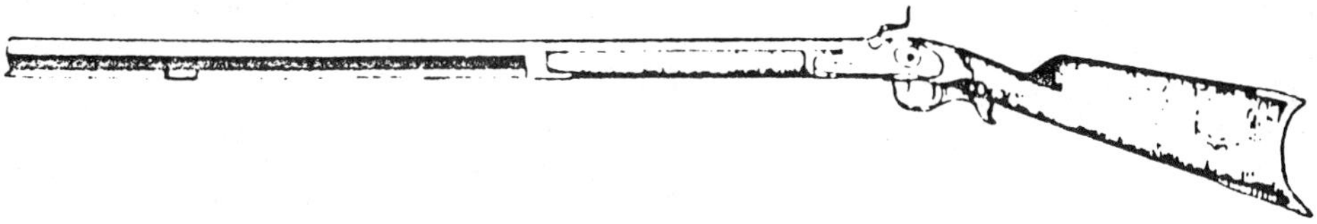

COLOPHON

"The Gold Seekers" was printed in the plant of the Statesman-Examiner, Inc., at Colville, Washington. The Statesman-Examiner, besides being Colville's weekly newspaper, is a small press book publisher, specializing in local histories. The Statesman-Examiner used a 4-unit Goss Community roll fed offset press to produce this work.

Cover drawing by Mike Somerlott.

Index

People, Places And Things
(Mines And Companies Pg. IV thru XVI)

NEW ADDITIONS TO AN OLD STORY

Hall Creek Mine building at Inchelium, Washington with Ora L. Richardson standing at doorway. Heavy output with little report of gain. (Courtesy of granddaughter Linda Terry of Colville).

Incorporated Under the Laws of the State of Washington

NUMBER SHARES

Hall Creek Mining & Milling Company

CAPITAL STOCK $1,000,000

FULLY PAID NON ASSESSABLE

THIS CERTIFIES THAT O. L. Richardson is the owner of

One Dollar each of the Capital Stock of

Hall Creek Mining & Milling Company

In Witness Whereof,

SHARES $1.00 EACH

A stock certificate from the little-known Hall Creek Mining and Milling Co., belonging to O.L. Richardson (courtesy granddaughter, Linda Terry of Colville). The mine at Inchelium, Washington may have been an off-shoot of activity at Nelson, B.C. — the great "Silver King", whose owners were of the Inchelium and Colville region. Refer to pages 40, 41, 43.

Hiram F. Smith "Okanogan Smith's" home. (Refer to Page 17, 134 and 135). Miner, legislator, and Alaska promoter; 1858-1894. (Courtesy Harry A. Sherling).

Power House on Similkameen River near Oroville. First power June 1906. (courtesy Harry A. Sherling). See page 135.

Molson, Washington 1906 — The "boom" and slow death of an 1896-1931 town (Pages 138-139). Courtesy of Harry A. Sherling.